Springer Atmospheric Sciences

More information about this series at http://www.springer.com/series/10176

Wenjie Dong • Jianbin Huang
Yan Guo • Fumin Ren

Atlas of Climate Change: Responsibility and Obligation of Human Society

Springer

Wenjie Dong
Beijing Normal University
Zhuhai, China

Jianbin Huang
Tsinghua University
Beijing, China

Yan Guo
Beijing Normal University
Beijing, China

Fumin Ren
China Meteorological Administration
Beijing, China

ISSN 2194-5217 ISSN 2194-5225 (electronic)
Springer Atmospheric Sciences
ISBN 978-3-662-48442-5 ISBN 978-3-662-48444-9 (eBook)
DOI 10.1007/978-3-662-48444-9

Library of Congress Control Number: 2015955135

Springer Berlin Heidelberg New York Dordrecht London

Printed on acid-free paper

Springer-Verlag GmbH Berlin Heidelberg is part of Springer Science+Business Media (www.springer.com)

There is a star twinkling forever in the holy and remote universe, which guides us towards an orderly human activity on future earth. The star is named Duzheng Ye. This book is dedicated to our great teacher and friend Duzheng Ye.

Foreword I

I am grateful to know that the publication of the atlas—*Climate Change: Responsibility and Obligation of Human Society*—and to write preface for it. Congratulations to the authors for their new achievements! The atlas presented the latest modelling outputs from the state-of-the-art earth system models to understand human's historical responsibility to climate change and obligation for future reduction of emission. Its experiments design is in light to Future Earth's themes, which try to use the earth system models to link human society's emission reduction commitment with climate change, giving meaningful and interesting results.

More than 20 years of my experiences participating in IPCC told me that IPCC is a great school. There are 259 scientists from the developed, developing and economic-transforming countries participated in IPCC AR5 WGI, among which 17 from China and Dr. Wenjie Dong is one among them. Chinese scientist actively got involved in IPCC activities and many of youth become the experts protecting climate and environment. IPCC is greatly contributing to growing up of young scientists from developing countries.

IPCC AR5 WGI confirms that it is extremely likely that more than half of temperature rising in our climate system since 1950s is due to human influence. Accordingly, there is a remarkable change in weather patterns and extreme events, such as heat waves, droughts and floods. Many of the extremes of the past decade were unprecedented. The decade 2001–2010 was the warmest on record, continuing a trend of pronounced global warming. More natural temperature records were broken than any other previous decade. The rate of temperature increase would have been even higher if it were not for the relative cooling influence of La Niña events and the role of the deep oceans in absorbing heat. Multiple lines of evidence confirm that the extra heat being trapped by greenhouse gases is warming the Earth's atmosphere, surface to record levels, heating and acidifying the oceans, raising sea levels, and melting sea ice, ice caps and glaciers.

The key findings of IPCC AR5, as expressed in its synthesis report, are: human influence on the climate system is clear; the more we disrupt our climate, the more we risk severe, pervasive and irreversible impacts; and we have the means to limit climate change and build a more prosperous, sustainable future. We must take actions as soon as possible to adapt to climate change and greatly to reduce emission so that they significantly mitigate climate change to avoid the worst effects and influences of climate change.

Based on CMIP5, the authors of the atlas designed new numerical experiments, with three CMIP5 models from developing country and one from developed country, to distinguish the historical responsibilities to the climate change of the developing and developed countries respectively and then evaluate the contributions of mitigation of climate change due to their emission reduction commitment. Additionally, the authors gave the climate change projection particularly for China and the East Asia. The atlas would be helpful for scientists, teachers, students and policy makers working on climate change.

State Key Laboratory of Cryosphere Sciences, CAS
Lanzhou, China
Oct 14, 2015

Dahe Qin

Foreword II

Climate change is one of the greatest issues which received increasingly more concerns worldwide. Its extensive impacts challenge human beings' survival and welfare. To research the causes and mechanism of climate change and predict/project the future climate change and its possible effects are of tremendous significance, and become hot scientific topics nowadays. In the past three decades, an amount of progresses have been achieved in climate change research, which have been systematically summarized by the Intergovernmental Panel on Climate Change (IPCC) Assessment Report periodically published every 4-7 years. In late 2013, the newest assessment report IPCC AR5 was released, in which several aspects have been concerned in comparison with the preceding reports, such as large-scale climate phenomena and their relevance to regional climate change, climate extreme, more detailed interaction processes between clouds and aerosols, and detailed detection and attribution of climate change, indicating a deeper and more sophisticated stage the climate change research is stepping into.

East Asia is the largest and representative monsoon zone in the world. Its climate variability is extremely complex under global warming background. Research of the climate change in East Asian region has great scientific values, and serves the sustainable development and diplomatic negotiation of our nation.

By my knowledge, this atlas provides some novel results in studying the responsibility and obligation of human society for historical climate change, which provides a potential solution to address the controversial issue of carbon reduction in international negotiations. In addition, it vividly shows the future climate change state over the globe and East Asia based on the advanced coupled climate models, which settles scientific basis for our national decision-making. Finally, it shows some informative results in applying the multi-model ensemble methods.

Overall, the atlas is informative and practical. I believe that the readers will benefit a lot from this atlas. It will offer immediate help for policy makers, scientists and the general public who are interested in climate change.

China Meteorological Administration
Beijing, China
Sep 1, 2015

Jifan Chou
丑纪范

Preface

This atlas provides the latest results in studying the responsibility and obligation of human society for historical climate change. It quantitatively shows how much responsibility should be taken by the developed and developing world for historical climate change in consideration of anthropogenic carbon and sulfur emissions as well as global carbon trade, and it can provide a potential solution to address the controversial issue of carbon emission reduction in the international climate negotiation. Since the quantitative attribution of historical climate change is conducted based on Coupled Model Inter-comparison Project phase 5 (CMIP5) model, the evaluation of the model's fidelity in reproducing the observed climate change is necessary. In addition to evaluation, future climate change based on CMIP5 models is also projected at the globe and China and the surrounding area in terms of surface air temperature, precipitation, atmospheric circulations, climate extremes, sea surface temperature, and Arctic sea ice. In addition, this atlas shows some comparison results among multi-model ensemble based on different ensemble methods and ensemble member sizes in order to obtain the most reliable estimate.

This atlas is supported by the project "Super-Ensemble Projection and Attribution of Climate Change Based on the CMIP5." Our main motivation for this atlas is to offer immediate help and guidance for decision-makers, researchers, and those who are interested in climate change.

This atlas is a collaborative achievement of the whole editorial group, and our sincere gratitude goes to Wenping Li, Haiqing Huang, and Libin Fan, who gave us selfless help and support during the process of editing and publication.

Beijing, China
August 31, 2014

Wenjie Dong

Contents

Contributors

Wenjie Dong State Key Laboratory of Earth Surface Processes and Resource Ecology, Zhuhai Joint Innovative Center for Climate-Environment-Ecosystem, Future Earth Research Institute, Beijing Normal University, Beijing/Zhuhai, China

Yan Guo State Key Laboratory of Earth Surface Processes and Resource Ecology, Zhuhai Joint Innovative Center for Climate-Environment-Ecosystem, Future Earth Research Institute, Beijing Normal University, Beijing, China

Chuanjiang Huang Key Laboratory of Marine Science and Numerical Modeling, First Institute of Oceanography, State Oceanic Administration, Qingdao, China

Jianbin Huang Ministry of Education Key Laboratory for Earth System Modeling, Center for Earth System Science, Tsinghua University, Beijing, China

Yunyun Liu National Climate Center, China Meteorological Administration, Beijing, China

Fumin Ren Chinese Academy of Meteorological Science, China Meteorological Administration, Beijing, China

Di Tian State Key Laboratory of Earth Surface Processes and Resource Ecology, Beijing Normal University, Beijing, China

Ting Wei Chinese Academy of Meteorological Science, China Meteorological Administration, Beijing, China

Chonghai Xu National Climate Center, China Meteorological Administration, Beijing, China

Ying Xu National Climate Center, China Meteorological Administration, Beijing, China

Dongdong Yan State Key Laboratory of Earth Surface Processes and Resource Ecology, Zhuhai Joint Innovative Center for Climate-Environment-Ecosystem, Future Earth Research Institute, Beijing Normal University, Beijing, China

Yao Yao Ministry of Education Key Laboratory for Earth System Modeling, Center for Earth System Science, Tsinghua University, Beijing, China

Wenping Yuan State Key Laboratory of Earth Surface Processes and Resource Ecology, Zhuhai Joint Innovative Center for Climate-Environment-Ecosystem, Future Earth Research Institute, Beijing Normal University, Beijing, China

Zongci Zhao Ministry of Education Key Laboratory for Earth System Modeling, Center for Earth System Science, Tsinghua University, Beijing, China

Jinqing Zuo National Climate Center, China Meteorological Administration, Beijing, China

Chapter 1
Data and Methods

Abstract This chapter introduces the data and methods utilized in the computation in the following chapters, as well as the experimental designs of numerical simulations in Chaps. 4, 5, and 6.

Keywords Data • Numerical simulation design

1.1 Data Introduction

In this atlas, Chaps. 2 and 3 are mainly based on coupled model numerical experiment data participating in Coupled Model Inter-comparison Project phase 5 (CMIP5), which is coordinated and released by the Program for Climate Model Diagnosis and Intercomparison (PCMDI, available at http://pcmdi9.llnl.gov/) and observations. Monthly data from the historical experiment (1850–2005) and three Representative Concentration Pathway (RCP) experiments (2006–2100), i.e., RCP2.6, RCP4.5, and RCP8.5 (Moss et al. 2010), are commonly used in these two chapters, except in the subsection of climate extremes in Chap. 2, where daily surface air temperature and precipitation are also used. The annual and seasonal (December–February, March–May, June–August, and September–November) changes in near-surface air temperature, precipitation, atmospheric circulations in the Asia-Pacific region, sea surface temperature in the South China Sea, and Arctic sea ice are mostly concerned.

In the process of analyzing the historical and future changes in surface air temperature and precipitation over the globe and China, atmospheric circulations in the Asia-Pacific region, climate extremes in China, and Arctic sea ice under three emission scenarios of RCP2.6, RCP4.5, and RCP8.5, 22 climate models in the CMIP5 program are consistently used (shown in Table 1.1). Another 21 climate models (shown in Table 1.2) in the CMIP5 program, in which 16 models are same as those in the other sections, are used to analyze the sea surface temperature in the South China Sea.

In order to evaluate the ability of the 22 models in the CMIP5 program in reproducing the observed historical variations of global climate, HadCRUT4 (Morice et al. 2012), GISSTEMP (Hansen et al. 2010), and ERA-Interim (Dee et al. 2011)

W. Dong et al., *Atlas of Climate Change: Responsibility and Obligation of Human Society*, Springer Atmospheric Sciences,
DOI 10.1007/978-3-662-48444-9_1

Table 1.1 Information of 22 CMIP5 models, which are used to analyze changes of temperature, precipitation, climate extremes, and Arctic sea ice

No.	Models	Nation	Horizontal resolution
1	BCC-CSM1-1	China	2.8°×2.8°
2	BNU-ESM	China	2.8°×2.8°
3	CanESM2	Canada	2.8°×2.8°
4	CCSM4	USA	1.25°×0.94°
5	CNRM-CM5	France	1.4°×1.4°
6	CSIRO-Mk3-6-0	Australia	1.875°×1.875°
7	FGOALS-g2	China	2.8°×2.8°
8	FIO-ESM	China	2.8°×2.8°
9	GFDL-CM3	USA	2.5°×2.0°
10	GFDL-ESM2G	USA	2.5°×2.0°
11	GFDL-ESM2M	USA	2.5°×2.0°
12	GISS-E2-H	USA	2.5°×2.0°
13	GISS-E2-R	USA	2.5°×2.0°
14	HadGEM2-AO	S. Korea	1.875°×1.25°
15	IPSL-CM5A-LR	France	3.75°×1.875°
16	IPSL-CM5A-MR	France	2.5°×1.25°
17	MIROC5	Japan	1.4°×1.4°
18	MIROC-ESM	Japan	2.8°×2.8°
19	MIROC-ESM-CHEM	Japan	2.8°×2.8°
20	MPI-ESM-LR	Germany	1.9°×1.9°
21	MRI-CGCM3	Japan	1.1°×1.1°
22	NorESM1-M	Norway	2.5°×1.875°

are used for global surface air temperature, while CRUTS3.1 (Harris et al. 2014), GPCC (Schneider et al. 2013), GPCP (Adler et al. 2003), CMAP (Xie and Arkin 1997), and 20CR (Compo et al. 2011) are employed for global precipitation. HadCRUT4 is a gridded dataset of monthly mean surface air temperature developed by Hadley Centre of the UK Met Office and the Climate Research Unit of the University of East Anglia (Morice et al. 2012), which covers the global land and ocean from 1850 to present on 5° latitude-longitude grids. GISSTEMP is also a gridded dataset of monthly mean surface air temperature developed by Goddard Institute for Space Studies in the National Aeronautics and Space Administration (Hansen et al. 2010), which covers the global land and ocean from 1880 to present on regular 2°×2° grids. ERA-Interim is a reanalysis dataset developed by the European Centre for Medium-Range Weather Forecasts (Dee et al. 2011), which covers the global land and ocean from 1979 to present on regular 0.75°×0.75° grids. GPCC (Global Precipitation Climatology Centre) precipitation dataset is a monthly mean product based on station observation (Schneider et al. 2013), which covers the global land from 1901 to present on regular 0.5°×0.5° grids. CRUTS3.1 monthly mean gridded precipitation dataset from the Climate Research Unit of the

Table 1.2 CMIP5 model details used in Sect. 2.5

No.	Models	Agency/nation	Model resolution (SCS, lat × lon)
1	BCC-CSM1-1	BCC-CMA/China	(0.33°~0.92°) × 1.0°
2	BCC-CSM1-1-m		
3	CanESM2	CCCMA/Canada	0.93° × 1.4°
4	CCSM4	NCAR/USA	(0.27°~0.56°) × 1.125°
5	CESM1-CAM5	NSF-DOE-NCAR/USA	(0.27°~0.56°) × 1.125°
6	CNRM-CM5	CNRM-CERFACS/France	(0.33°~0.92°) × 1.0°
7	CSIRO-Mk3-6-0	CSIRO-QCCCE/Australia	0.93° × 1.875°
8	FIO-ESM	FIO-SOA/China	(0.27°~0.56°) × 1.125°
9	GFDL-CM3	NOAA-GFDL/USA	(0.375°~1.0°) × 1.0°
10	GFDL-ESM2G		
11	GFDL-ESM2M		
12	HadGEM2-AO	NMR/KMA/S. Korea	(0.34°~0.94°) × 1.0°
13	HadGEM2-ES	MOHC/UK	(0.34°~0.94°) × 1.0°
14	IPSL-CM5A-LR	IPSL/France	(0.50°~1.89°) × 2.0°
15	IPSL-CM5A-MR		
16	MIROC5	MIROC/Japan	(0.57°~1.62°) × (~1.65°)
17	MPI-ESM-LR	MPI-M/Germany	~(1.56° × 1.60°)
18	MPI-ESM-MR		~(0.36° × 0.45°)
19	MRI-CGCM3	MRI/Japan	0.5° × 1.0°
20	NorESM1-M	NCC/Norway	(0.27°~0.56°) × 1.125°
21	NorESM1-ME		
Counts	21	15	

University of East Anglia covers global land from 1901 to 2009 and on regular 0.5° × 0.5° grids (Harris et al. 2014). The 20CR is a reanalysis dataset developed by the Physical Sciences Division (PSD) of the Earth System Research Laboratory (ESRL) in the National Oceanic and Atmospheric Administration (NOAA) and Cooperative Institute for Research in Environmental Sciences (CIRES) at the University of Colorado (Compo et al. 2011), which spans from 1871 to 2012 on a regular grid. GPCP is a monthly mean global precipitation dataset developed and computed by the NASA/Goddard Space Flight Center's Laboratory for Atmospheres as a contribution to the GEWEX Global Precipitation Climatology Project (Adler et al. 2003), which combines observations and satellite precipitation since 1979 into 2.5° × 2.5° global grids. CMAP is a global precipitation dataset which merges satellite estimates since 1979 into 2.5° × 2.5° global grids (Xie and Arkin 1997).

When evaluating the performance of 22 models in simulating climate change over China, four observed surface air temperature data are used, including CRUTS3.1 dataset derived from the Climate Research Unit (on a 0.5° latitude-longitude grid and be available at http://badc.nerc.ac.uk/view/badc.nerc.ac.uk_ATOM_ACTIVITY_0c08abfc-f2d5-11e2-a948-00163e251233), the gridded daily temperature data on a 0.5° latitude-longitude grid over China (1961–2005) developed by the China Meteorological Administration/National Climate Center (referred to as NCC) (Xu et al. 2009), the time series of the annual and seasonal mean temperature anomalies relative to climate state 1971–2000 developed by Tang Guoli (referred to as Tang) (Tang et al. 2009), and the time series of the annual mean temperature anomalies relative to climate state 1961–1990 developed by Wang Shaowu (referred to as Wang, available at http://adrem.org.cn/Faculty/GongDY/data/Tempdata.htm); two observed precipitation data are used, including CRUTS3.21 dataset derived from the Climate Research Unit (on a 0.5° latitude-longitude grid, available at http://badc.nerc.ac.uk/view/badc.nerc.ac.uk_ATOM_ACTIVITY_0c08abfc-f2d5-11e2-a948-00163e251233) and the gridded daily precipitation data on a 0.5° latitude-longitude grid over East Asia (1961–2005) developed by Xie et al. (2007).

The subsection of atmospheric circulations in the Asia-Pacific region includes the East Asian Summer Monsoon (EASM) and Western Pacific Subtropical High (WPSH) area and strength indices, which are calculated using the National Centers for Environmental Prediction/National Center for Atmospheric Research (NCEP/NCAR) reanalysis dataset (Kalnay et al. 1996). The Arctic Oscillation (AO) and Siberian High (SH) indices are calculated using the HadSLP2 data (Allan and Ansell 2006).

In the subsection of sea surface temperature (SST) in the South China Sea and Arctic sea ice, the monthly mean SST and sea ice concentration (SIC), provided by the Met Office Hadley Centre (Rayner et al. 2003) with a horizontal resolution of $1° \times 1°$, are used to evaluate the model performance.

1.2 Numerical Simulation Design

Chapters 4, 5, and 6 are concerned on quantifying the contributions of man-made carbon emissions and sulfur emissions from developed and developing countries as well as global carbon trade to historical global climate change. This quantification was performed with numerical model NCAR/CESM1_0_2, whose detailed description document and forcing data, as well as this model, are available at http://www.cesm.ucar.edu/models/cesm1.0. The temporal resolution of model output is one month.

The experimental design to attribute the contribution of anthropogenic carbon emissions from developed and developing countries to historical climate change (Chap. 4) is shown in Fig. 1.1. In detail, the CESM1_0_2 model runs for 162 years (1850–2011) with three sets of different human emissions of CO_2 flux data (available at http://cdiac.ornl.gov/, accessed 5 July 2011) and other observed forcing data in the twentieth century (available at http://www.cesm.ucar.edu/models/cesm1.0/, accessed 5 July 2011). The three sets of numerical experiment are listed as follows:

- ALL experiment: all countries release man-made CO_2 in accordance with the true historical level.

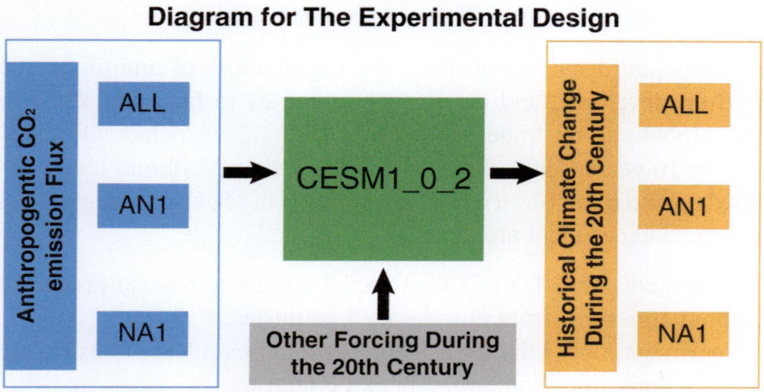

Fig. 1.1 Diagram of the experimental design to attribute historical contribution of the man-made carbon emissions to climate change from developed/developing countries

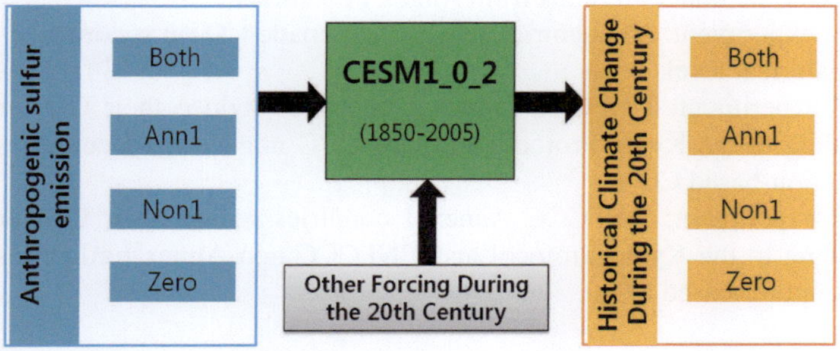

Fig. 1.2 Diagram of the experimental design to attribute historical contribution of the man-made sulfur emissions to climate change from developed/developing countries

- AN1 experiment: only UNFCCC Annex I countries release CO_2 without emissions from non-Annex I countries.
- NA1 experiment: only UNFCCC non-Annex I countries release CO_2 without emissions from Annex I countries.

The experiment design to attribute the contribution of anthropogenic sulfur emissions from developed and developing countries to historical climate change (Chap. 5) is shown in Fig. 1.2. In detail, the CESM1_0_2 model runs for 156 years (1850–2005) with four sets of different human emission of sulfur flux data and other observed forcing data in the twentieth century. The four sets of numerical experiment are listed as follows:

- Both experiment: all countries release man-made sulfur in accordance with the true historical level.
- Ann1 experiment: only UNFCCC Annex I countries release sulfur without emissions from non-Annex I countries.
- Non1 experiment: only UNFCCC non-Annex I countries release sulfur without emissions from Annex I countries.

– Zero experiment: no country releases man-made sulfur.

The experimental design to attribute the contribution of anthropogenic carbon emissions from developed and developing countries to historical climate change considering global carbon trade is shown in Fig. 1.3. In detail, the CESM1_0_2 model runs for 16 years (1990–2005) with seven sets of different human emissions of CO_2 flux data and other observed forcing data in the twentieth century. The seven sets of numerical experiment are listed as follows:

– IPN0 experiment: only UNFCCC Annex I countries release production-based CO_2 without emissions from non-Annex I countries.
– ICN0 experiment: only UNFCCC Annex I countries release consumption-based CO_2 without emissions from non-Annex I countries.
– I0NP experiment: only UNFCCC non-Annex I countries release production-based CO_2 without emissions from Annex I countries.
– I0NC experiment: only UNFCCC non-Annex I countries release consumption-based CO_2 without emissions from Annex I countries.
– Global experiment: all countries release man-made CO_2 in accordance with the true historical level.
– IKNP experiment: UNFCCC Annex I countries reduce their CO_2 emissions according to the Kyoto Protocol and UNFCCC non-Annex I countries release production-based CO_2.
– IKNC experiment: UNFCCC Annex I countries reduce their CO_2 emissions according to the Kyoto Protocol and UNFCCC non-Annex I countries release consumption-based CO_2.

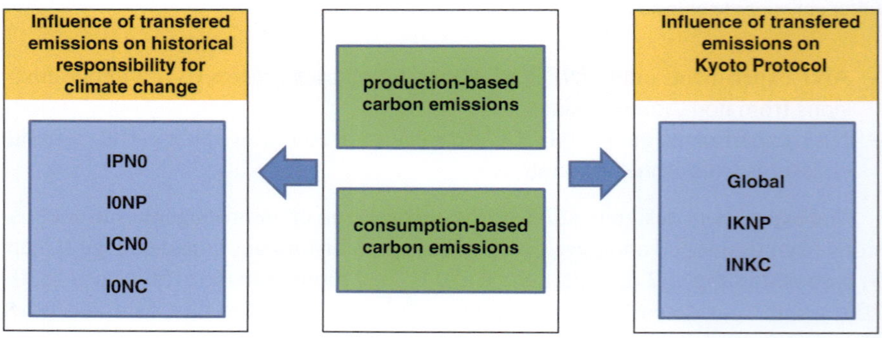

Fig. 1.3 Diagram of the experimental design to research influence of transferred emissions on attribution of historical responsibility from developed/developing countries' carbon emissions and effect of the Kyoto Protocol

References

Adler RF et al (2003) The version 2 Global Precipitation Climatology Project (GPCP) monthly precipitation analysis (1979–present). J Hydrometeorol 4:1147–1167

Allan R, Ansell T (2006) A new globally complete monthly historical gridded mean sea level pressure dataset (HadSLP2): 1850–2004. J Clim 19:5816–5842

Compo GP et al (2011) The twentieth century reanalysis project. Q J R Meteorol Soc 137:1–28. doi:10.1002/qj.776

Dee DP et al (2011) The ERA-Interim reanalysis: configuration and performance of the data assimilation system. Q J R Meteorol Soc 137:553–597. doi:10.1002/qj.828

Hansen J et al (2010) Global surface temperature change. Rev Geophys 48:RG4004. doi:10.1029/2010RG000345

Harris I et al (2014) Updated high-resolution grids of monthly climatic observations – the CRU TS3.10 Dataset. Int J Climatol 34:623–642. doi:10.1002/joc.3711

Kalnay E et al (1996) The NCEP/NCAR 40–year reanalysis project. Bull Am Meteorol Soc 77:437–472

Morice CP et al (2012) Quantifying uncertainties in global and regional temperature change using an ensemble of observational estimates: the HadCRUT4 dataset. J Geophys Res 117:D08101. doi:10.1029/2011JD017187

Moss RH et al (2010) The next generation of scenarios for climate change research and assessment. Nature 463:747–756

Rayner NA et al (2003) Global analyses of sea surface temperature, sea ice, and night marine air temperature since the late nineteenth century. J Geophys Res 108(D14):4407. doi:10.1029/2002JD002670

Schneider U et al (2013) GPCC's new land surface precipitation climatology based on quality-controlled in situ data and its role in quantifying the global water cycle. Theor Appl Climatol 115:15–40. doi:10.1007/s00704-013-0860-x

Tang G et al (2009) Comparative analysis of the time series of surface air temperature over China for the last 100 years. Adv Clim Change Res 5:71–78

Xie P, Arkin PA (1997) Global precipitation: a 17-year monthly analysis based on gauge observations, satellite estimates, and numerical model outputs. Bull Am Meteorol Soc 78:2539–2558

Xie P et al (2007) A gauge-based analysis of daily precipitation over East Asia. J Hydrometeorol 8:607–626

Xu Y et al (2009) A daily temperature dataset over China and its application in validating a RCM simulation. Adv Atmos Sci 26:763–772

Chapter 2
Climate Change: Historical Simulations and Projections

Abstract Sections 2.1 and 2.2 present the evaluations of CMIP5 models in reproducing the historical climate change and projections of future climate change under three scenarios of RCP2.6, RCP4.5, and RCP8.5 in terms of surface air temperature and precipitation over the globe and China; Sect. 2.3 presents the atmospheric circulations over the Asia-Pacific region in terms of the East Asian Summer Monsoon (EASM) (Guo, Acta Geograph Sin 38(3):208–217, 1983), Western Pacific Subtropical High (WPSH) area and strength (Liu et al., J Appl Meteorol Sci 23(4):414–423, 2012), Arctic Oscillation (AO) (Thompson and Wallace, Geophys Res Lett 25:1297–1300, 1998), and Siberian High (SH) (Wu and Wang, Geophys Res Lett 29(19):1897, 2002); and Sect. 2.4 presents the climate extremes in China. The analyzed climate extreme elements include surface air maximum temperature and surface air minimum temperature, consecutive dry days – maximum number of consecutive days with less than 1 mm of precipitation (CDD) – and simple daily precipitation intensity (SDII) (Frisch et al., Clim Res 19:193–212, 2002). The historical climatology is set to long-term mean over 1986–2005. In projection, the uncertainty range is given described by one inter-model standard deviation.

In Sect. 2.5, the simulation ability and future change of the sea surface temperature (SST) in the South China Sea (SCS) are analyzed by using 21 CMIP5 models. Finally, Sect. 2.6 provides the simulated historical Northern Hemisphere sea ice concentration (SIC) and sea ice area (SIA) by 22 coupled climate models from CMIP5. The SIA is defined as the total area with at least 15 % sea ice in Northern Hemisphere. The projected future changes of Northern Hemisphere SIC and SIA under RCP4.5 and RCP8.5 are also shown.

Keywords Historical • RCPs • Surface air temperature • Precipitation • Atmospheric circulation • Climate extremes • Sea surface temperature • Arctic sea ice

W. Dong et al., *Atlas of Climate Change: Responsibility and Obligation of Human Society*, Springer Atmospheric Sciences,
DOI 10.1007/978-3-662-48444-9_2

2.1 Surface Air Temperature

2.1.1 Global Scale

2.1.1.1 Historical Global (GH) Temperature (Figs. 2.1, 2.2, 2.3, 2.4, 2.5, 2.6, 2.7, 2.8, 2.9, and 2.10)

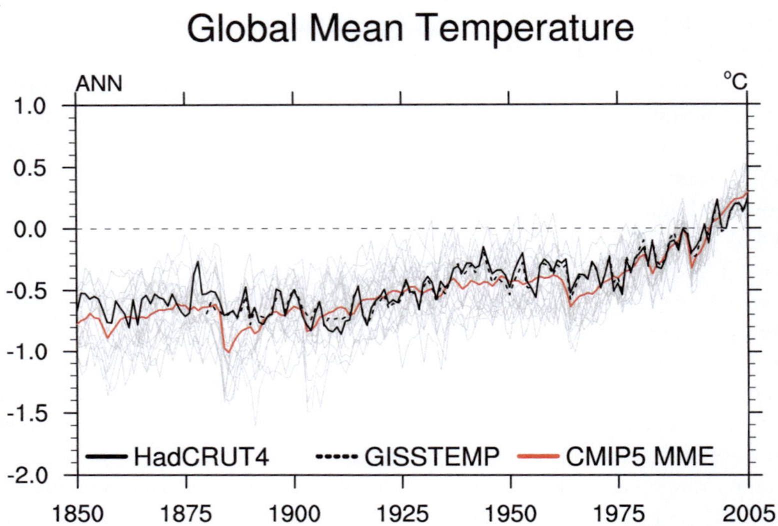

Fig. 2.1 Global annual mean surface air temperature anomalies, relative to 1986–2005 climatology from two observations and 22 CMIP5 models (*gray lines*) and their ensemble mean (*red line*) (unit: °C)

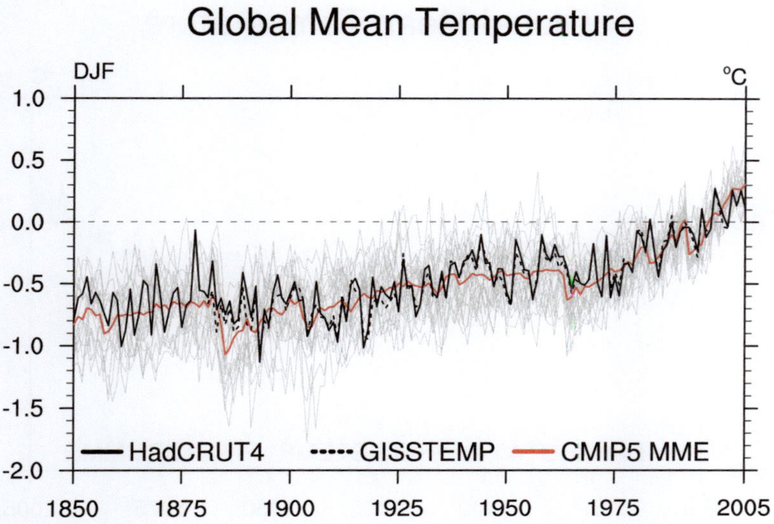

Fig. 2.2 Global mean surface air temperature anomalies for December–February, relative to 1986–2005 climatology from two observations and 22 CMIP5 models (*gray lines*) and their ensemble mean (*red line*) (unit: °C)

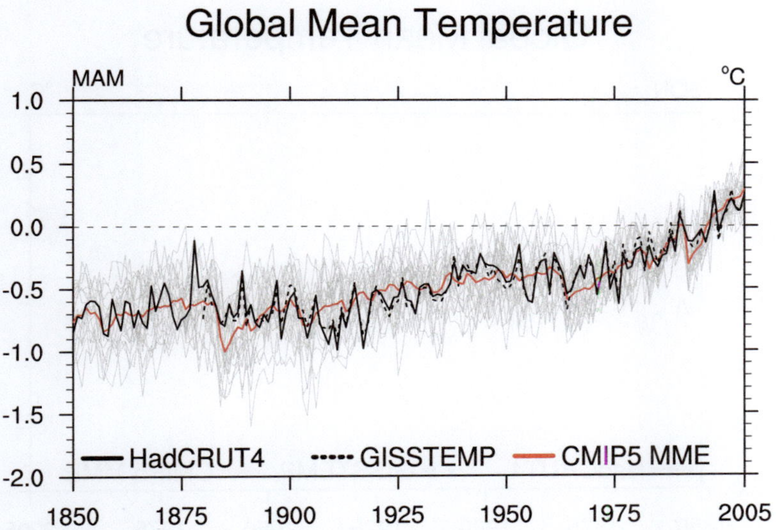

Fig. 2.3 Global mean surface air temperature anomalies for March–May, relative to 1986–2005 climatology from two observations and 22 CMIP5 models (*gray lines*) and their ensemble mean (*red line*) (unit: °C)

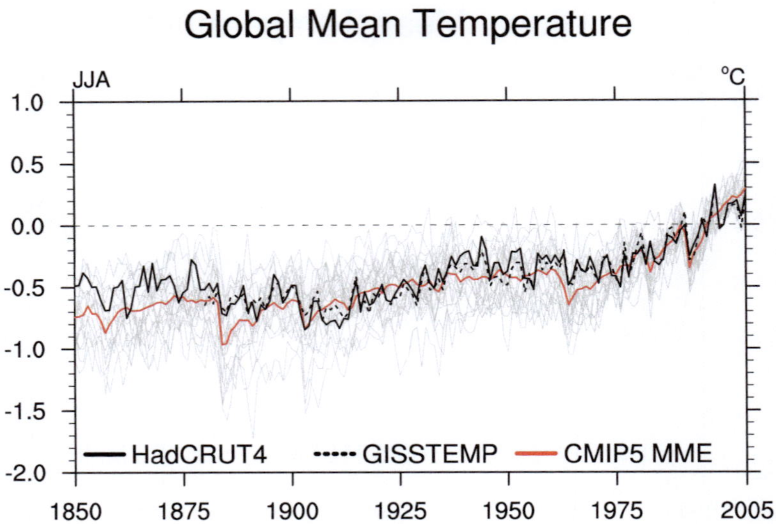

Fig. 2.4 Global mean surface air temperature anomalies for June–August, relative to 1986–2005 climatology from two observations and 22 CMIP5 models (*gray lines*) and their ensemble mean (*red line*) (unit: °C)

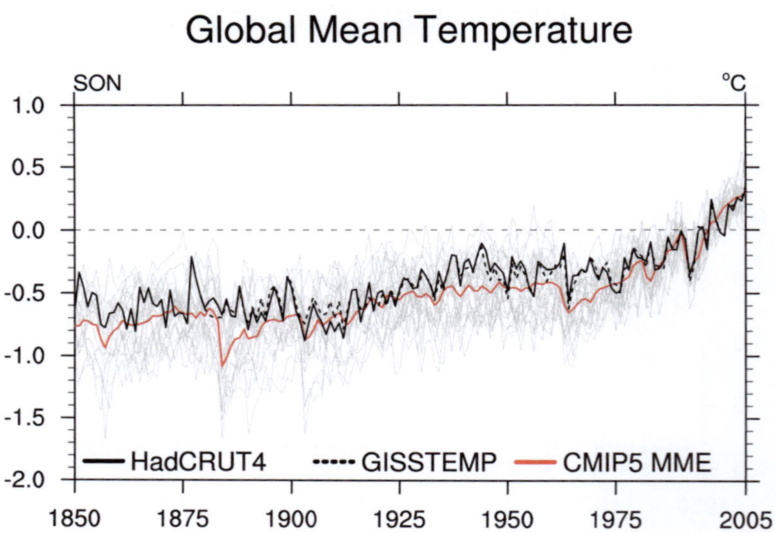

Fig. 2.5 Global mean surface air temperature anomalies for September–November, relative to 1986–2005 climatology from two observations and 22 CMIP5 models (*gray lines*) and their ensemble mean (*red line*) (unit: °C)

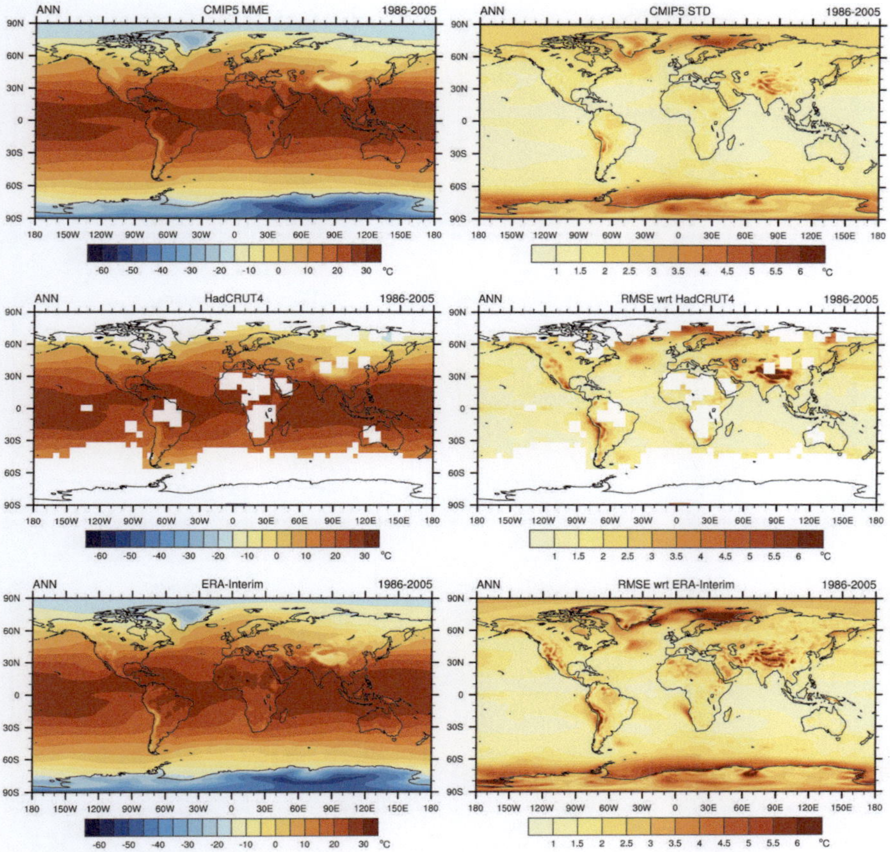

Fig. 2.6 Global annual mean surface air temperature over 1986–2005: multi-model average (*top left*), HadCRUT4 (*middle left*), ERA-Interim (*bottom left*), standard deviation of CMIP5 (*top right*), RMSE with regard to HadCRUT4 (*middle right*), and RMSE with regard to ERA-Interim (*bottom right*) (unit: °C)

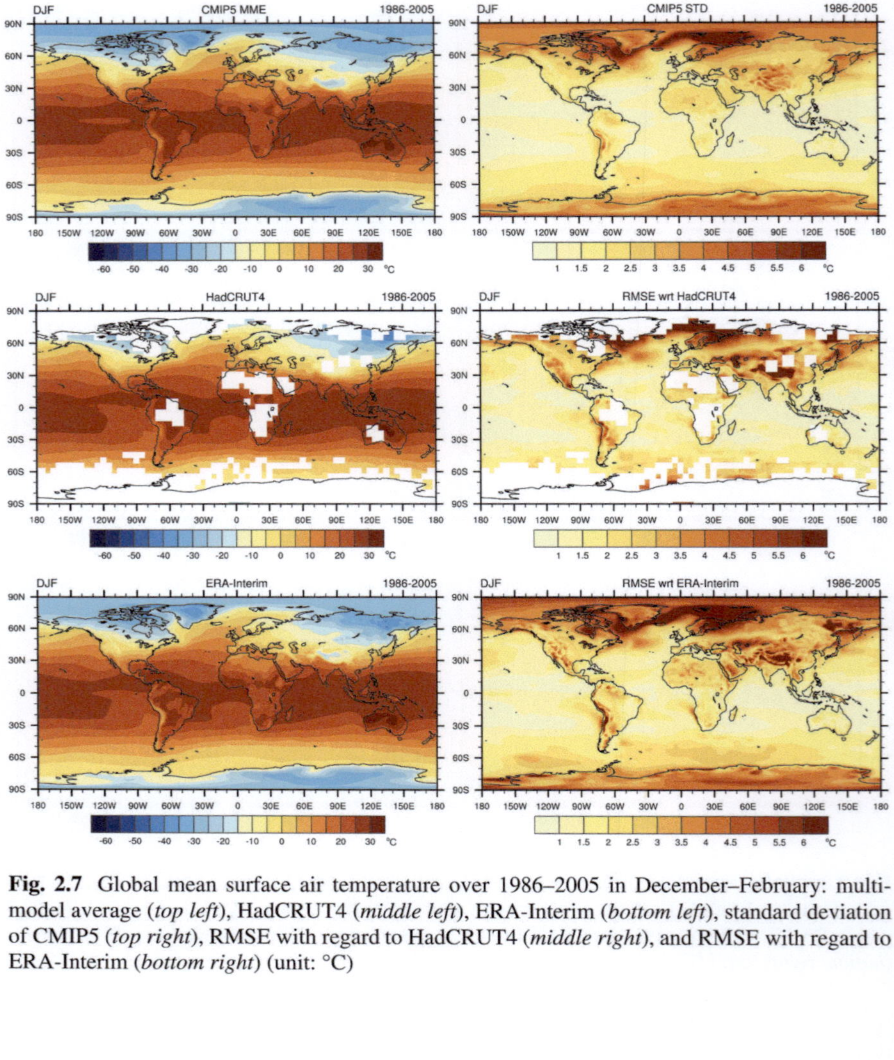

Fig. 2.7 Global mean surface air temperature over 1986–2005 in December–February: multi-model average (*top left*), HadCRUT4 (*middle left*), ERA-Interim (*bottom left*), standard deviation of CMIP5 (*top right*), RMSE with regard to HadCRUT4 (*middle right*), and RMSE with regard to ERA-Interim (*bottom right*) (unit: °C)

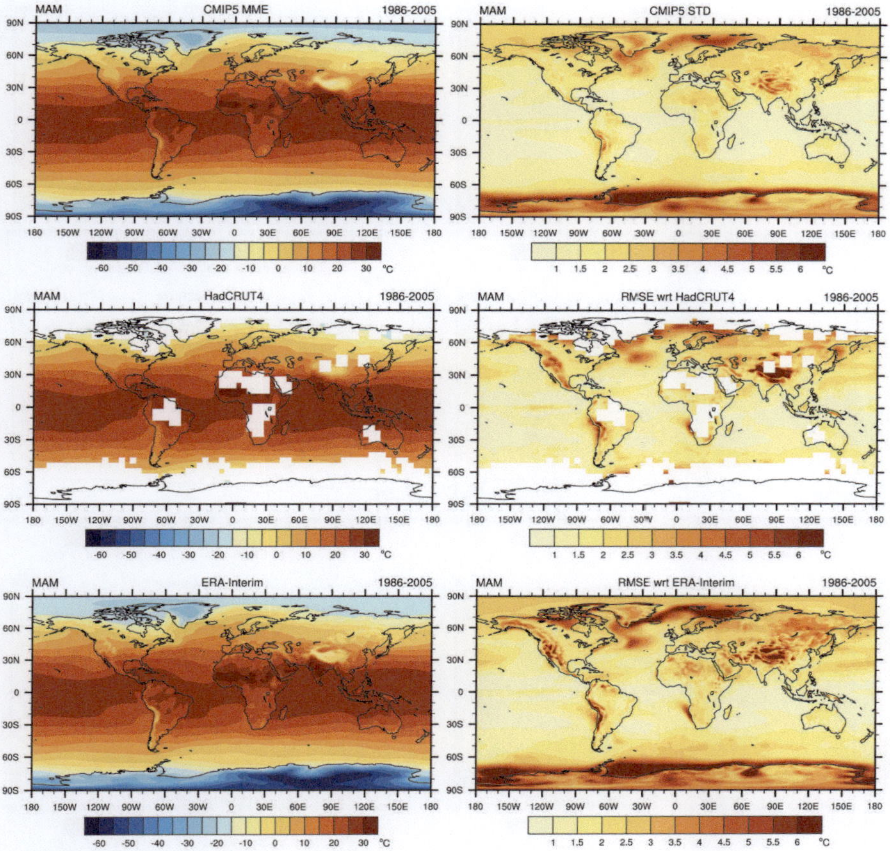

Fig. 2.8 Global mean surface air temperature over 1986–2005 in March–May: multi-model average (*top left*), HadCRUT4 (*middle left*), ERA-Interim (*bottom left*), standard deviation of CMIP5 (*top right*), RMSE with regard to HadCRUT4 (*middle right*), and RMSE with regard to ERA-Interim (*bottom right*) (unit: °C)

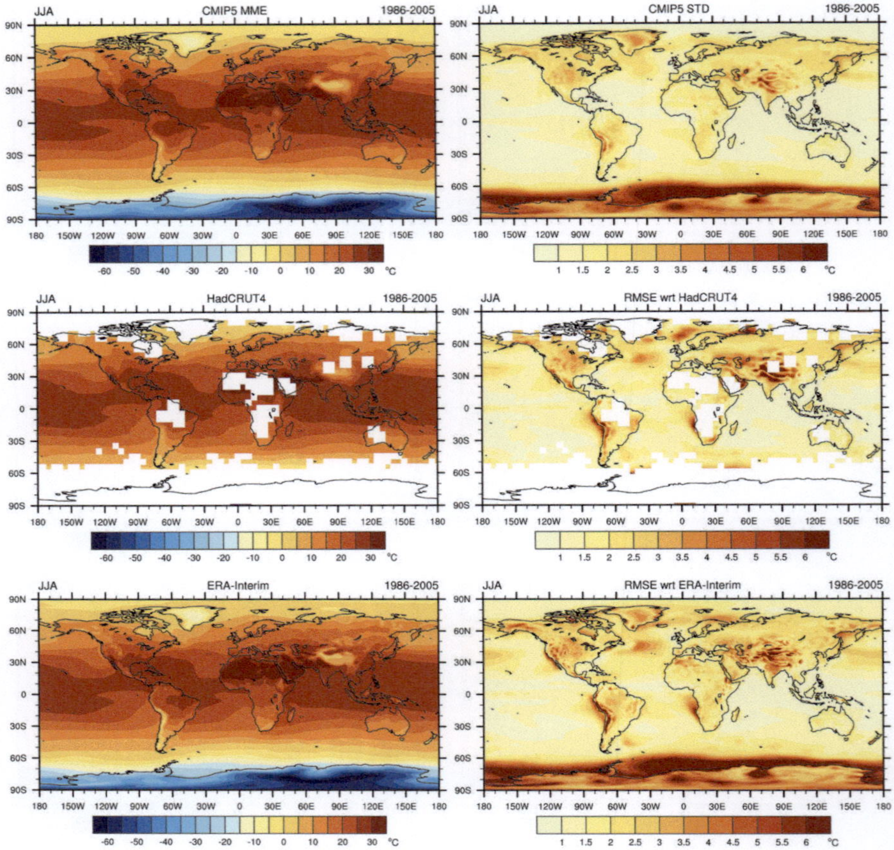

Fig. 2.9 Global mean surface air temperature over 1986–2005 in June–August: multi-model average (*top left*), HadCRUT4 (*middle left*), ERA-Interim (*bottom left*), standard deviation of CMIP5 (*top right*), RMSE with regard to HadCRUT4 (*middle right*), and RMSE with regard to ERA-Interim (*bottom right*) (unit: °C)

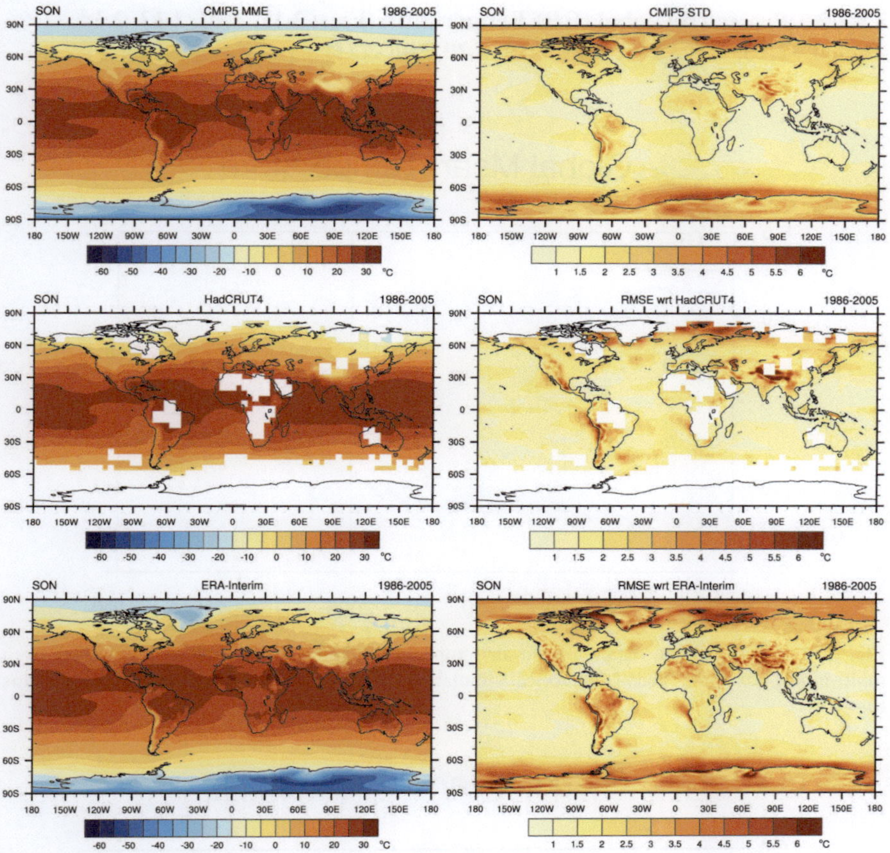

Fig. 2.10 Global mean surface air temperature over 1986–2005 in September–November: multi-model average (*top left*), HadCRUT4 (*middle left*), ERA-Interim (*bottom left*), standard deviation of CMIP5 (*top right*), RMSE with regard to HadCRUT4 (*middle right*), and RMSE with regard to ERA-Interim (*bottom right*) (unit: °C)

2.1.1.2 Projected Global (GP) Temperature (Figs. 2.11, 2.12, 2.13, 2.14, 2.15, 2.16, 2.17, 2.18, 2.19, 2.20, and 2.21)

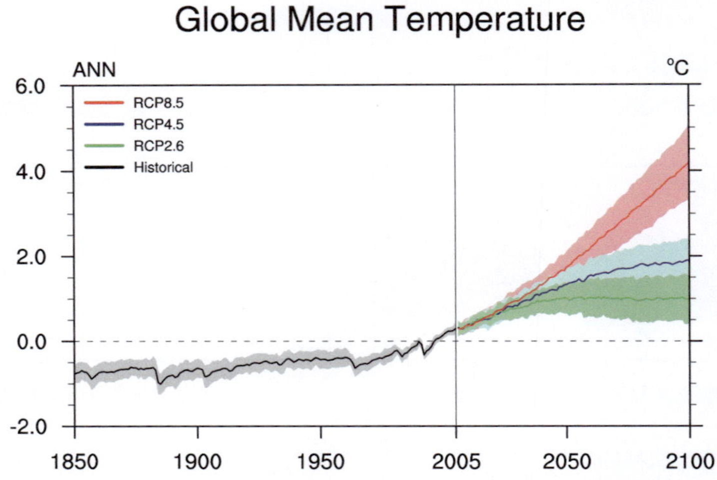

Fig. 2.11 Global annual mean surface air temperature anomalies relative to 1986–2005 from multi-model ensemble mean of historical simulations and projections under three RCP scenarios (unit: °C)

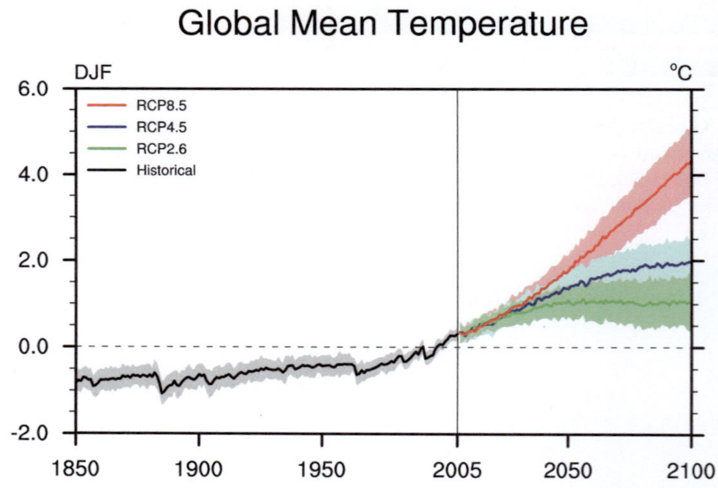

Fig. 2.12 Global mean surface air temperature anomalies relative to 1986–2005 in December–February from multi-model ensemble mean of historical simulations and projections under three RCP scenarios (unit: °C)

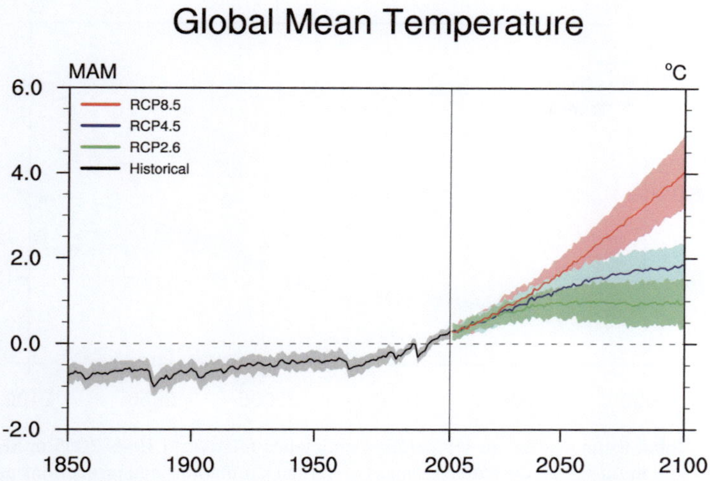

Fig. 2.13 Global mean surface air temperature anomalies relative to 1986–2005 in March–May from multi-model ensemble mean of historical simulations and projections under three RCP scenarios (unit: °C)

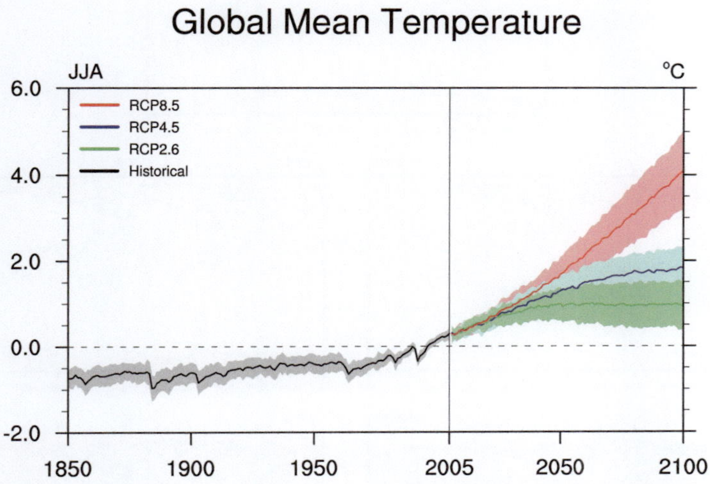

Fig. 2.14 Global mean surface air temperature anomalies relative to 1986–2005 in June–August from multi-model ensemble mean of historical simulations and projections under three RCP scenarios (unit: °C)

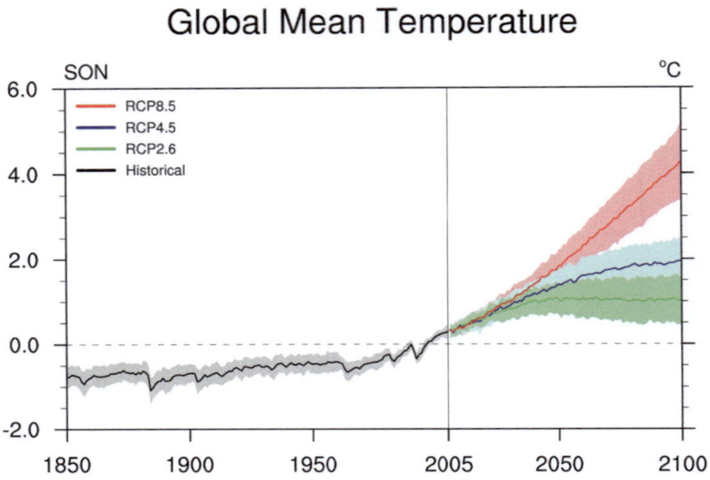

Fig. 2.15 Global mean surface air temperature anomalies relative to 1986–2005 in September–November from multi-model ensemble mean of historical simulations and projections under three RCP scenarios (unit: °C)

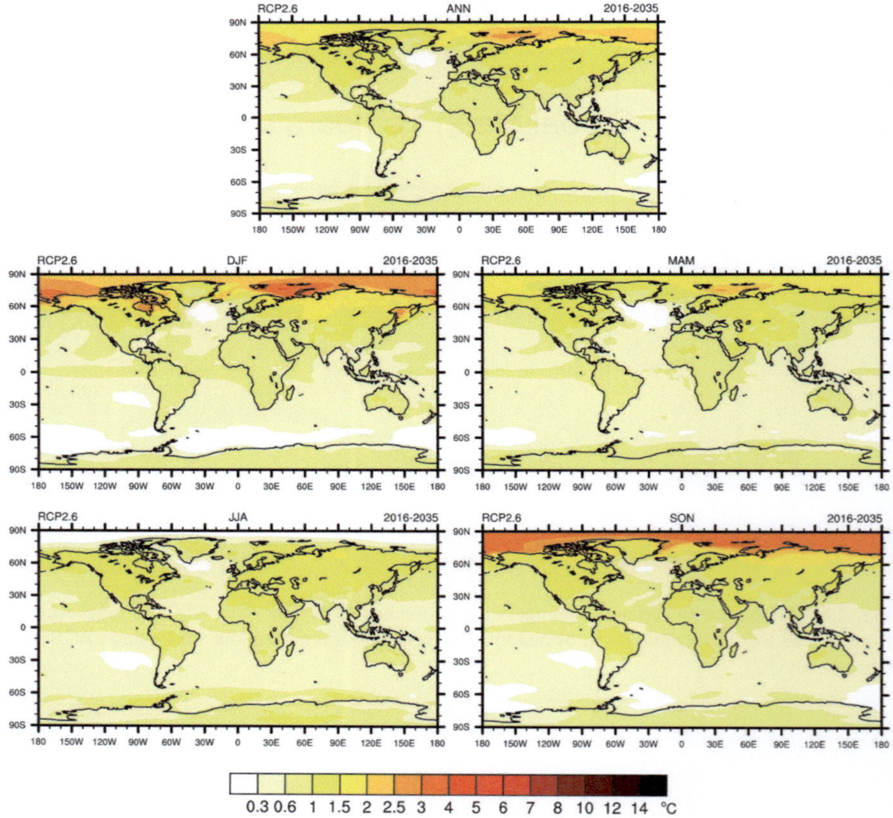

Fig. 2.16 Changes in surface air temperature in 2016–2035 under RCP2.6 relative to the climatology in 1986–2005: annual mean (*top*), December–February mean (*middle left*), March–May mean (*middle right*), June–August mean (*bottom left*), and September–November mean (*bottom right*) (unit: °C)

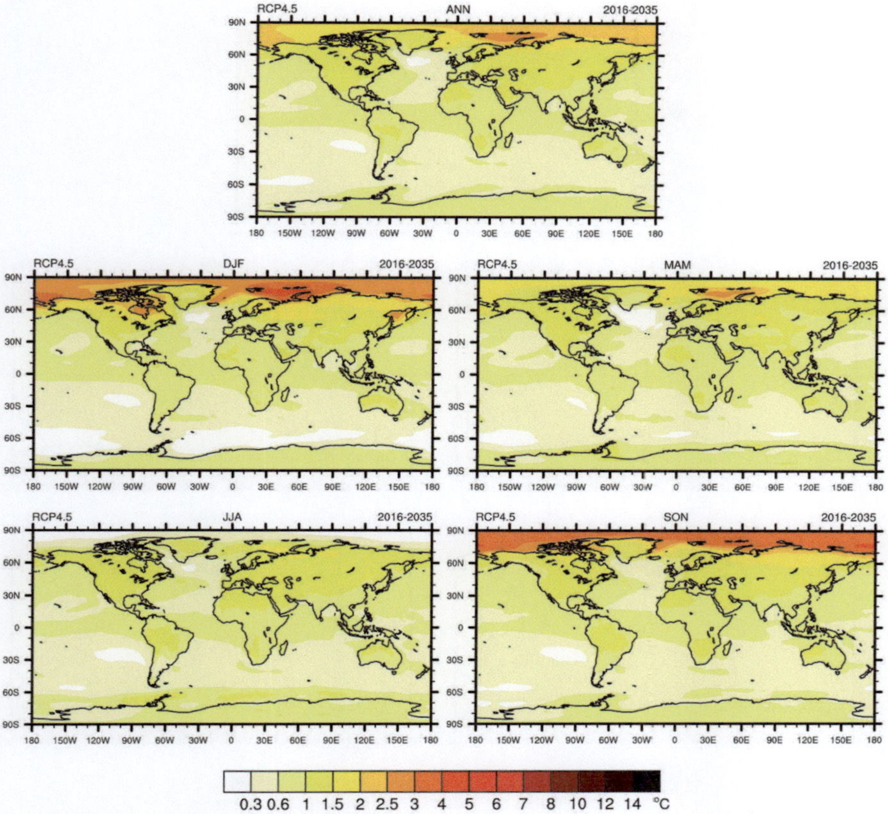

Fig. 2.17 Changes in surface air temperature in 2016–2035 under RCP4.5 relative to the climatology in 1986–2005: annual mean (*top*), December–February mean (*middle left*), March–May mean (*middle right*), June–August mean (*bottom left*), and September–November mean (*bottom right*) (unit: °C)

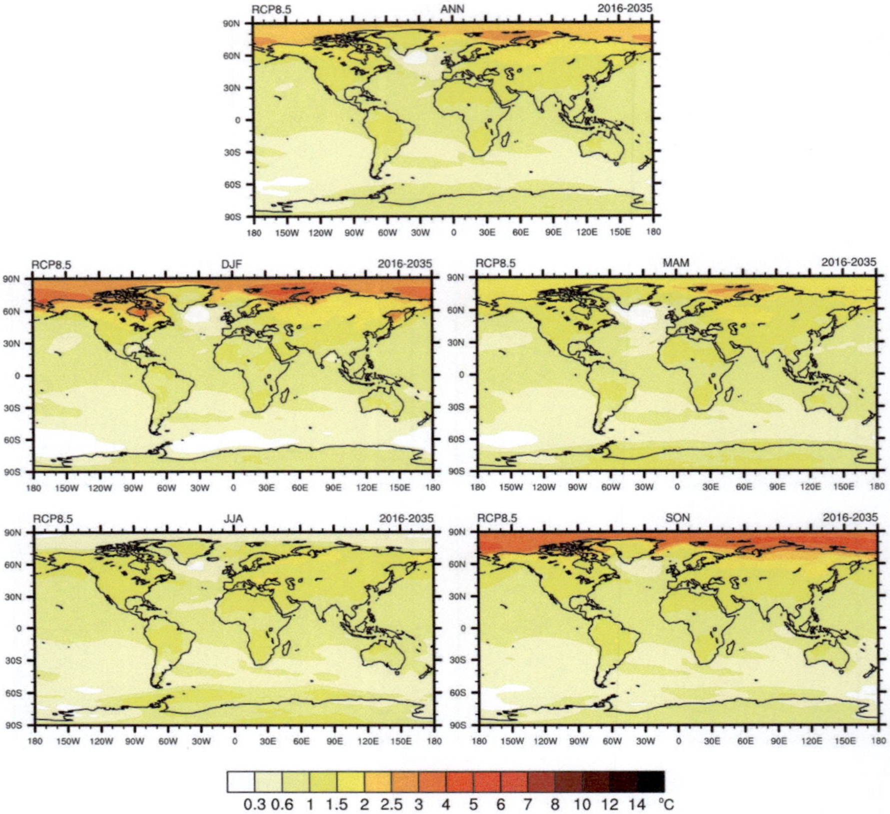

Fig. 2.18 Changes in surface air temperature in 2016–2035 under RCP8.5 relative to the climatology in 1986–2005: annual mean (*top*), December–February mean (*middle left*), March–May mean (*middle right*), June–August mean (*bottom left*), and September–November mean (*bottom right*) (unit: °C)

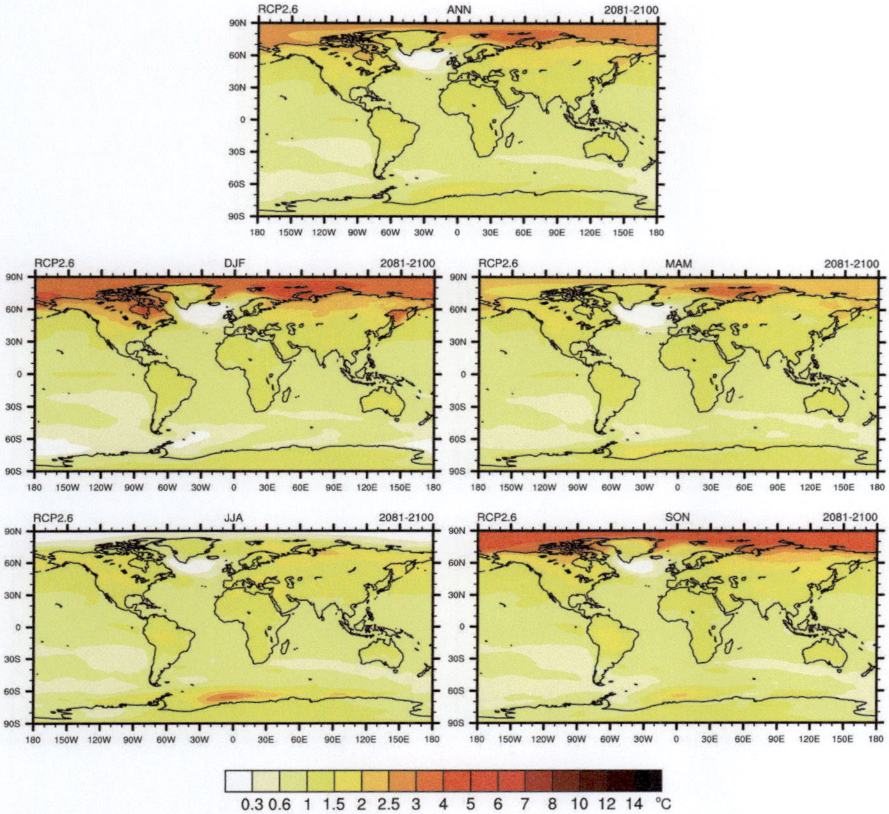

Fig. 2.19 Changes in surface air temperature in 2081–2100 under RCP2.6 relative to the climatology in 1986–2005: annual mean (*top*), December–February mean (*middle left*), March–May mean (*middle right*), June–August mean (*bottom left*), and September–November mean (*bottom right*) (unit: °C)

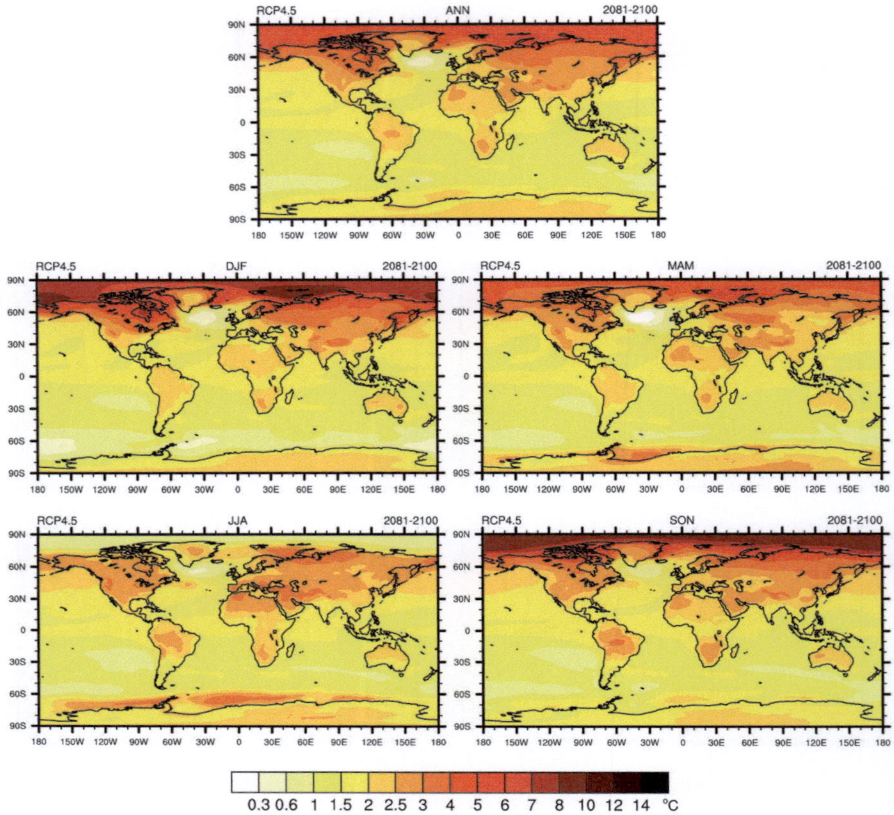

Fig. 2.20 Changes in surface air temperature in 2081–2100 under RCP4.5 relative to the climatology in 1986–2005: annual mean (*top*), December–February mean (*middle left*), March–May mean (*middle right*), June–August mean (*bottom left*), and September–November mean (*bottom right*) (unit: °C)

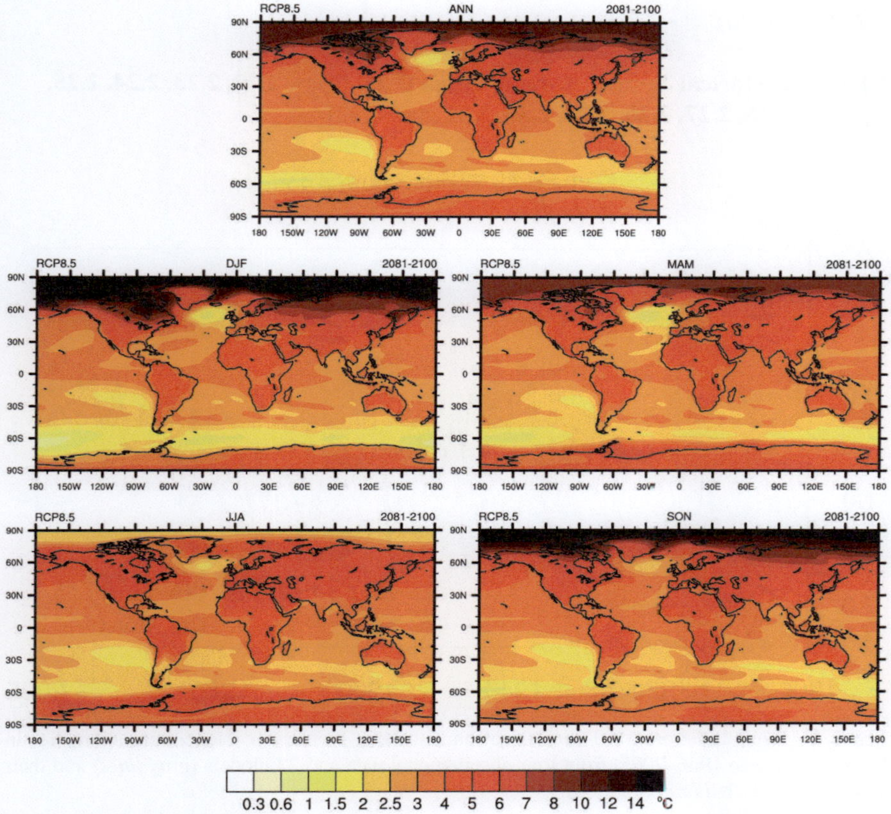

Fig. 2.21 Changes in surface air temperature in 2081–2100 under RCP8.5 relative to the climatology in 1986–2005: annual mean (*top*), December–February mean (*middle left*), March–May mean (*middle right*), June–August mean (*bottom left*), and September–November mean (*bottom right*) (unit: °C)

2.1.2 China

2.1.2.1 Historical Temperature in China (CH) (Figs. 2.22, 2.23, 2.24, 2.25, 2.26, 2.27, 2.28, 2.29, 2.30, 2.31, and 2.32)

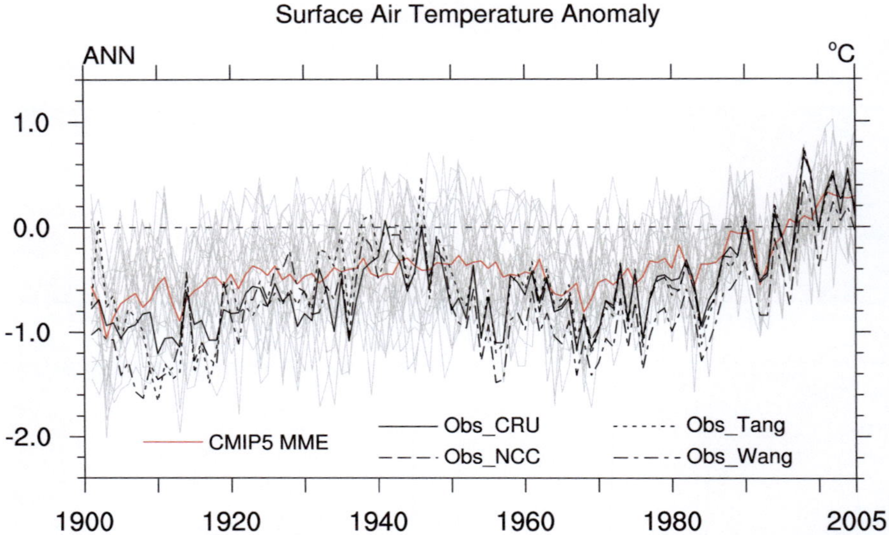

Fig. 2.22 Evolution from 1901 to 2005 of annual mean surface air temperature anomalies in China, relative to 1986–2005, from four observational data and 22 models (*gray lines*) and their ensemble mean (*red line*) (unit: °C)

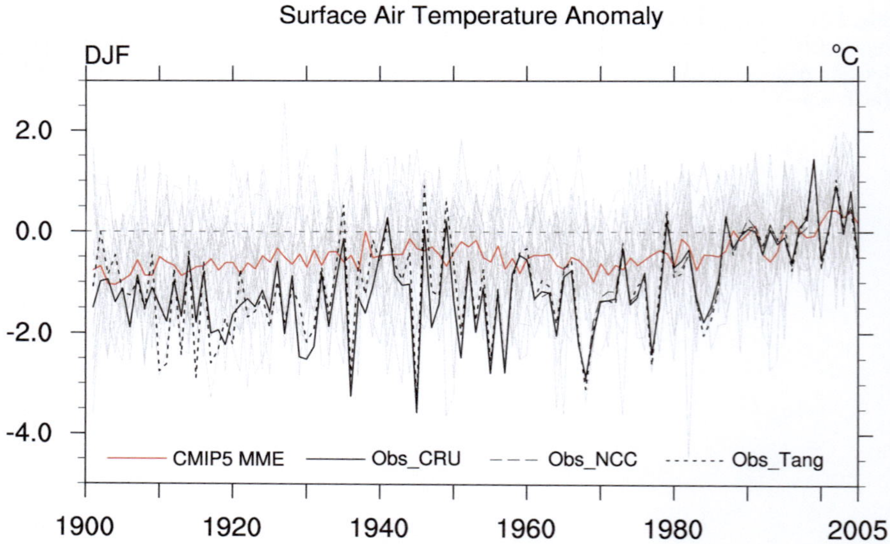

Fig. 2.23 Evolution from 1901 to 2005 of winter (December–February) surface air temperature anomalies in China, relative to 1986–2005, from three observational data and 22 models (*gray lines*) and their ensemble mean (*red line*) (unit: °C)

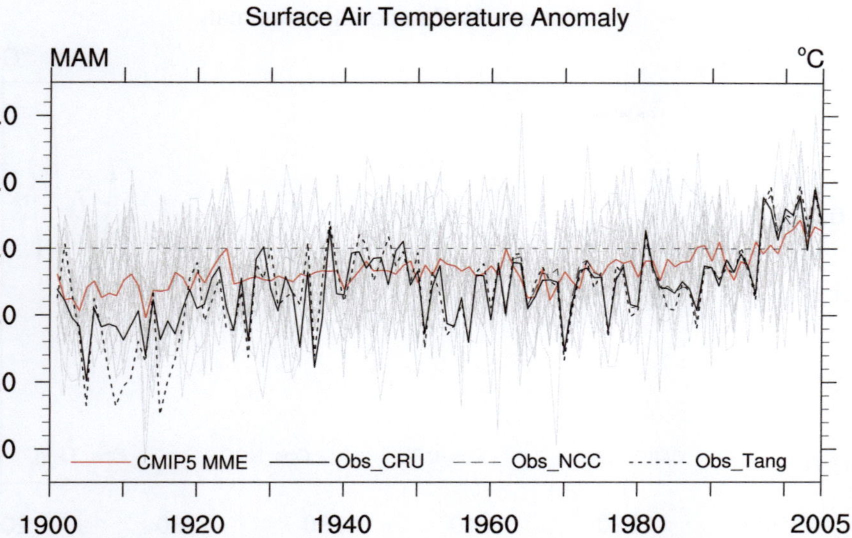

Fig. 2.24 Evolution from 1901 to 2005 of spring (March–May) surface air temperature anomalies in China, relative to 1986–2005, from three observational data and 22 models (gray lines) and their ensemble mean (red line) (unit: °C)

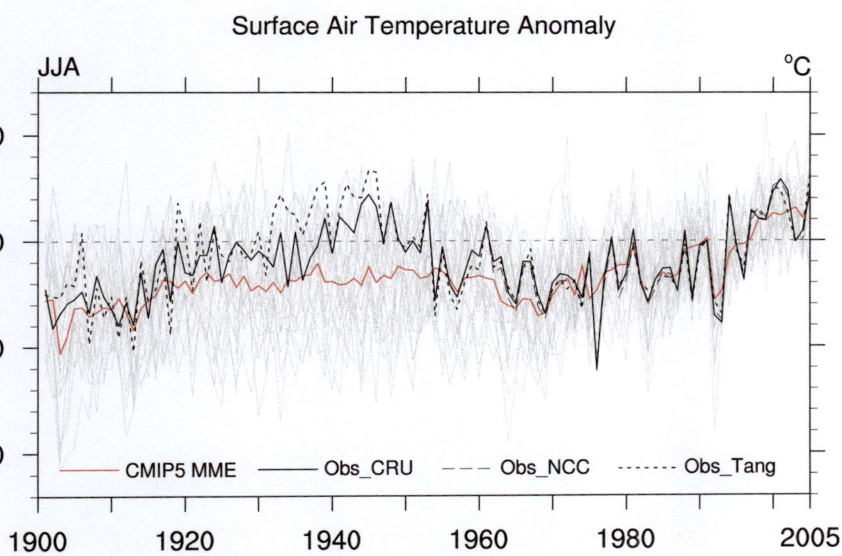

Fig. 2.25 Evolution from 1901 to 2005 of summer (June–August) surface air temperature anomalies in China, relative to 1986–2005, from three observational data and 22 models (*gray lines*) and their ensemble mean (*red line*) (unit: °C)

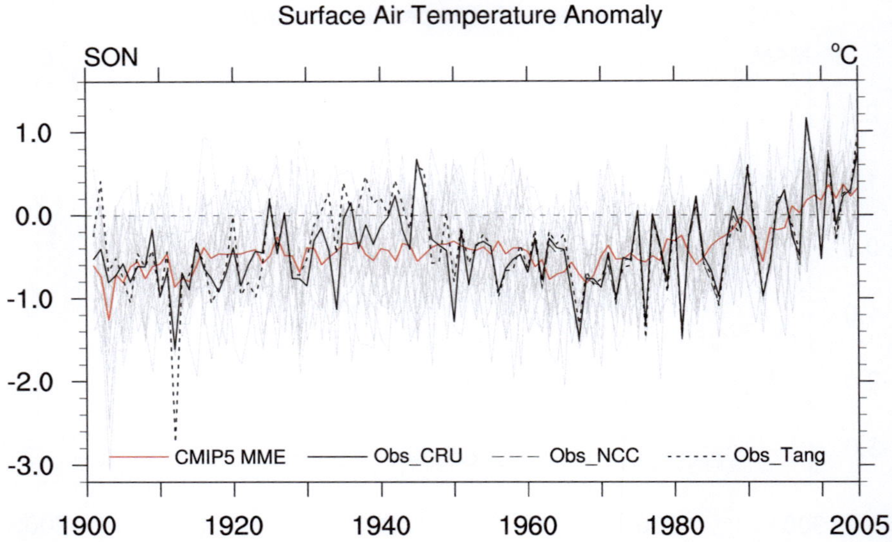

Fig. 2.26 Evolution from 1901 to 2005 of autumn (September–November) surface air temperature anomalies in China, relative to 1986–2005, from three observational data and 22 models (*gray lines*) and their ensemble mean (*red line*) (unit: °C)

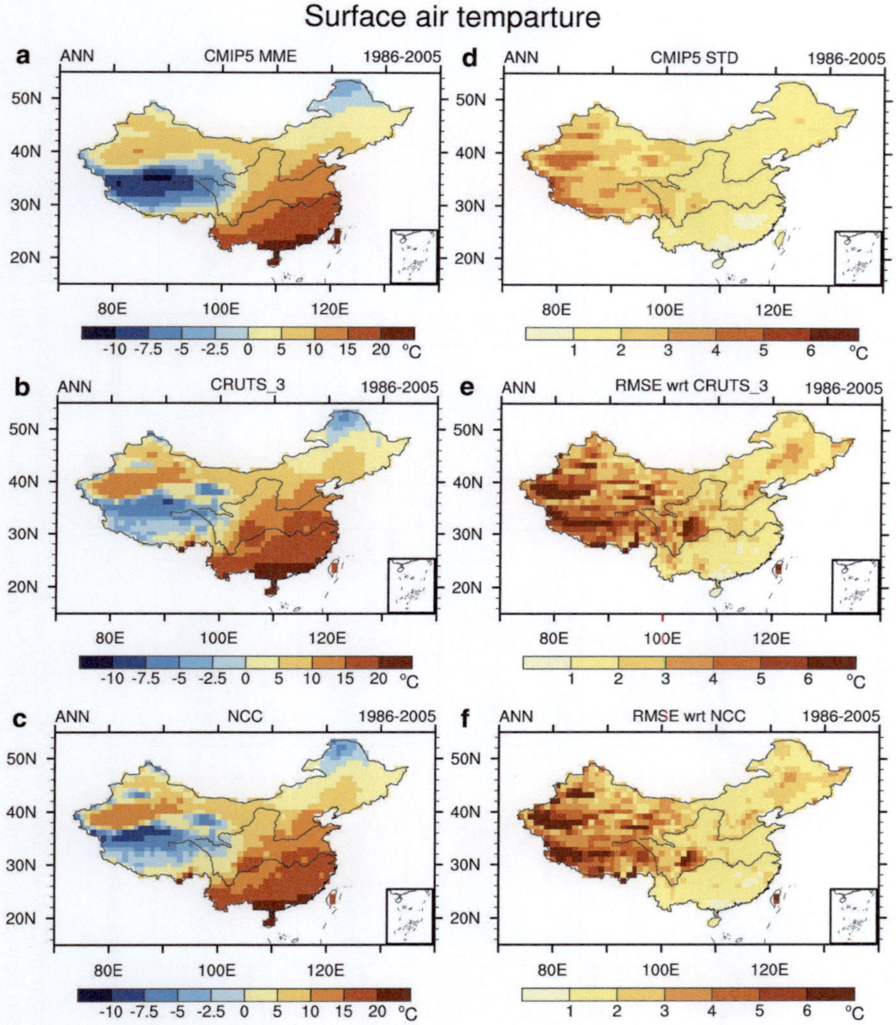

Fig. 2.27 Annual mean surface air temperature in China averaged over 1986–2005: (**a**) multi-model ensemble mean, (**b**) observation (CRUTS_3), (**c**) observation (NCC). (**d**) Inter-model standard deviation among 22 models. RMSE between model simulations and observation (**e**) CRUTS_3, (**f**) NCC (unit: °C)

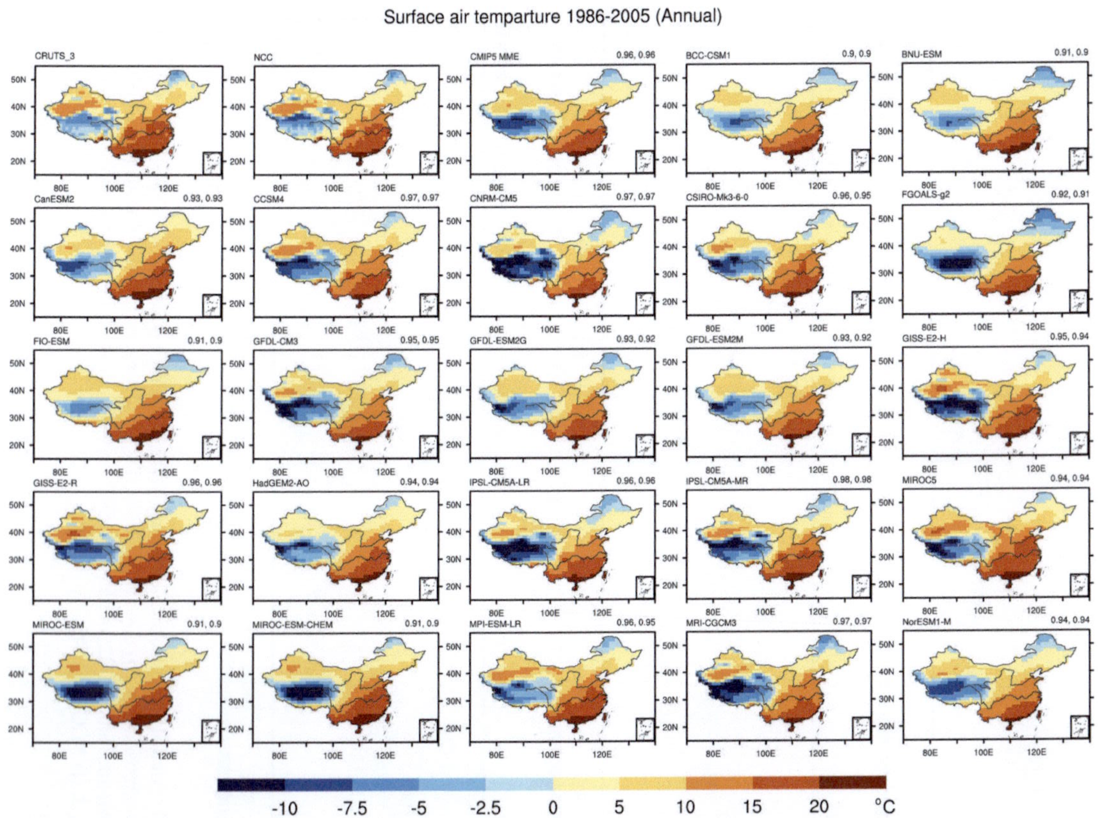

Fig. 2.28 Annual mean surface air temperature in China averaged over 1986–2005 from observations, multi-model ensemble mean, and 22 individual models (unit: °C). The spatial correlation coefficients between observations (CRUTS_3, NCC) and model simulation are listed at the top right corner

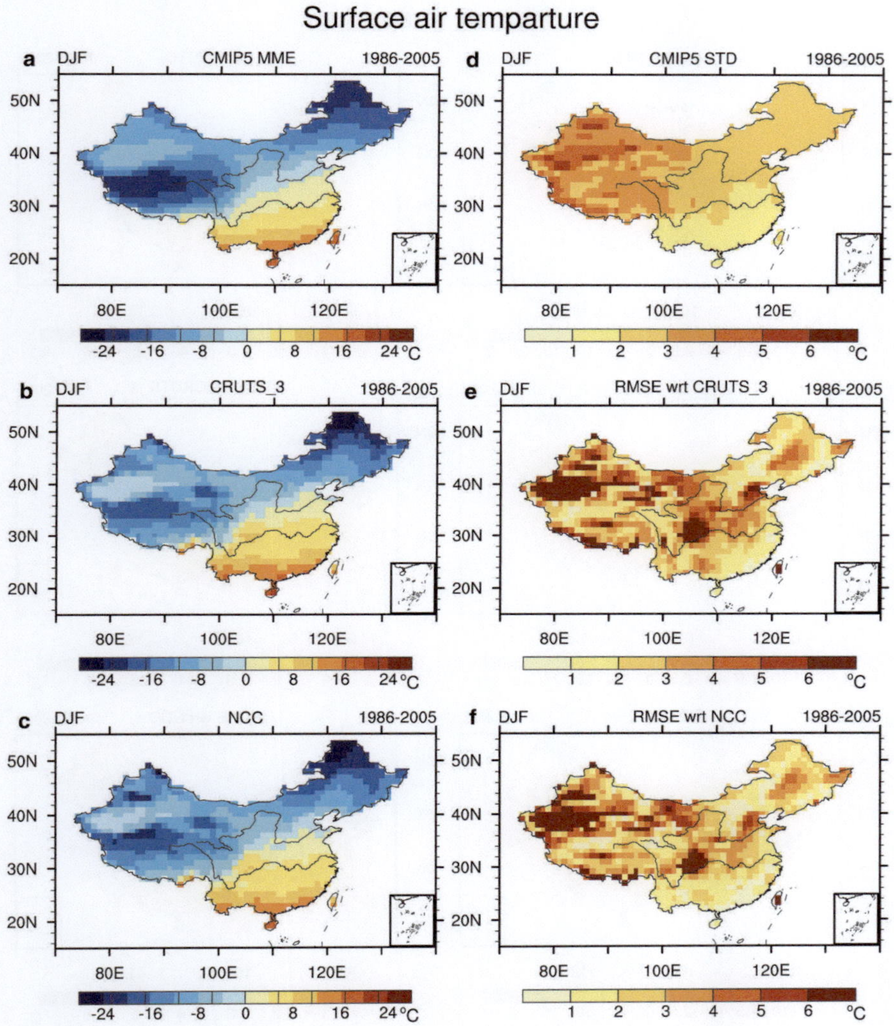

Fig. 2.29 Surface air temperature in China for December–February averaged over 1986–2005: (**a**) multi-model ensemble mean, (**b**) observation (CRU), (**c**) observation (NCC). (**d**) Inter-model standard deviation among 22 models. RMSE between model simulations and observation (**e**) CRUTS_3, (**f**) NCC (unit: °C)

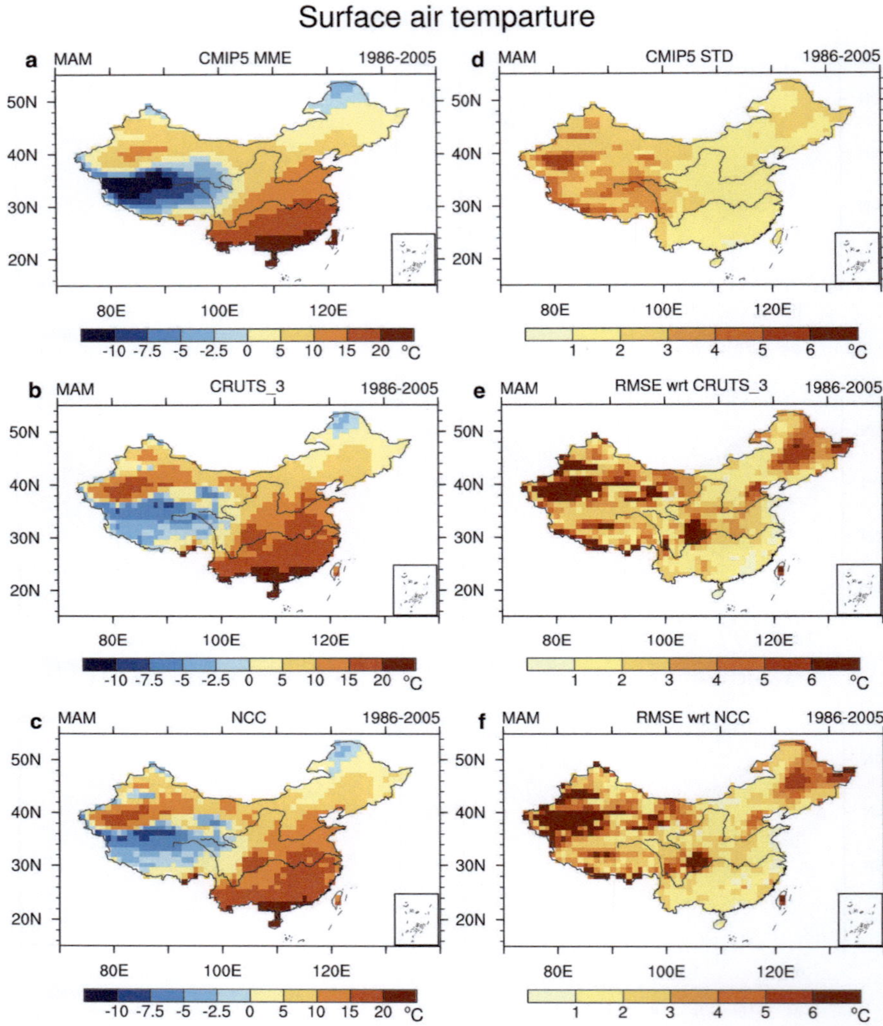

Fig. 2.30 Surface air temperature in China for March–May averaged over 1986–2005: (**a**) multi-model ensemble mean, (**b**) observation (CRUTS_3), (**c**) observation (NCC). (**d**) Inter-model standard deviation among 22 models. RMSE between model simulations and observation (**e**) CRUTS_3, (**f**) NCC (unit: °C)

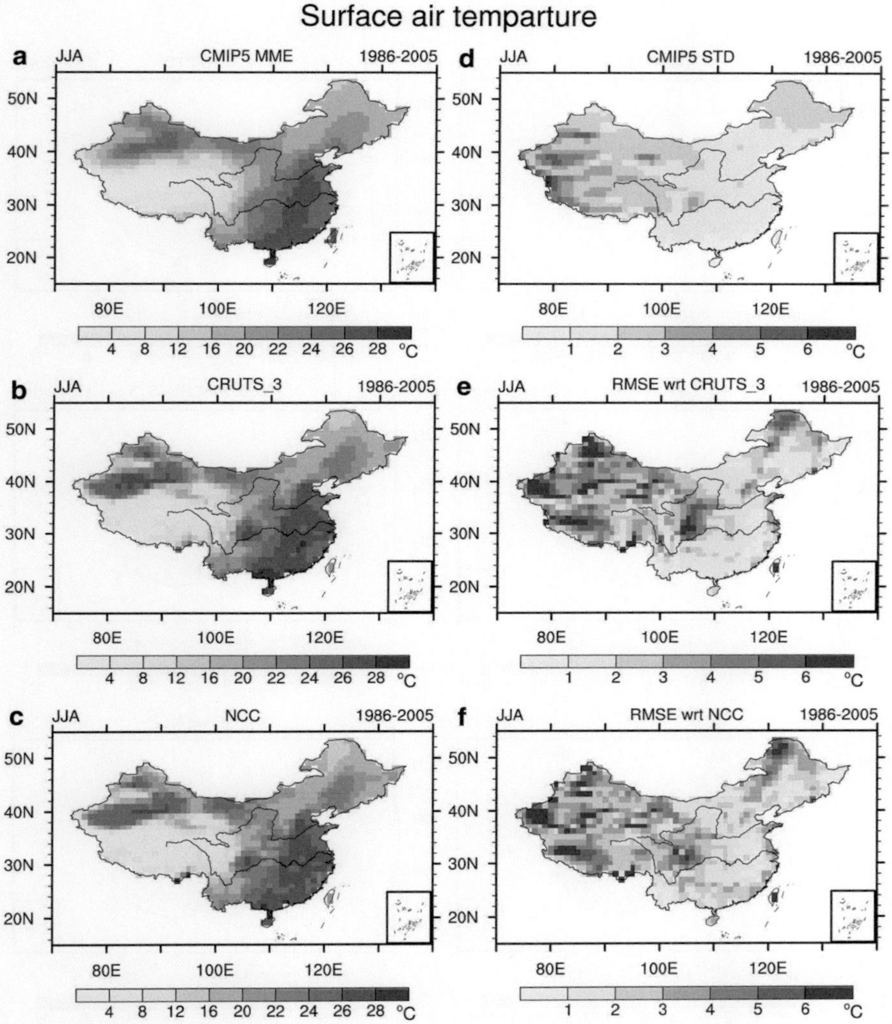

Fig. 2.31 Surface air temperature in China for June–August averaged over 1986–2005: (**a**) multi-model ensemble mean, (**b**) observation (CRUTS_3), (**c**) observation (NCC). (**d**) Inter-model standard deviation among 22 models. RMSE between model simulations and observation (**e**) CRUTS_3, (**f**) NCC (unit: °C)

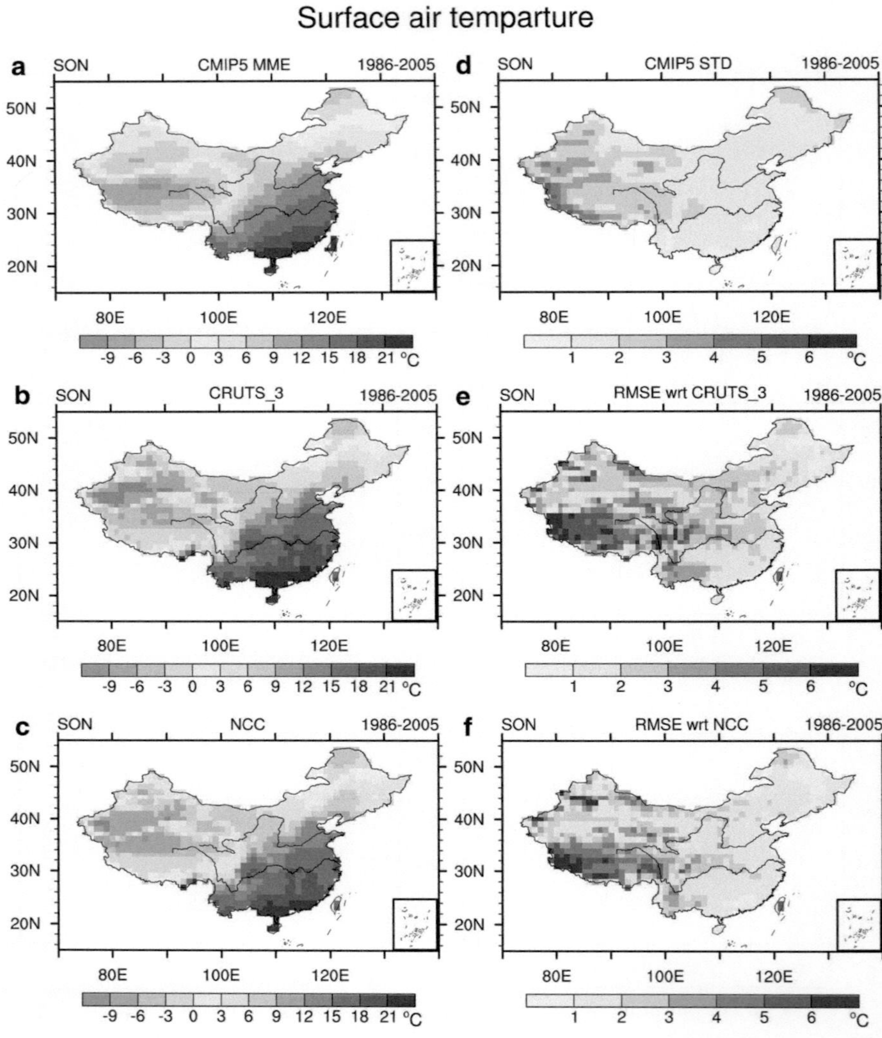

Fig. 2.32 Surface air temperature in China for September–November averaged over 1986–2005: (**a**) multi-model ensemble mean, (**b**) observation (CRUTS_3), (**c**) observation (NCC). (**d**) Inter-model standard deviation among 22 models. RMSE between model simulations and observation (**e**) CRUTS_3, (**f**) NCC (unit: °C)

2.1.2.2 Projected Temperature in China (CP) (Figs. 2.33, 2.34, 2.35, 2.36, 2.37, 2.38, 2.39, 2.40, 2.41, 2.42, 2.43, 2.44, 2.45, 2.46, and 2.47)

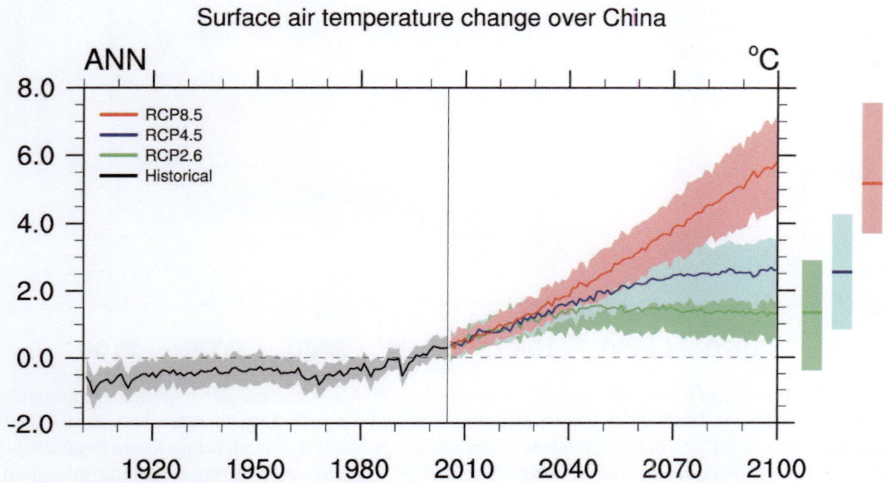

Fig. 2.33 Evolution from 1901 to 2100 of annual mean surface air temperature anomalies in China relative to 1986–2005, from multi-model ensemble mean. *Shadow* represents inter-model standard deviation. The band in the box (*right side*) represents the mean temperature over 2081–2100 from multi-model ensemble mean. The *bottom* (*top*) of the box represents the minimum (maximum) mean temperature over 2081–2100 among multi-model (unit: °C)

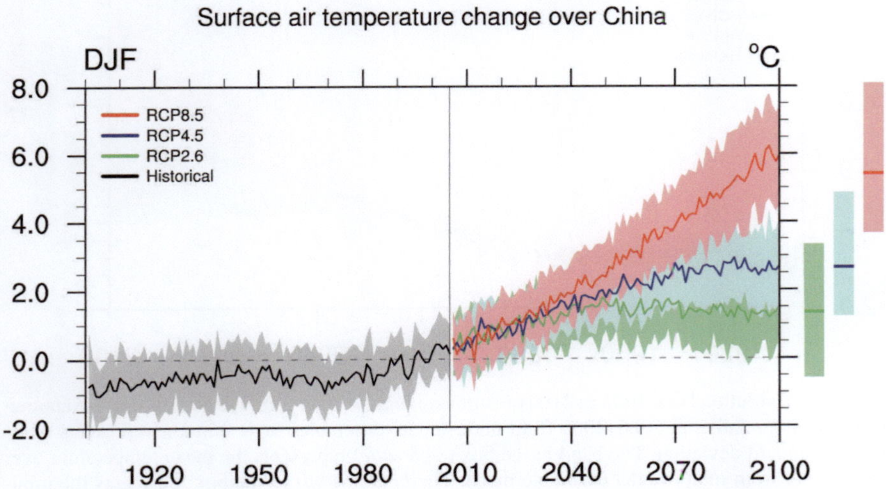

Fig. 2.34 Evolution from 1901 to 2100 of winter (December–February) surface air temperature anomalies in China relative to 1986–2005, from multi-model ensemble mean. *Shadow* represents inter-model standard deviation. The band in the box (*right side*) represents the mean temperature over 2081–2100 from multi-model ensemble mean. The *bottom* (*top*) of the box represents the minimum (maximum) mean temperature over 2081–2100 among multi-model (unit: °C)

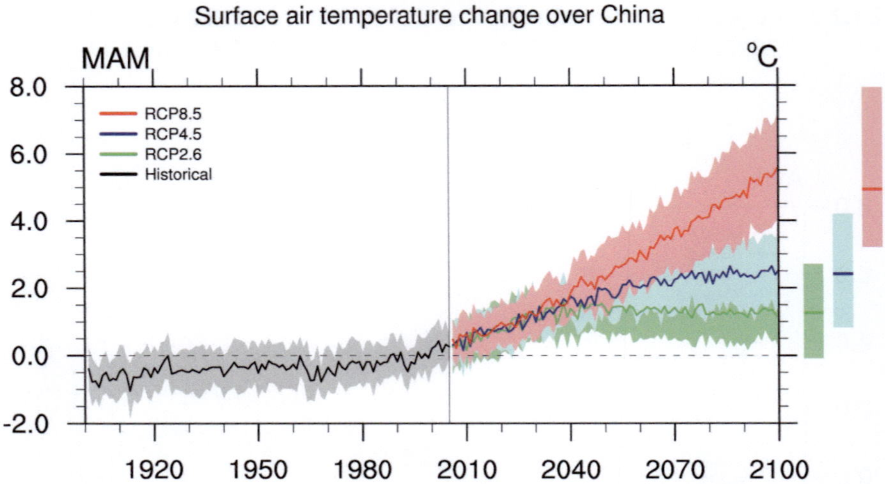

Fig. 2.35 Evolution from 1901 to 2100 of spring (March–May) surface air temperature anomalies in China relative to 1986–2005 from multi-model ensemble mean. *Shadow* represents inter-model standard deviation. The band in the box (*right side*) represents the mean temperature over 2081–2100 from multi-model ensemble mean. The *bottom* (*top*) of the box represents the minimum (maximum) mean temperature over 2081–2100 among multi-model (unit: °C)

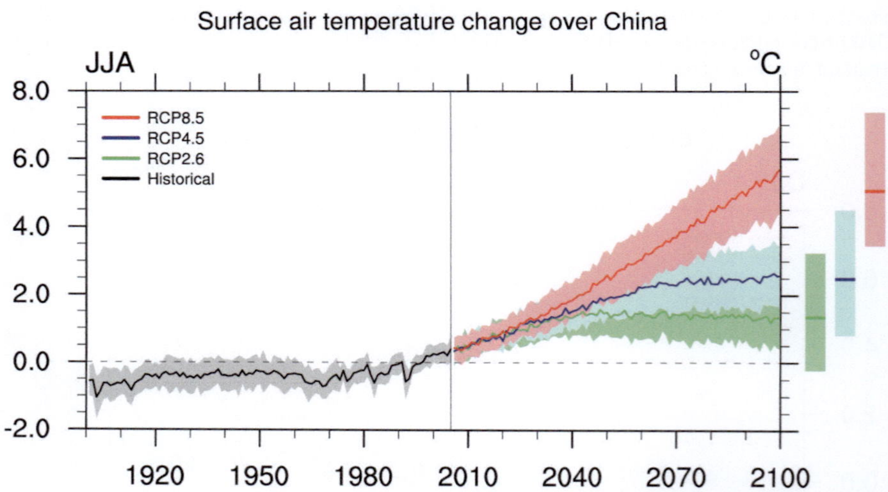

Fig. 2.36 Evolution from 1901 to 2100 of summer (June–August) surface air temperature anomalies in China relative to 1986–2005 from multi-model ensemble mean. *Shadow* represents inter-model standard deviation. The band in the box (*right side*) represents the mean temperature over 2081–2100 from multi-model ensemble mean. The *bottom* (*top*) of the box represents the minimum (maximum) mean temperature over 2081–2100 among multi-model (unit: °C)

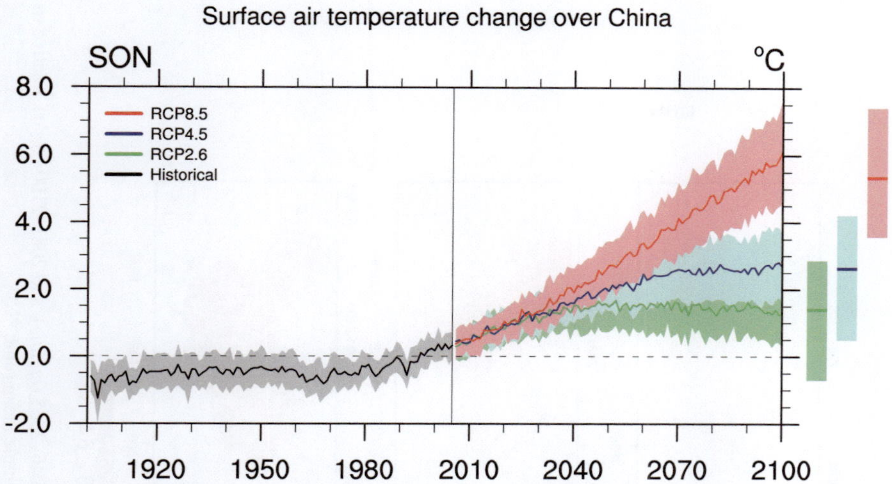

Fig. 2.37 Evolution from 1901 to 2100 of autumn (September–November) surface air temperature anomalies in China relative to 1986–2005 from multi-model ensemble mean. *Shadow* represents inter-model standard deviation. The band in the box (*right side*) represents the mean temperature over 2081–2100 from multi-model ensemble mean. The *bottom* (*top*) of the box represents the minimum (maximum) mean temperature over 2081–2100 among multi-model (unit: °C)

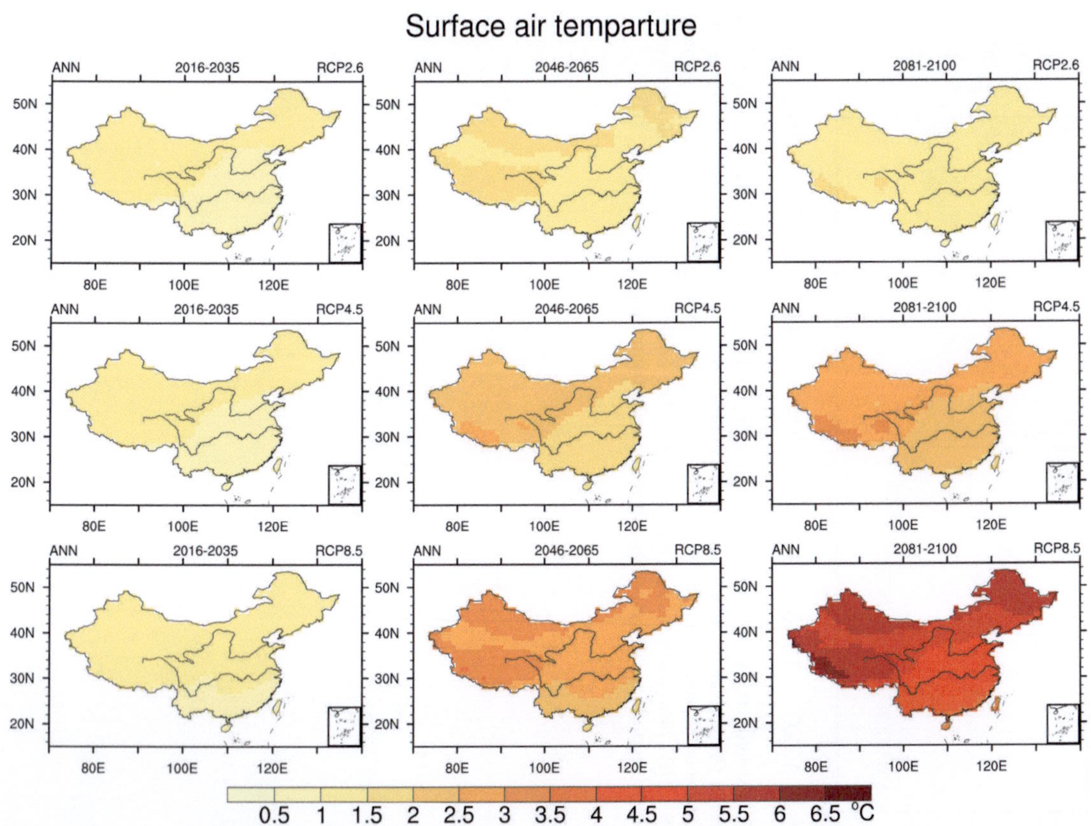

Fig. 2.38 Changes in annual mean surface air temperature in China averaged over 2016–2035 (*left*), 2046–2065 (*middle*), and 2081–2100 (*right*), relative to 1986–2005, for RCP2.6 (*top*), RCP4.5 (*middle*), and RCP8.5 (*bottom*) scenarios from multi-model ensemble mean (unit: °C)

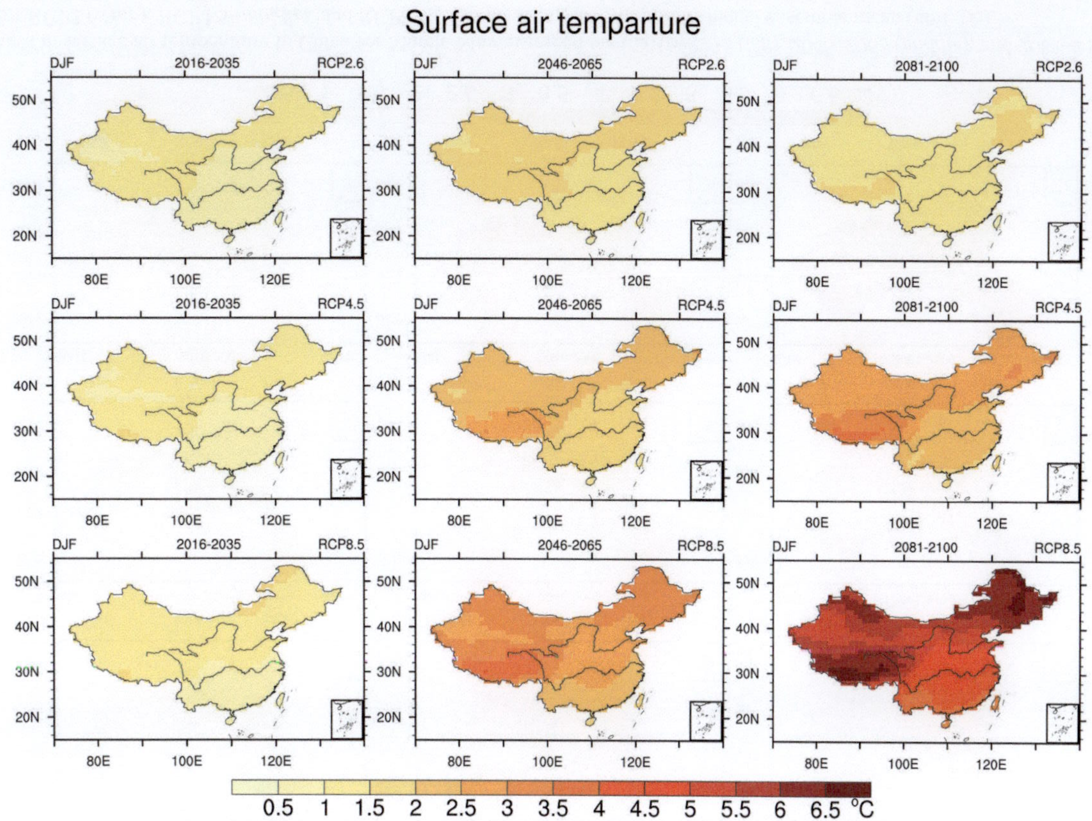

Fig. 2.39 Changes in surface air temperature in China for December–February averaged over 2016–2035 (*left*), 2046–2065 (*middle*), and 2081–2100 (*right*), relative to 1986–2005, for RCP2.6 (*top*), RCP4.5 (*middle*), and RCP8.5 (*bottom*) scenarios from multi-model ensemble mean (unit: °C)

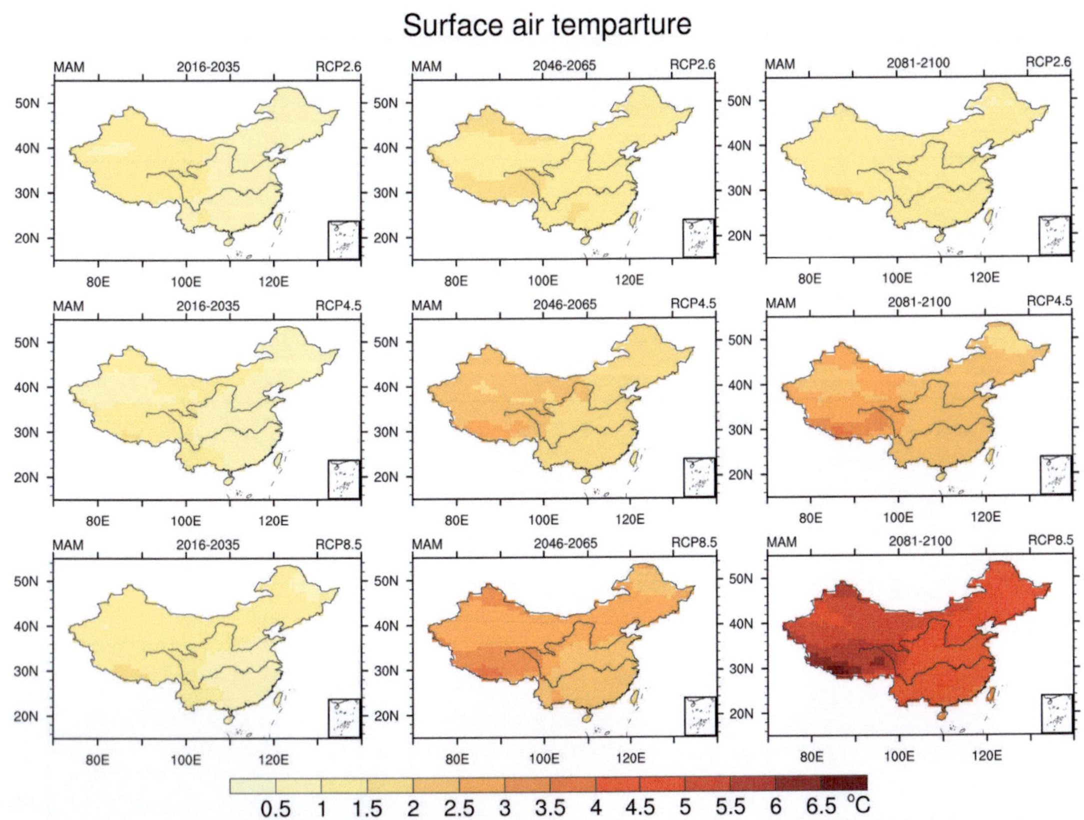

Fig. 2.40 Changes in surface air temperature in China for March–May averaged over 2016–2035 (*left*), 2046–2065 (*middle*), and 2081–2100 (*right*), relative to 1986–2005, for RCP2.6 (*top*), RCP4.5 (*middle*), and RCP8.5 (*bottom*) scenarios from multi-model ensemble mean (unit: °C)

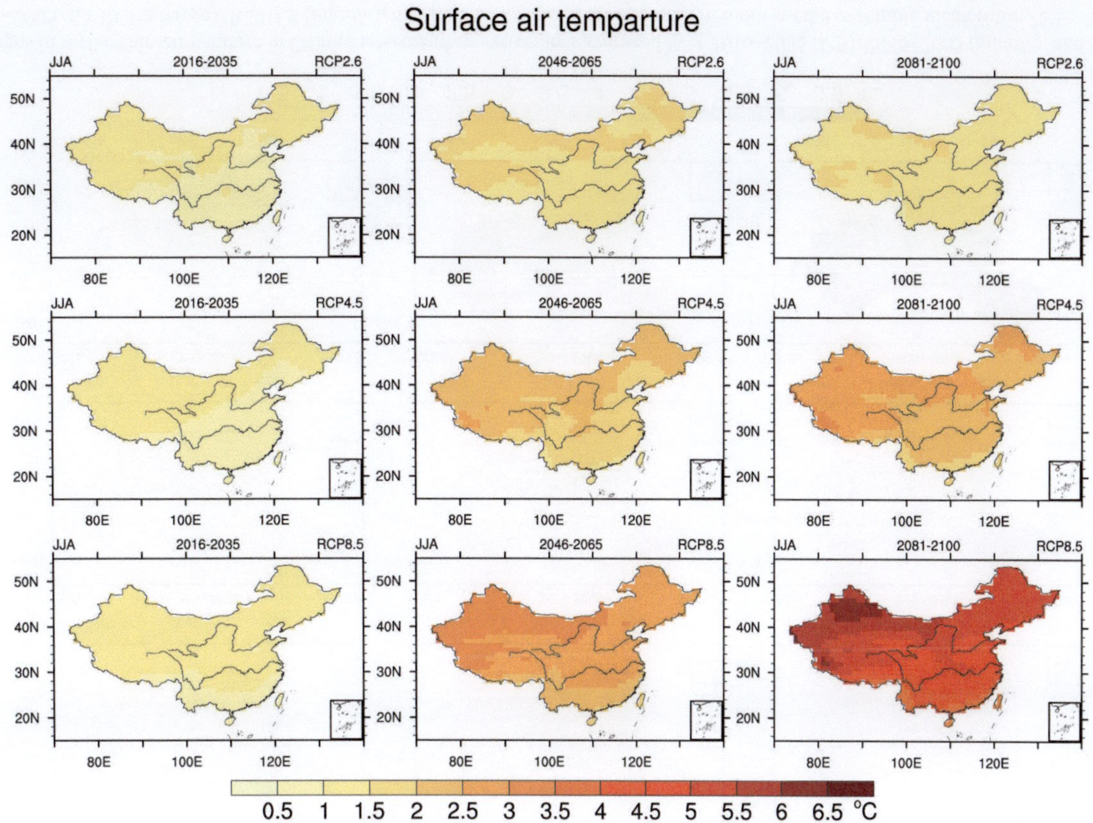

Fig. 2.41 Changes in surface air temperature in China for June–August averaged over 2016–2035 (*left*), 2046–2065 (*middle*), and 2081–2100 (*right*), relative to 1986–2005, for RCP2.6 (*top*), RCP4.5 (*middle*), and RCP8.5 (*bottom*) scenarios from multi-model ensemble mean (unit: °C)

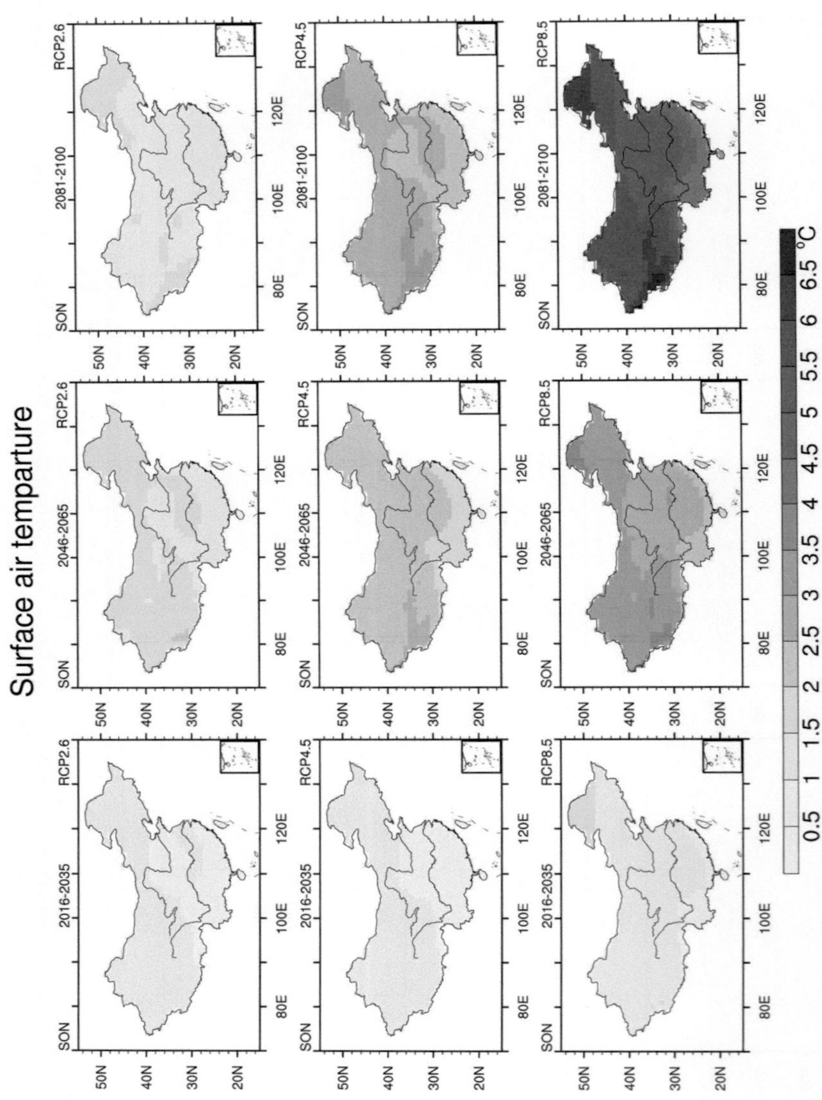

Fig. 2.42 Changes in surface air temperature in China for September–November averaged over 2016–2035 (*left*), 2046–2065 (*middle*), and 2081–2100 (*right*), relative to 1986–2005, for RCP2.6 (*top*), RCP4.5 (*middle*), and RCP8.5 (*bottom*) scenarios from multi-model ensemble mean (unit: °C)

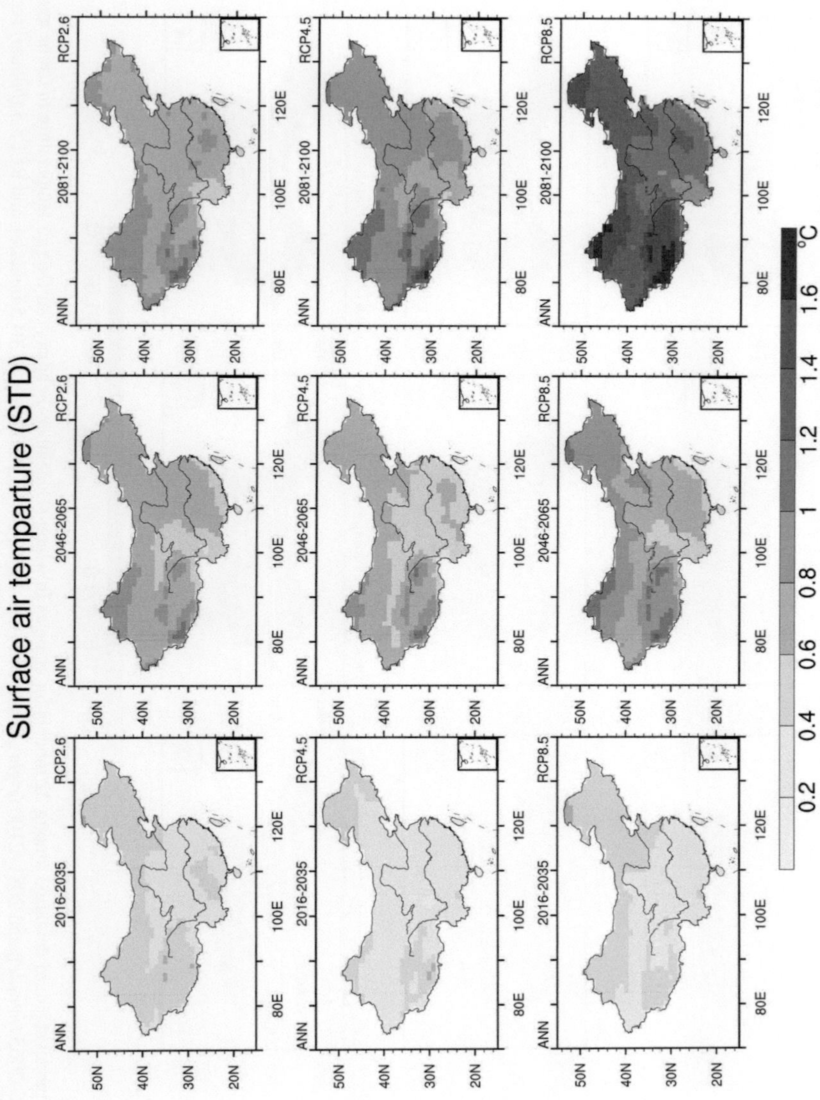

Fig. 2.43 Inter-model standard deviation among 22 models for changes in annual mean surface air temperature in China averaged over 2016–2035 (*left*), 2046–2065 (*middle*), and 2081–2100 (*right*), relative to 1986–2005, for RCP2.6 (*top*), RCP4.5 (*middle*), and RCP8.5 (*bottom*) scenarios (unit: °C)

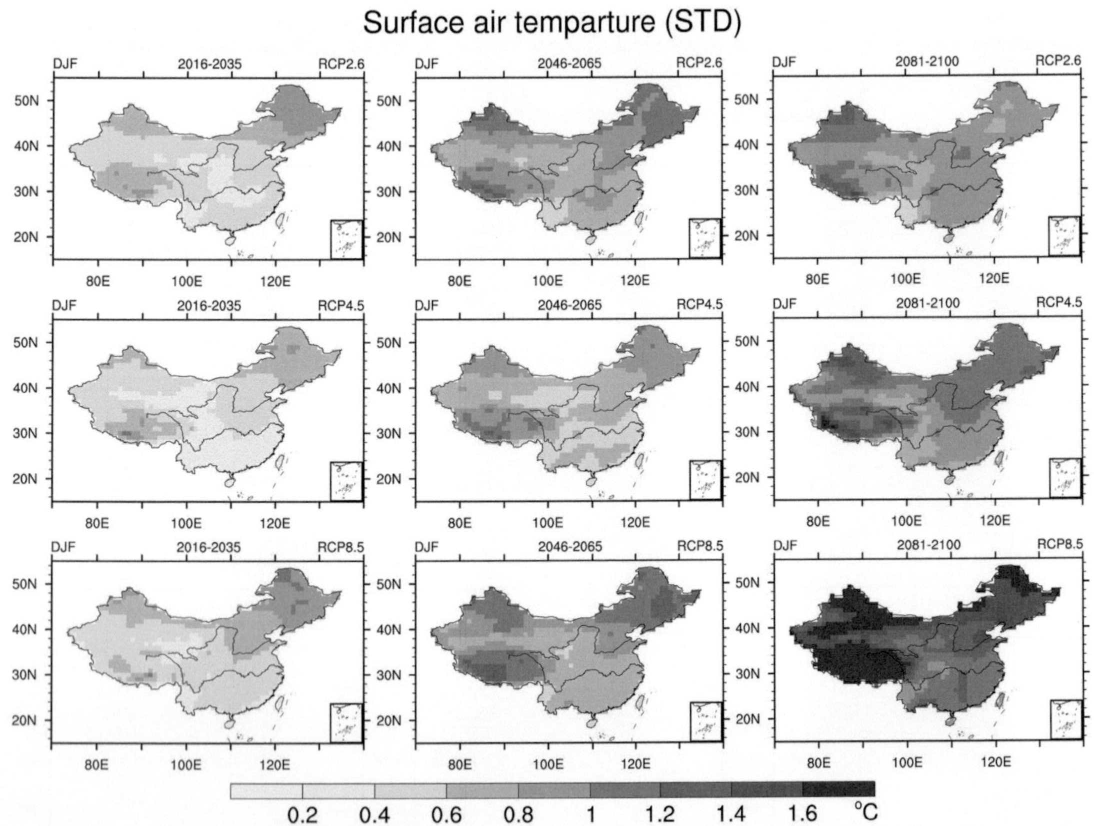

Fig. 2.44 Inter-model standard deviation among 22 models for changes in winter (December–February) surface air temperature in China averaged over 2016–2035 (*left*), 2046–2065 (*middle*), and 2081–2100 (*right*), relative to 1986–2005, for RCP2.6 (*top*), RCP4.5 (*middle*), and RCP8.5 (*bottom*) scenarios (unit: °C)

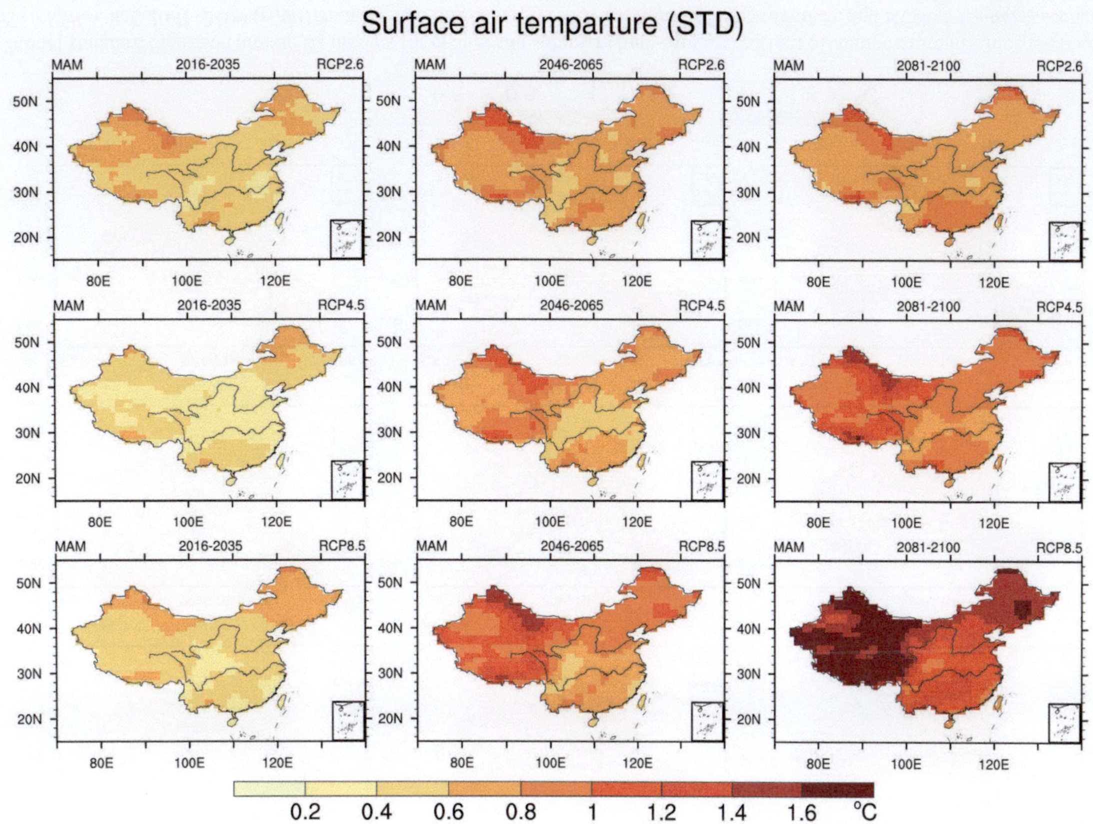

Fig. 2.45 Inter-model standard deviation among 22 models for changes in spring (March–May) surface air temperature in China averaged over 2016–2035 (*left*), 2046–2065 (*middle*), and 2081–2100 (*right*), relative to 1986–2005, for RCP2.6 (*top*), RCP4.5 (*middle*), and RCP8.5 (*bottom*) scenarios (unit: °C)

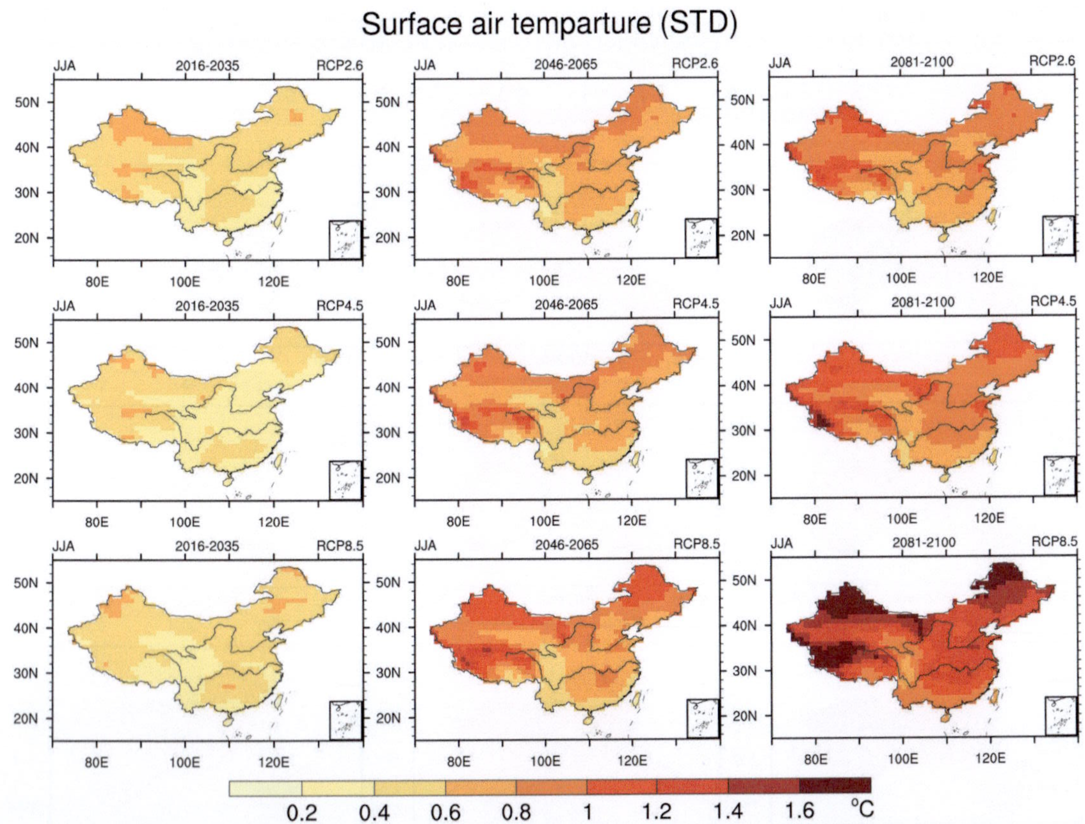

Fig. 2.46 Inter-model standard deviation among 22 models for changes in summer (June–August) surface air temperature in China averaged over 2016–2035 (*left*), 2046–2065 (*middle*), and 2081–2100 (*right*), relative to 1986–2005, for RCP2.6 (*top*), RCP4.5 (*middle*), and RCP8.5 (*bottom*) scenarios (unit: °C)

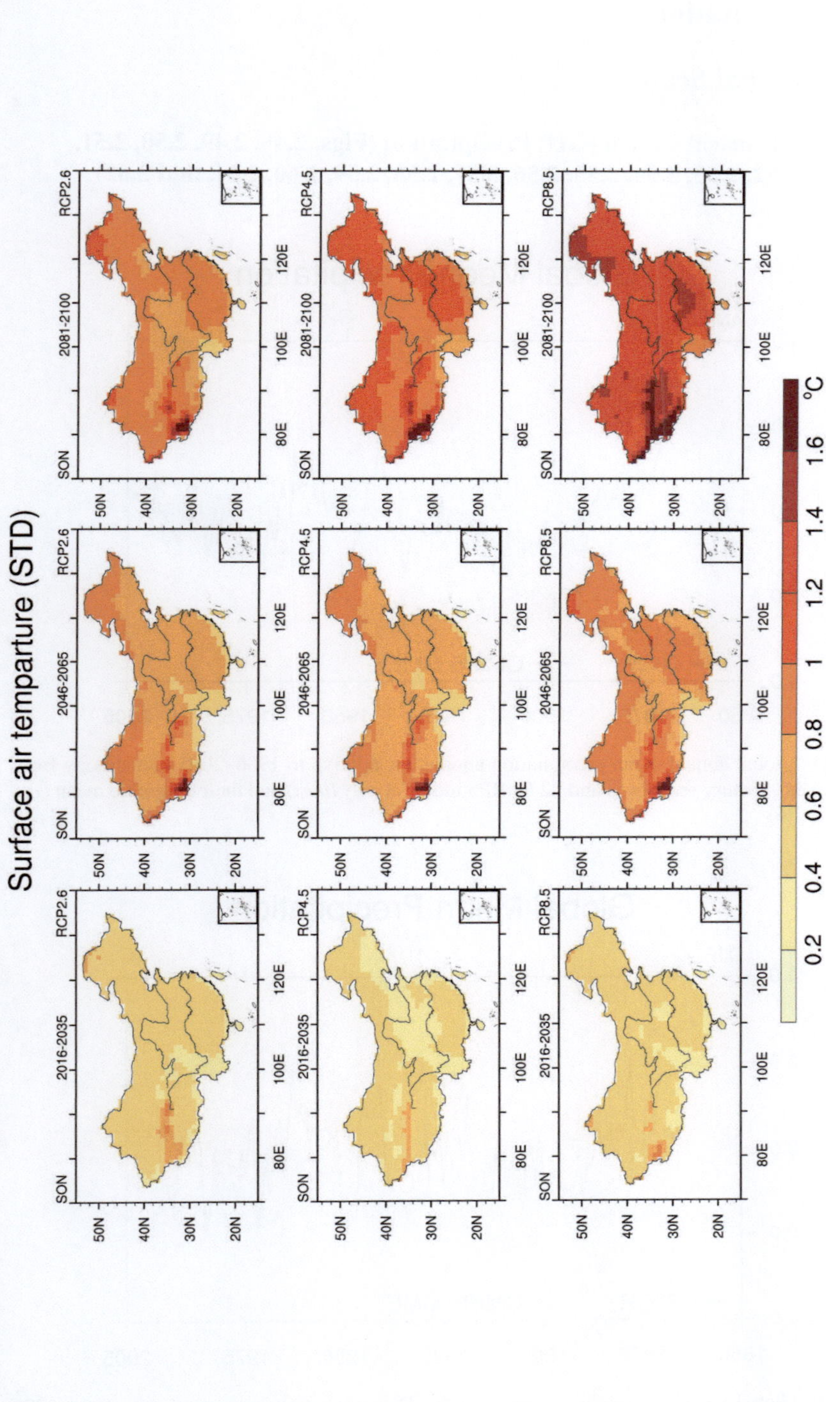

Fig. 2.47 Inter-model standard deviation among 22 models for changes in autumn (September–November) surface air temperature in China averaged over 2016–2035 (*left*), 2046–2065 (*middle*), and 2081–2100 (*right*), relative to 1986–2005, for RCP2.6 (*top*), RCP4.5 (*middle*), and RCP8.5 (*bottom*) scenarios (unit: °C)

2.2 Precipitation

2.2.1 Global Scale

2.2.1.1 Historical Global (GH) Precipitation (Figs. 2.48, 2.49, 2.50, 2.51, 2.52, 2.53, 2.54, 2.55, 2.56, 2.57, 2.58, 2.59, 2.60, 2.61, and 2.62)

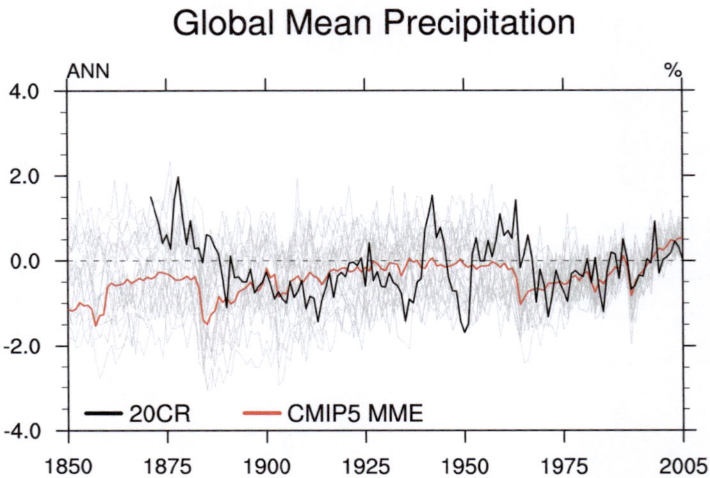

Fig. 2.48 Global annual mean precipitation anomalies, relative to 1986–2005 climatology from the twentieth-century reanalysis and 22 CMIP5 models (*gray lines*) and their ensemble mean (*red line*) (unit: %)

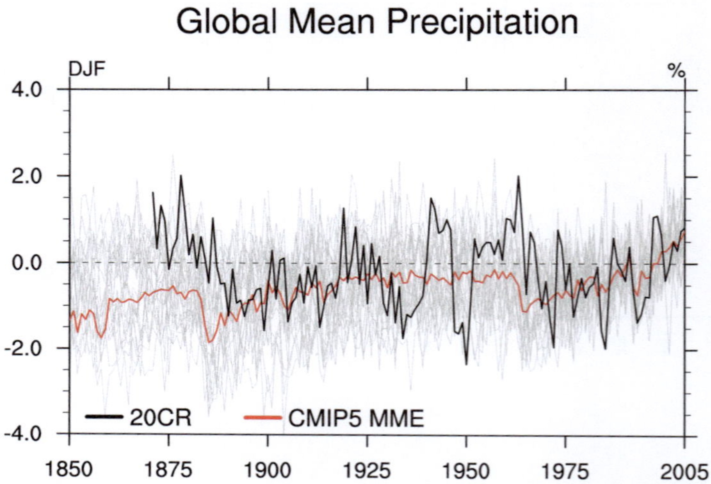

Fig. 2.49 Global mean precipitation anomalies for December–February, relative to 1986–2005 climatology from the twentieth-century reanalysis and 22 CMIP5 models (*gray lines*) and their ensemble mean (*red line*) (unit: %)

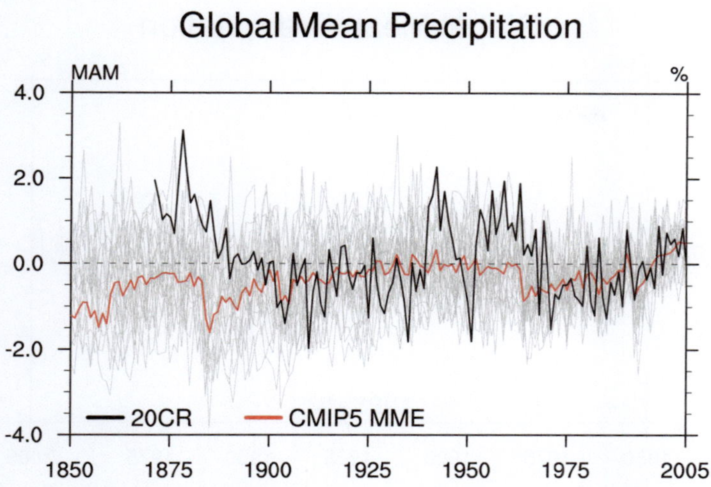

Fig. 2.50 Global mean precipitation anomalies for March–May, relative to 1986–2005 climatology from the twentieth-century reanalysis and 22 CMIP5 models (*gray lines*) and their ensemble mean (*red line*) (unit: %)

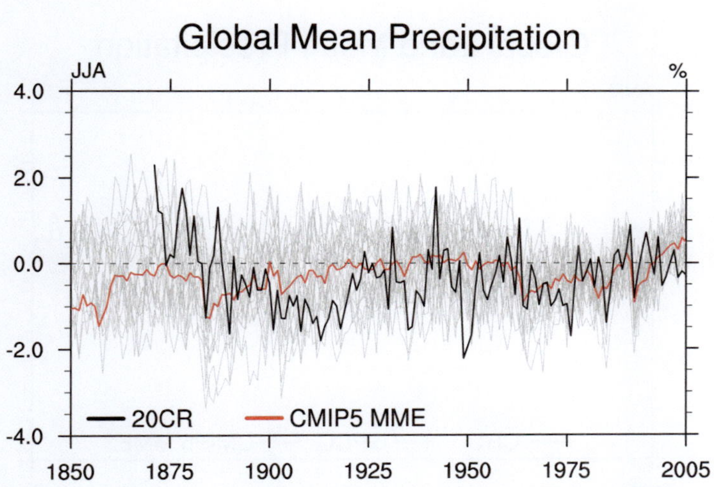

Fig. 2.51 Global mean precipitation anomalies for June–August, relative to 1986–2005 climatology from the twentieth-century reanalysis and 22 CMIP5 models (*gray lines*) and their ensemble mean (*red line*) (unit: %)

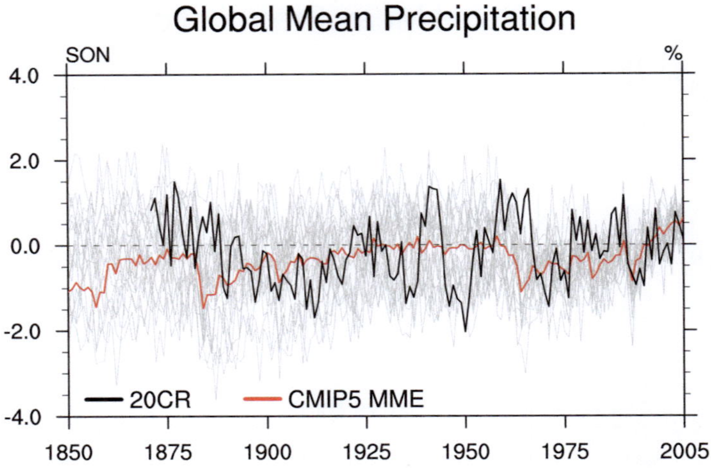

Fig. 2.52 Global mean precipitation anomalies for September–November, relative to 1986–2005 climatology from the twentieth-century reanalysis and 22 CMIP5 models (*gray lines*) and their ensemble mean (*red line*) (unit: %)

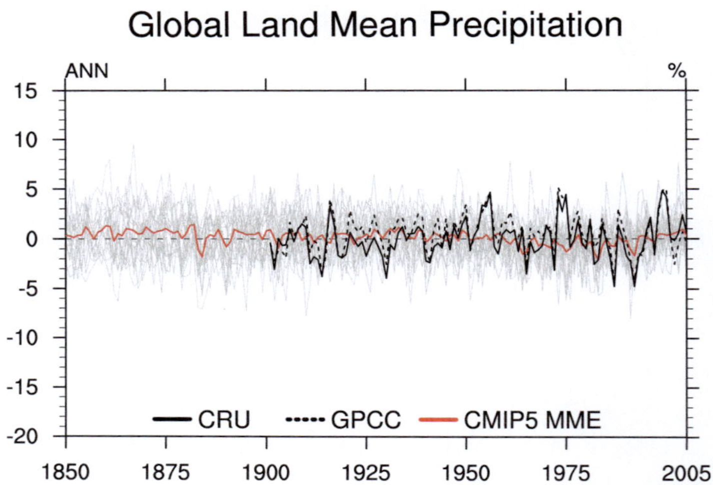

Fig. 2.53 Global land mean annual precipitation anomalies, relative to 1986–2005 climatology from two observations and 22 CMIP5 models (*gray lines*) and their ensemble mean (*red line*) (unit: %)

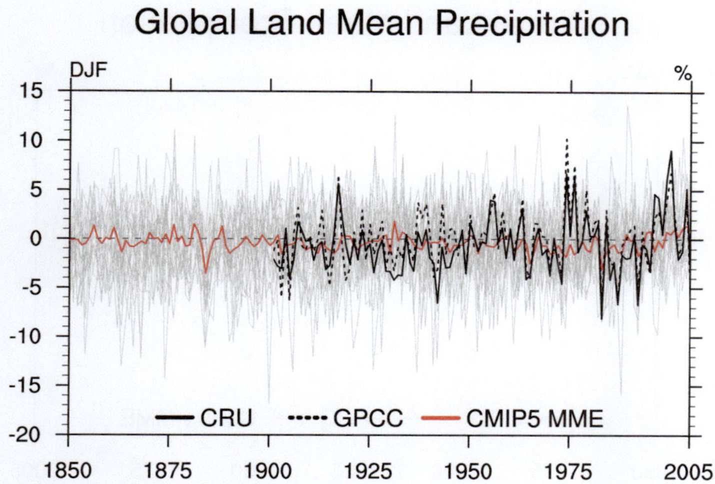

Fig. 2.54 Global land mean precipitation anomalies for December–February, relative to 1986–2005 climatology from two observations and 22 CMIP5 models (*gray lines*) and their ensemble mean (*red line*) (unit: %)

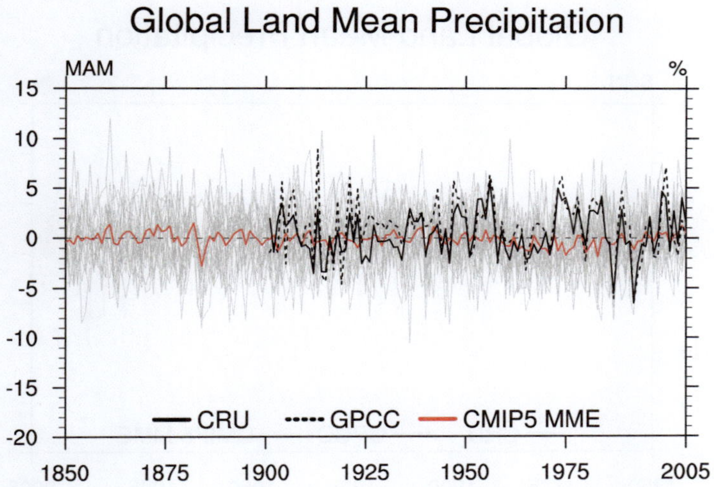

Fig. 2.55 Global land mean precipitation anomalies for March–May, relative to 1986–2005 climatology from two observations and 22 CMIP5 models (*gray lines*) and their ensemble mean (*red line*) (unit: %)

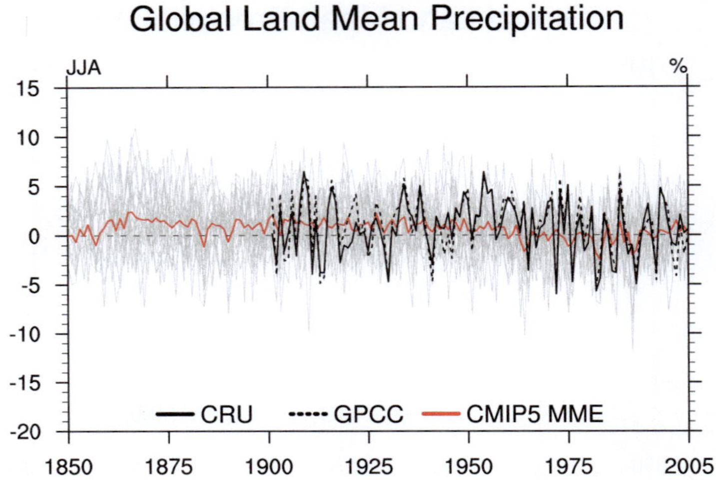

Fig. 2.56 Global land mean precipitation anomalies for June–August, relative to 1986–2005 climatology from two observations and 22 CMIP5 models (*gray lines*) and their ensemble mean (*red line*) (unit: %)

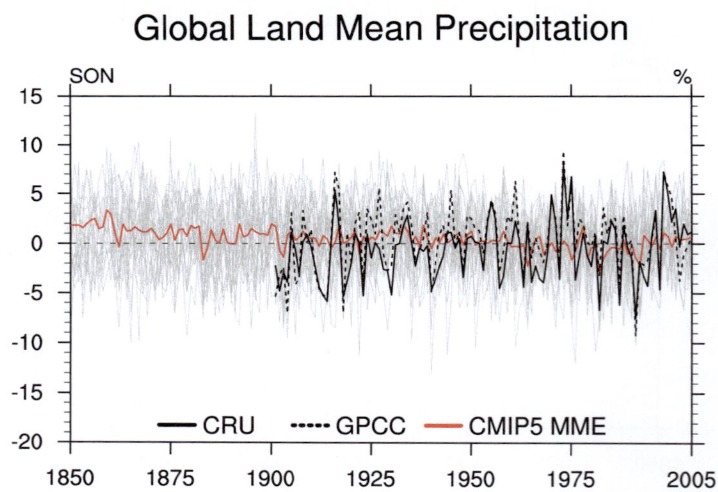

Fig. 2.57 Global land mean precipitation anomalies for September–November, relative to 1986–2005 climatology from two observations and 22 CMIP5 models (*gray lines*) and their ensemble mean (*red line*) (unit: %)

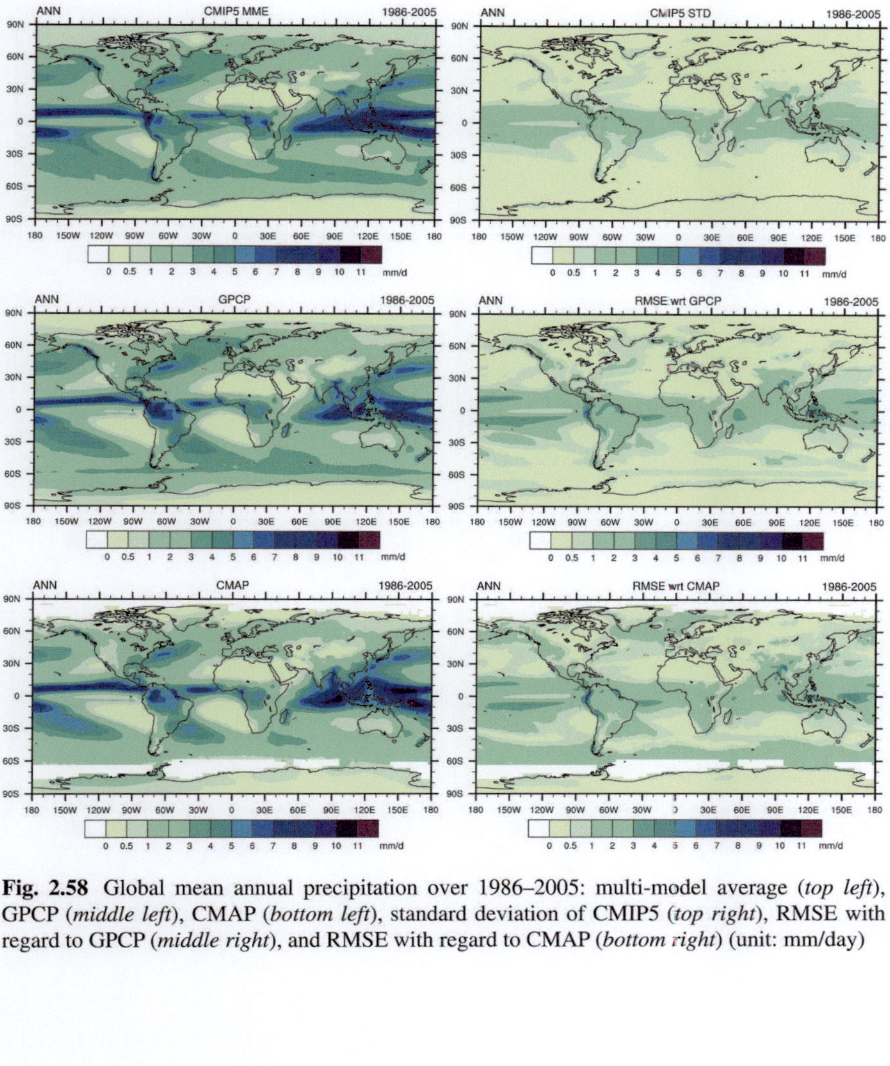

Fig. 2.58 Global mean annual precipitation over 1986–2005: multi-model average (*top left*), GPCP (*middle left*), CMAP (*bottom left*), standard deviation of CMIP5 (*top right*), RMSE with regard to GPCP (*middle right*), and RMSE with regard to CMAP (*bottom right*) (unit: mm/day)

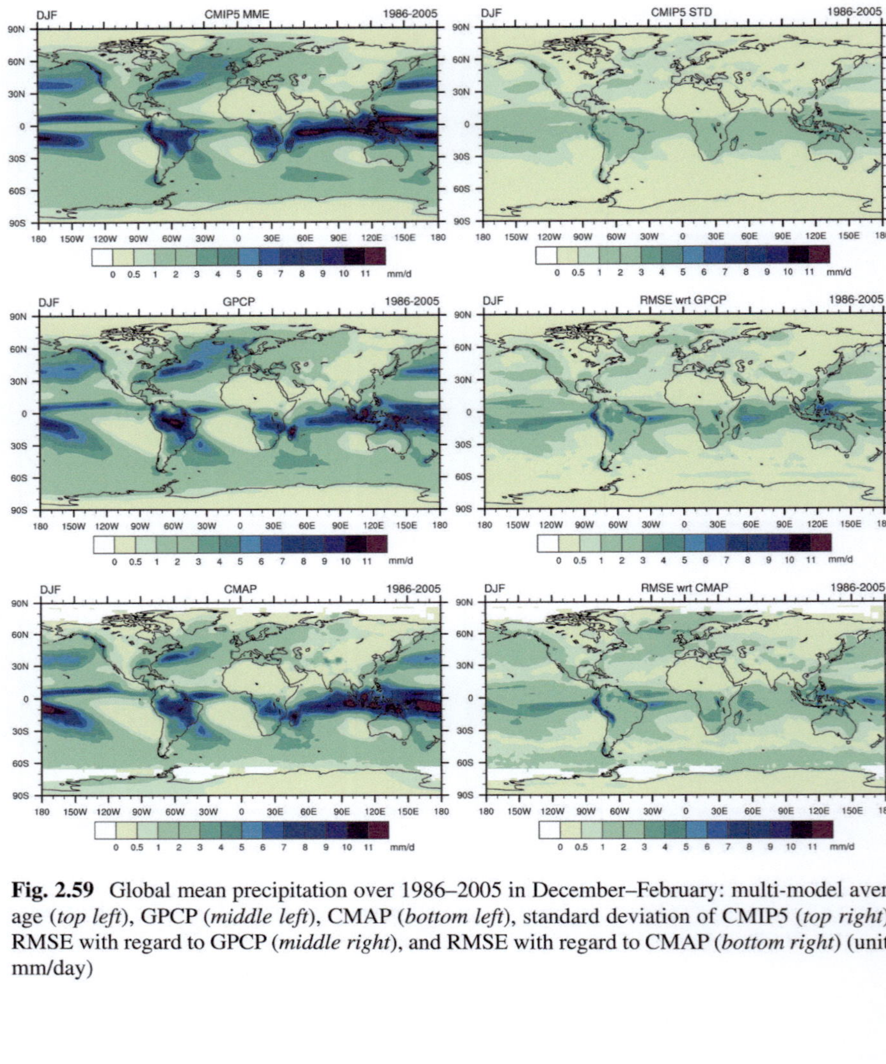

Fig. 2.59 Global mean precipitation over 1986–2005 in December–February: multi-model average (*top left*), GPCP (*middle left*), CMAP (*bottom left*), standard deviation of CMIP5 (*top right*), RMSE with regard to GPCP (*middle right*), and RMSE with regard to CMAP (*bottom right*) (unit: mm/day)

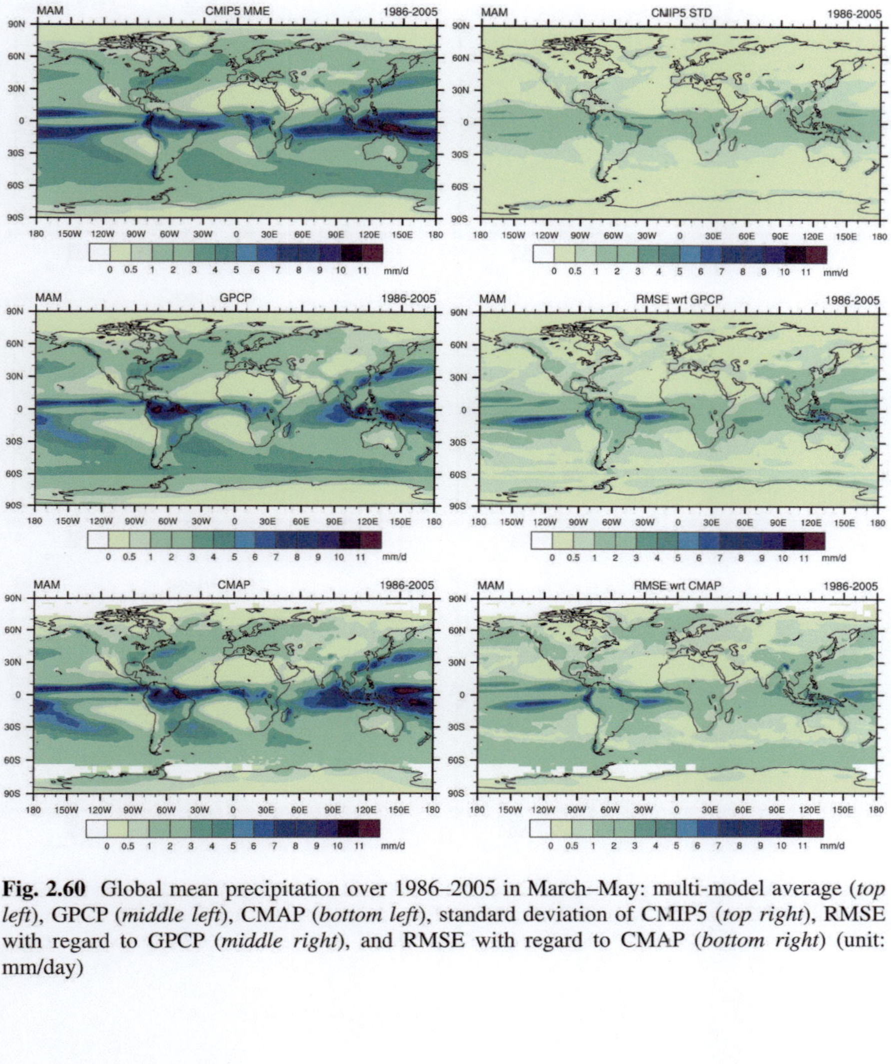

Fig. 2.60 Global mean precipitation over 1986–2005 in March–May: multi-model average (*top left*), GPCP (*middle left*), CMAP (*bottom left*), standard deviation of CMIP5 (*top right*), RMSE with regard to GPCP (*middle right*), and RMSE with regard to CMAP (*bottom right*) (unit: mm/day)

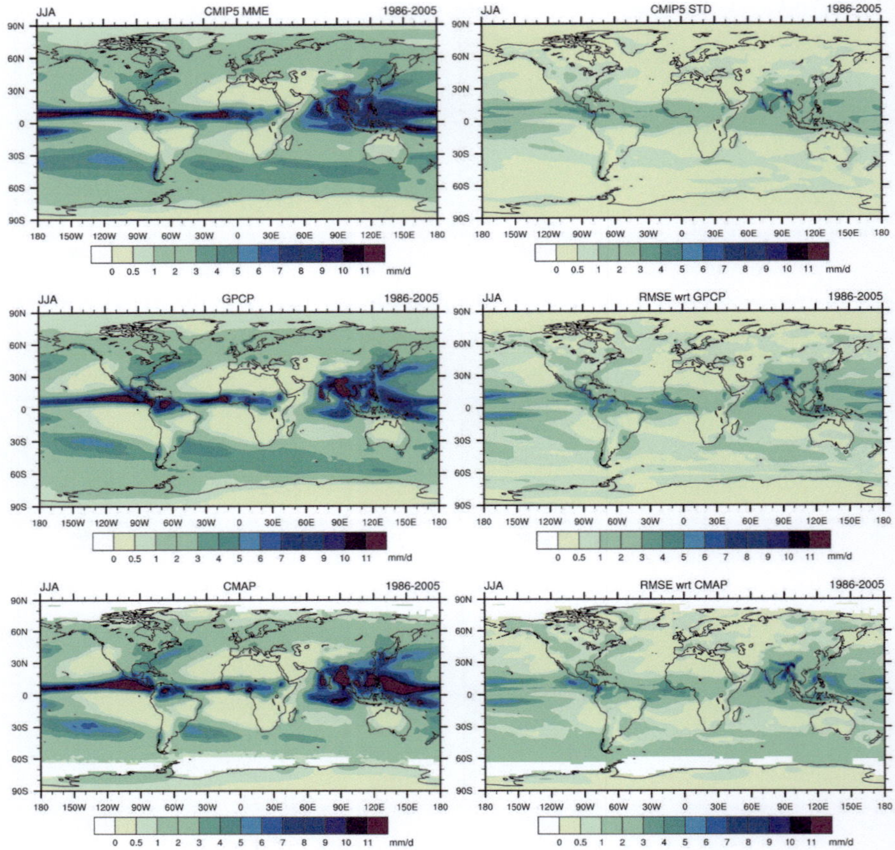

Fig. 2.61 Global mean precipitation over 1986–2005 in June–August: multi-model average (*top left*), GPCP (*middle left*), CMAP (*bottom left*), standard deviation of CMIP5 (*top right*), RMSE with regard to GPCP (*middle right*), and RMSE with regard to CMAP (*bottom right*) (unit: mm/day)

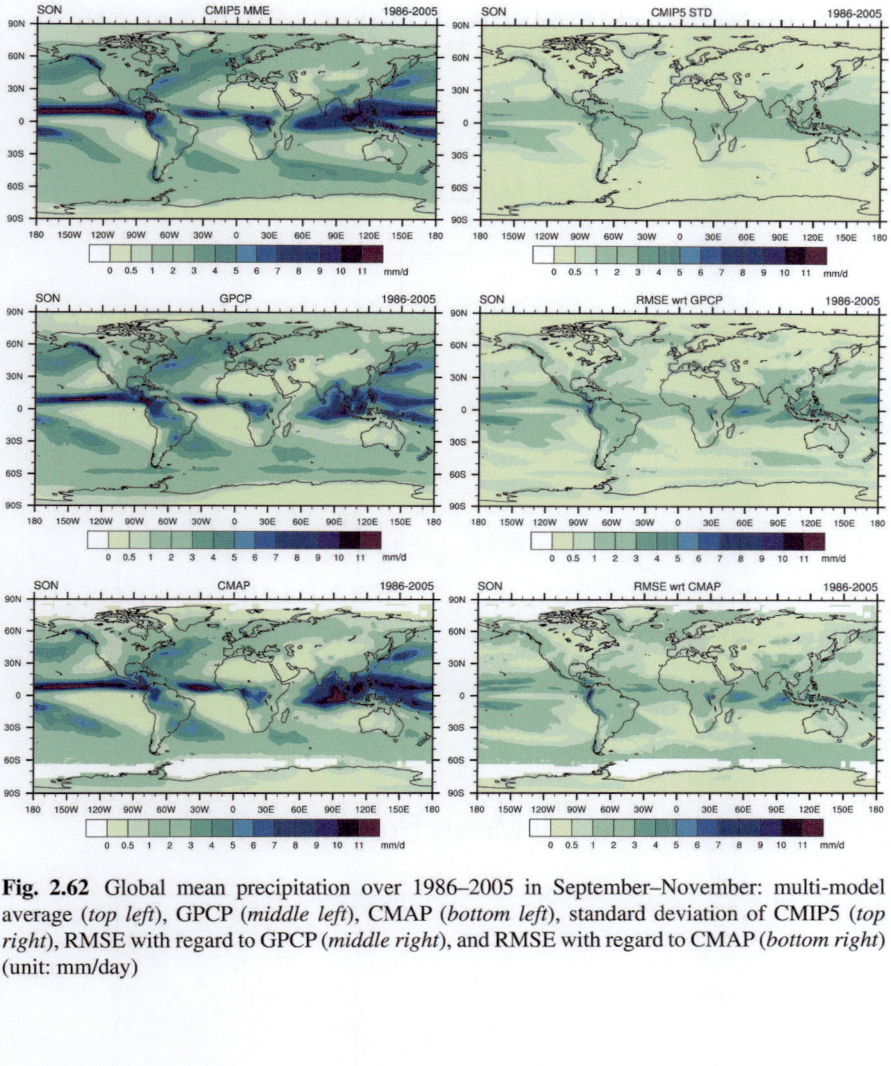

Fig. 2.62 Global mean precipitation over 1986–2005 in September–November: multi-model average (*top left*), GPCP (*middle left*), CMAP (*bottom left*), standard deviation of CMIP5 (*top right*), RMSE with regard to GPCP (*middle right*), and RMSE with regard to CMAP (*bottom right*) (unit: mm/day)

2.2.1.2 Projected Global Precipitation (GP) (Figs. 2.63, 2.64, 2.65, 2.66, 2.67, 2.68, 2.69, 2.70, 2.71, 2.72, and 2.73)

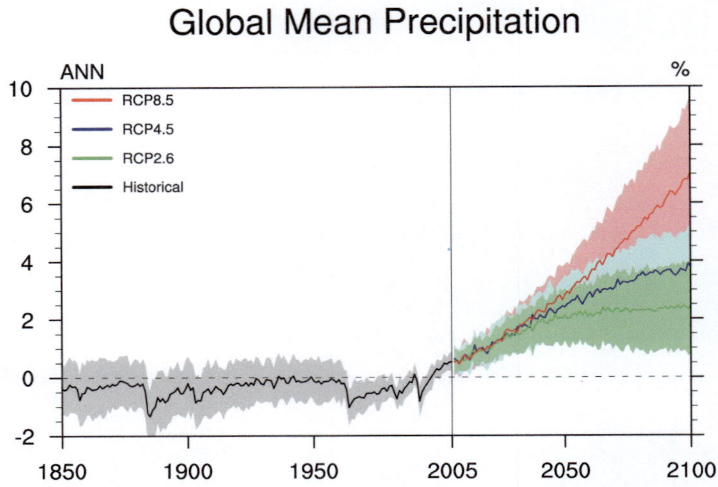

Fig. 2.63 Global annual mean precipitation anomalies relative to 1986–2005 from multi-model ensemble mean of historical simulations and projections under three RCP scenarios (unit: %)

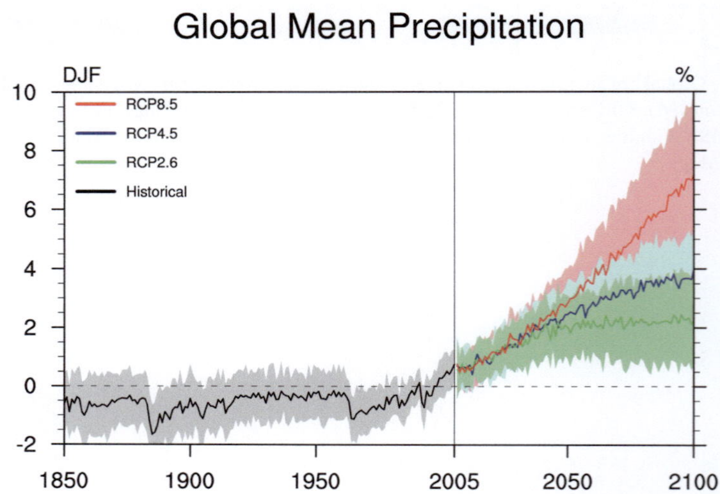

Fig. 2.64 Global mean precipitation anomalies relative to 1986–2005 in December–February from multi-model ensemble mean of historical simulations and projections under three RCP scenarios (unit: %)

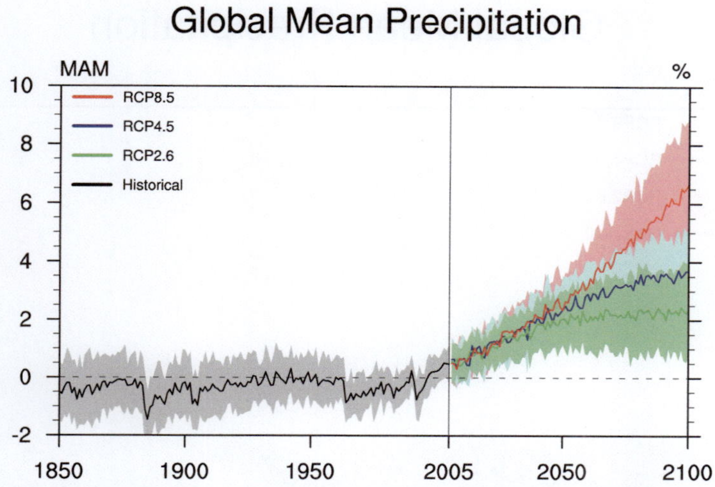

Fig. 2.65 Global mean precipitation anomalies relative to 1986–2005 in March–May from multi-model ensemble mean of historical simulations and projections under three RCP scenarios (unit: %)

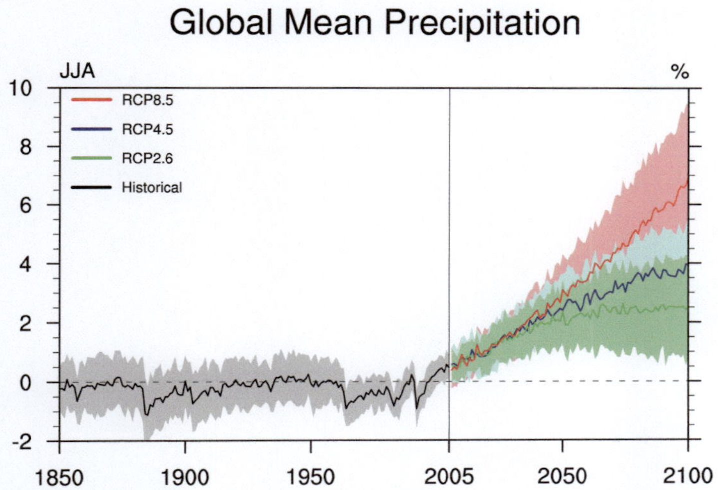

Fig. 2.66 Global mean precipitation anomalies relative to 1986–2005 in June–August from multi-model ensemble mean of historical simulations and projections under three RCP scenarios (unit: %)

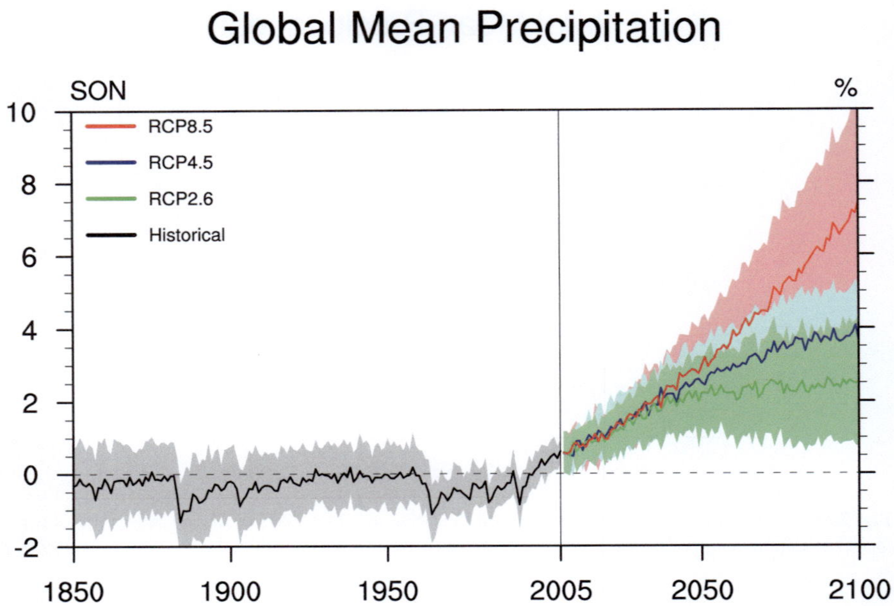

Fig. 2.67 Global mean precipitation anomalies relative to 1986–2005 in September–November from multi-model ensemble mean of historical simulations and projections under three RCP scenarios (unit: %)

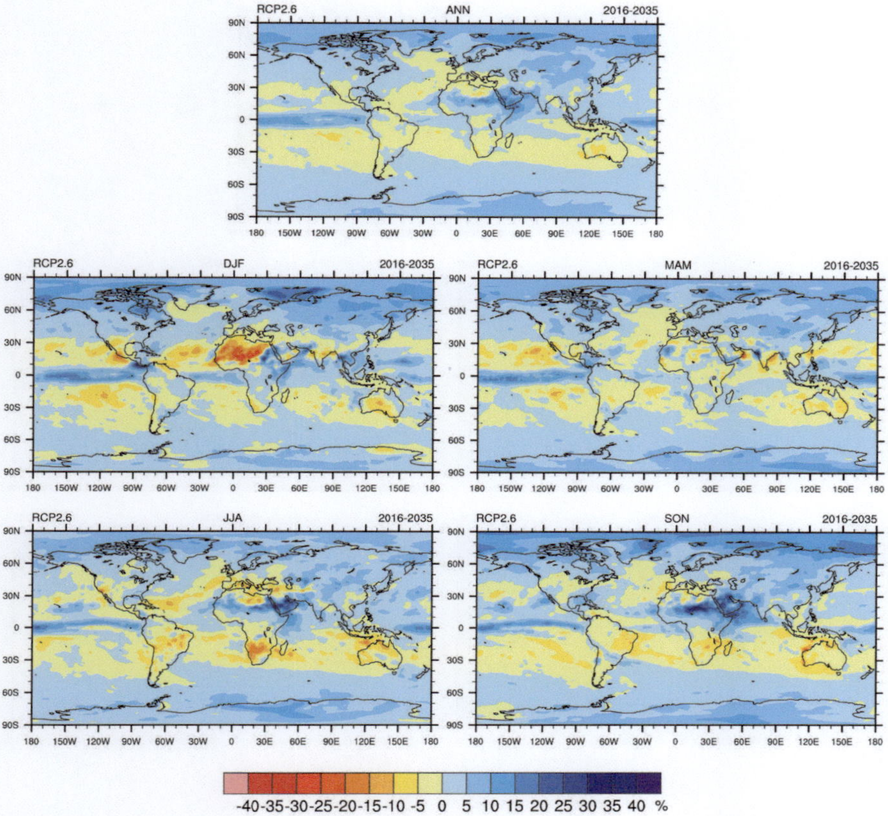

Fig. 2.68 Changes in precipitation in 2016–2035 under RCP2.6 relative to the climatology in 1986–2005: annual mean (*top*), December–February mean (*middle left*), March–May mean (*middle right*), June–August mean (*bottom left*), and September–November mean (*bottom right*) (unit: %)

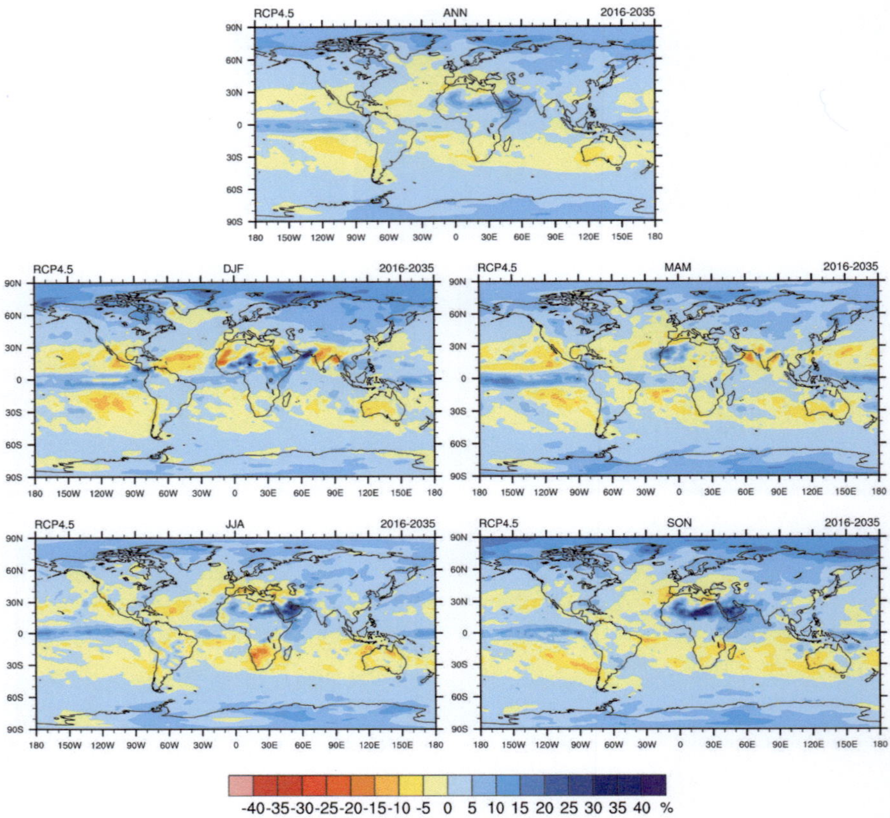

Fig. 2.69 Changes in precipitation in 2016–2035 under RCP4.5 relative to the climatology in 1986–2005: annual mean (*top*), December–February mean (*middle left*), March–May mean (*middle right*), June–August mean (*bottom left*), and September–November mean (*bottom right*) (unit: %)

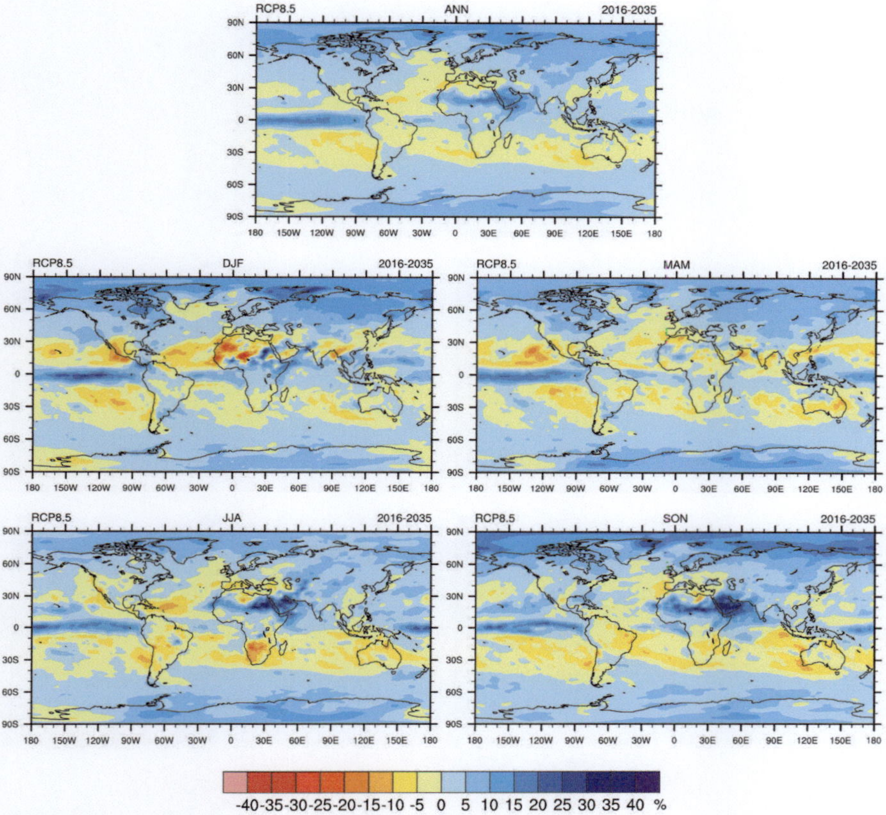

Fig. 2.70 Changes in precipitation in 2016–2035 under RCP8.5 relative to the climatology in 1986–2005: annual mean (*top*), December–February mean (*middle left*), March–May mean (*middle right*), June–August mean (*bottom left*), and September–November mean (*bottom right*) (unit: %)

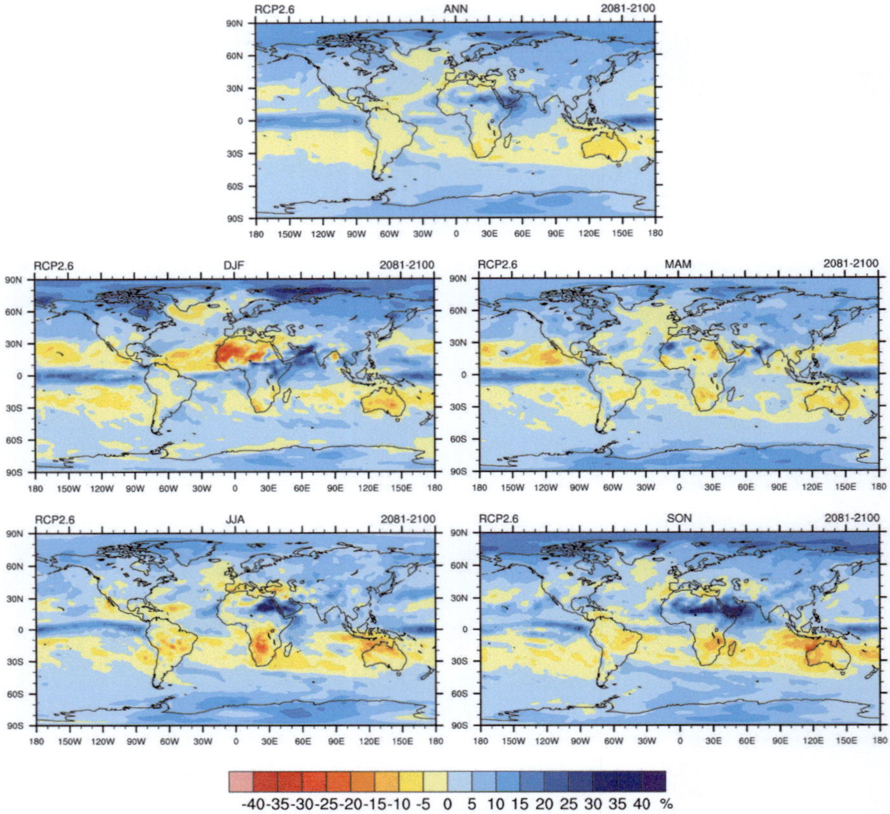

Fig. 2.71 Changes in precipitation in 2081–2100 under RCP2.6 relative to the climatology in 1986–2005: annual mean (*top*), December–February mean (*middle left*), March–May mean (*middle right*), June–August mean (*bottom left*), and September–November mean (*bottom right*) (unit: %)

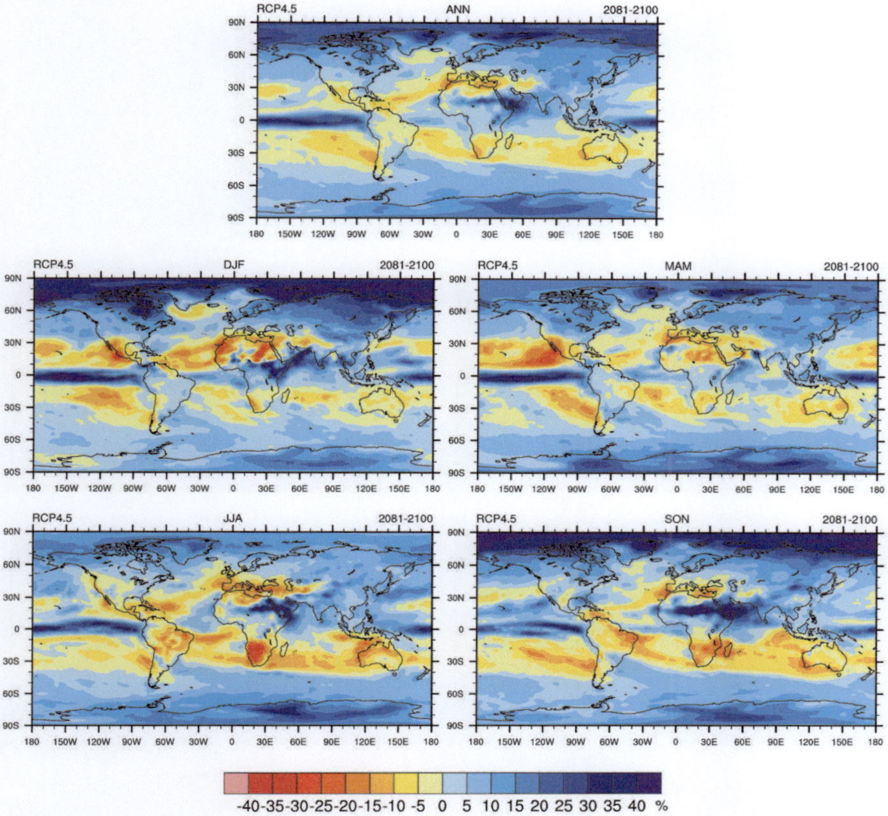

Fig. 2.72 Changes in precipitation in 2081–2100 under RCP4.5 relative to the climatology in 1986–2005: annual mean (*top*), December–February mean (*middle left*), March–May mean (*middle right*), June–August mean (*bottom left*), and September–November mean (*bottom right*) (unit: %)

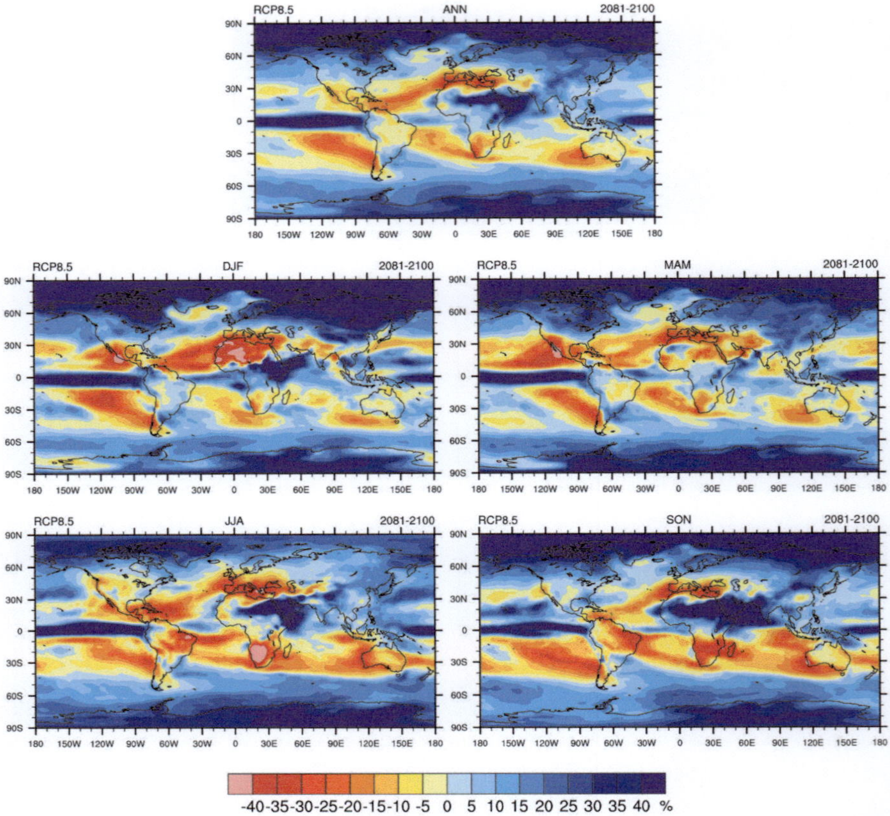

Fig. 2.73 Changes in precipitation in 2081–2100 under RCP8.5 relative to the climatology in 1986–2005: annual mean (*top*), December–February mean (*middle left*), March–May mean (*middle right*), June–August mean (*bottom left*), and September–November mean (*bottom right*) (unit: %)

2.2.2 China

2.2.2.1 Historical Precipitation in China (CH) (Figs. 2.74, 2.75, 2.76, 2.77, 2.78, 2.79, 2.80, 2.81, 2.82, 2.83, and 2.84)

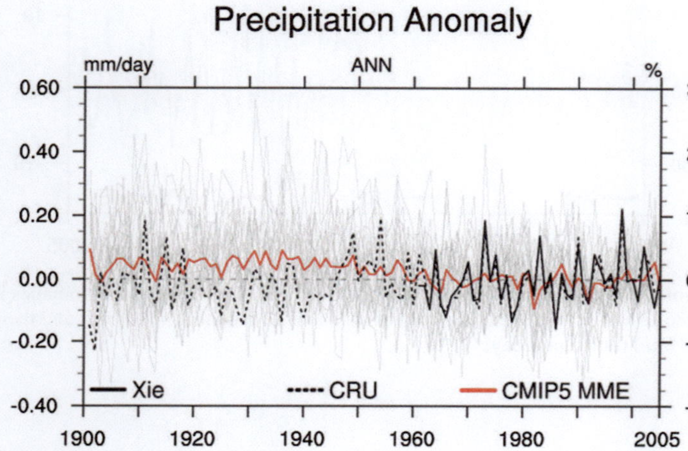

Fig. 2.74 Evolution from 1901 to 2005 of annual average precipitation anomalies in China, relative to 1986–2005, from four observational data and 22 models (*gray lines*) and their ensemble mean (*red line*) (units: mm/day; %)

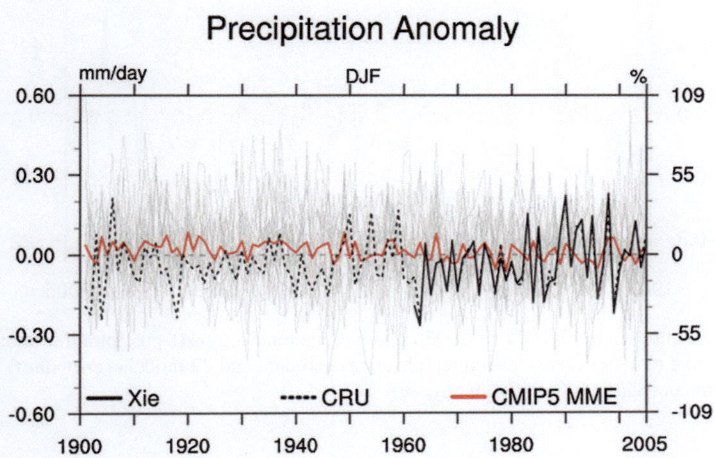

Fig. 2.75 Evolution from 1901 to 2005 of winter (December–February) precipitation anomalies in China, relative to 1986–2005, from three observational data and 22 models (*gray lines*) and their ensemble mean (*red line*) (units: mm/day; %)

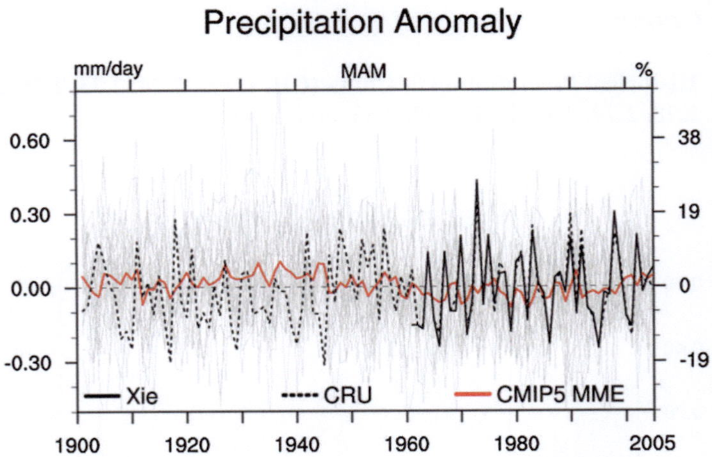

Fig. 2.76 Evolution from 1901 to 2005 of spring (March–May) precipitation anomalies in China, relative to 1986–2005, from three observational data and 22 models (*gray lines*) and their ensemble mean (*red line*) (units: mm/day; %)

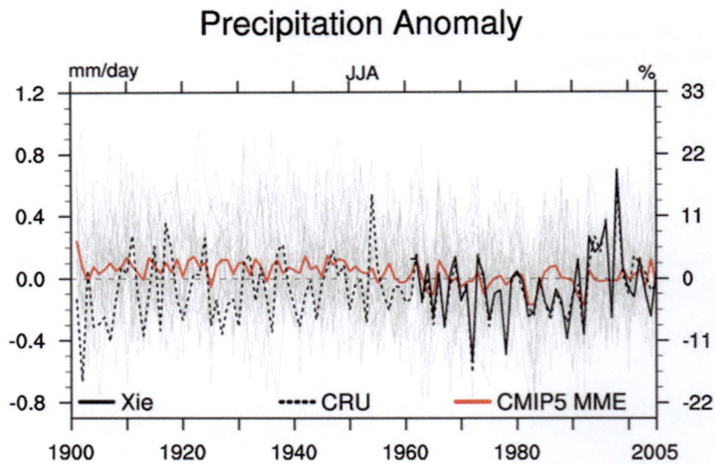

Fig. 2.77 Evolution from 1901 to 2005 of summer (June–August) precipitation anomalies in China, relative to 1986–2005, from three observational data and 22 models (*gray lines*) and their ensemble mean (*red line*) (units: mm/day; %)

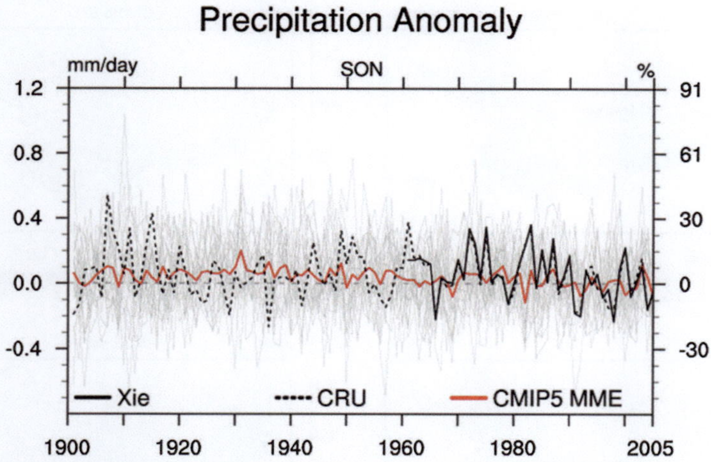

Fig. 2.78 Evolution from 1901 to 2005 of autumn (September–November) precipitation anomalies in China, relative to 1986–2005, from three observational data and 22 models (*gray lines*) and their ensemble mean (*red line*) (units: mm/day; %)

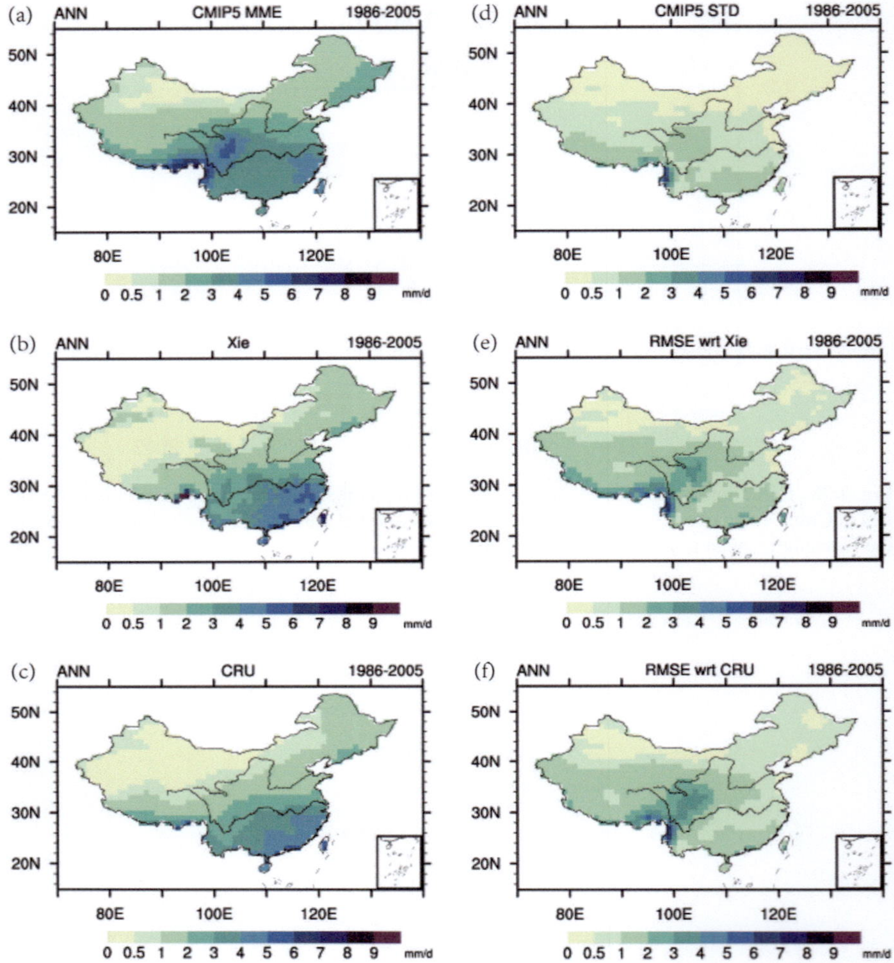

Fig. 2.79 Annual mean precipitation in China averaged over 1986–2005: (**a**) multi-model ensemble mean, (**b**) observation (Xie), (**c**) observation (CRU). (**d**) Inter-model standard deviation among 22 models. RMSE between model simulations and observation (**e**) Xie, (f) CRU (unit: mm/day)

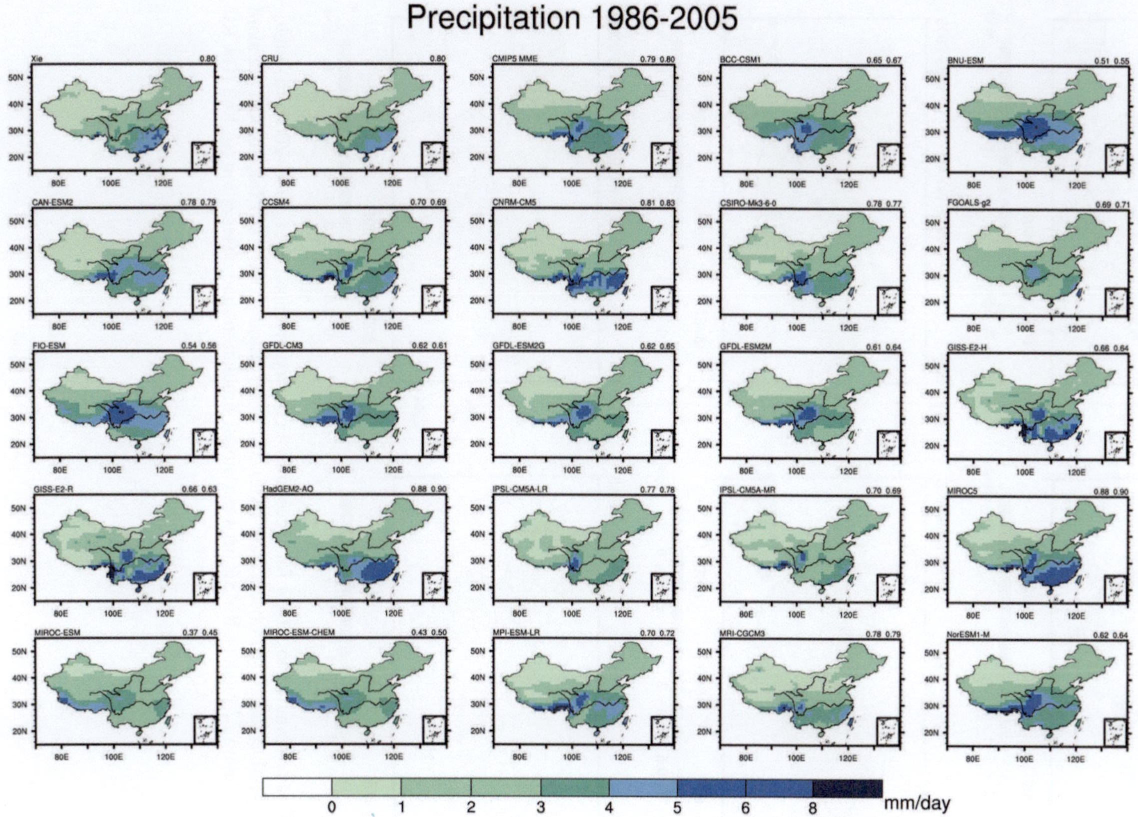

Fig. 2.80 Annual mean precipitation in China averaged over 1986–2005 from observations, multi-model ensemble mean, and 22 individual models (unit: mm/day). The spatial correlation coefficients between observations (Xie, CRU) and model simulation are listed at the top right corner

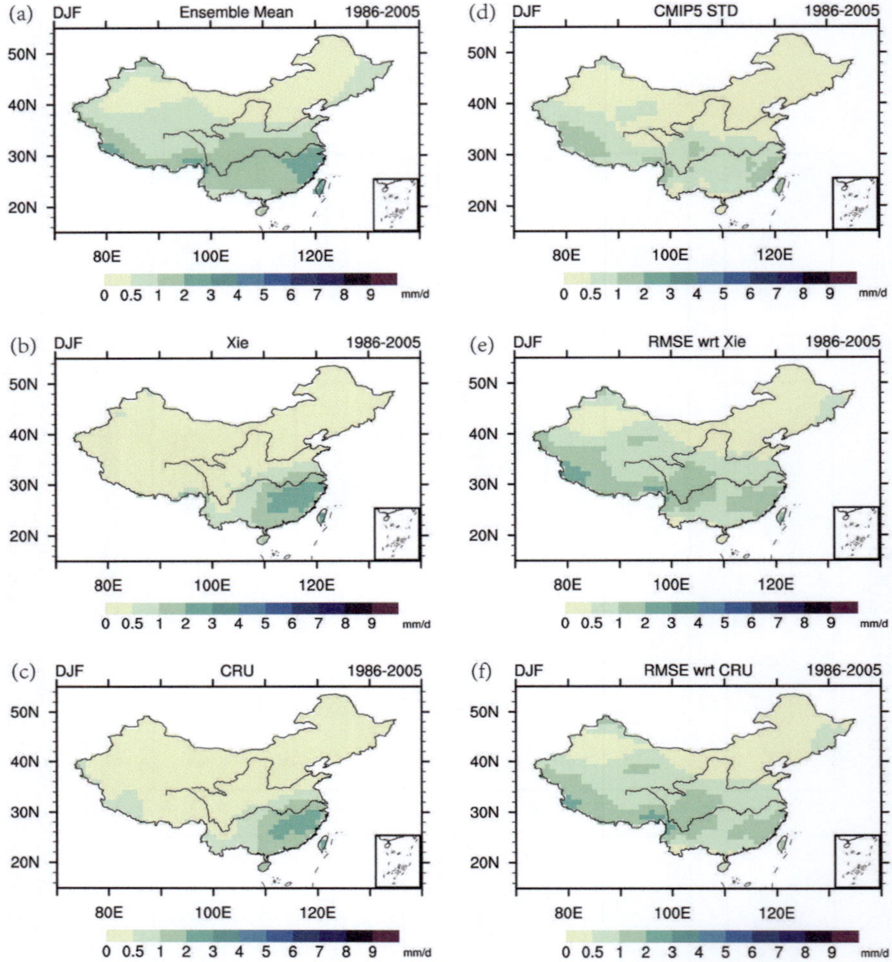

Fig. 2.81 Precipitation in China for December–February averaged over 1986–2005: (**a**) multi-model ensemble mean, (**b**) observation (Xie), (**c**) observation (CRU). (**d**) Inter-model standard deviation among 22 models. RMSE between model simulations and observation (**e**) Xie, (**f**) CRU (unit: mm/day)

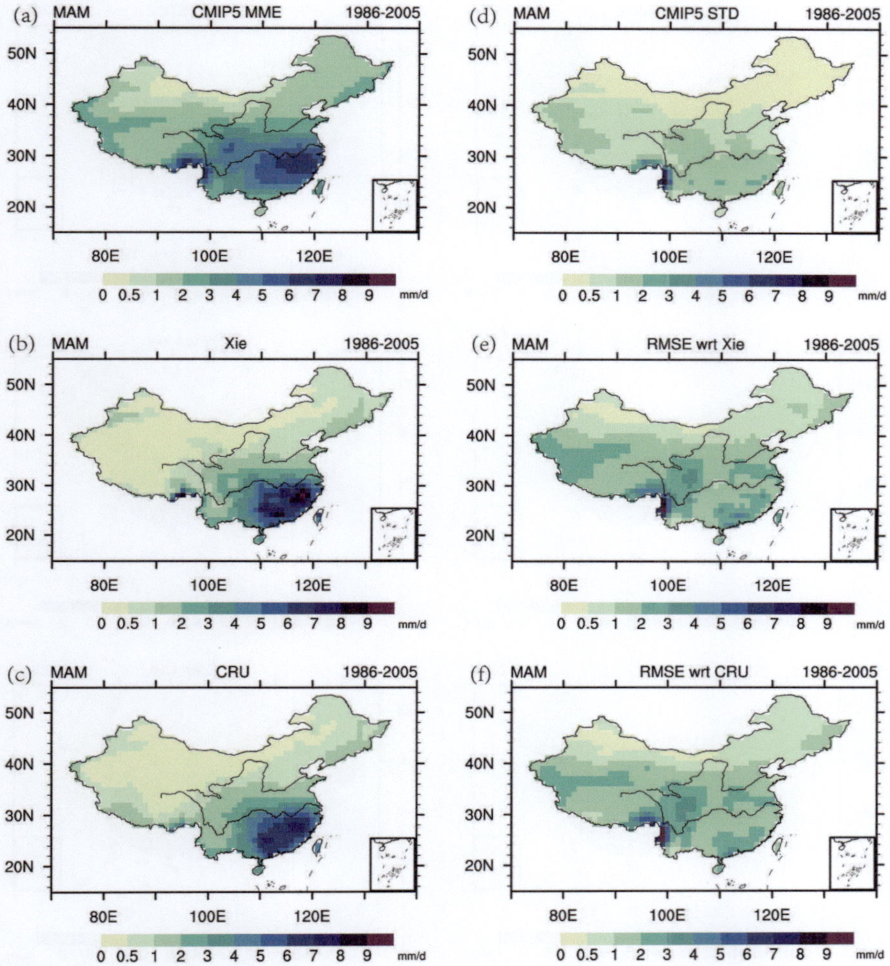

Fig. 2.82 Precipitation in China for March–May averaged over 1986–2005: (**a**) multi-model ensemble mean, (**b**) observation (Xie), (**c**) observation (CRU). (**d**) Inter-model standard deviation among 22 models. RMSE between model simulations and observation (**e**) Xie, (**f**) CRU (unit: mm/day)

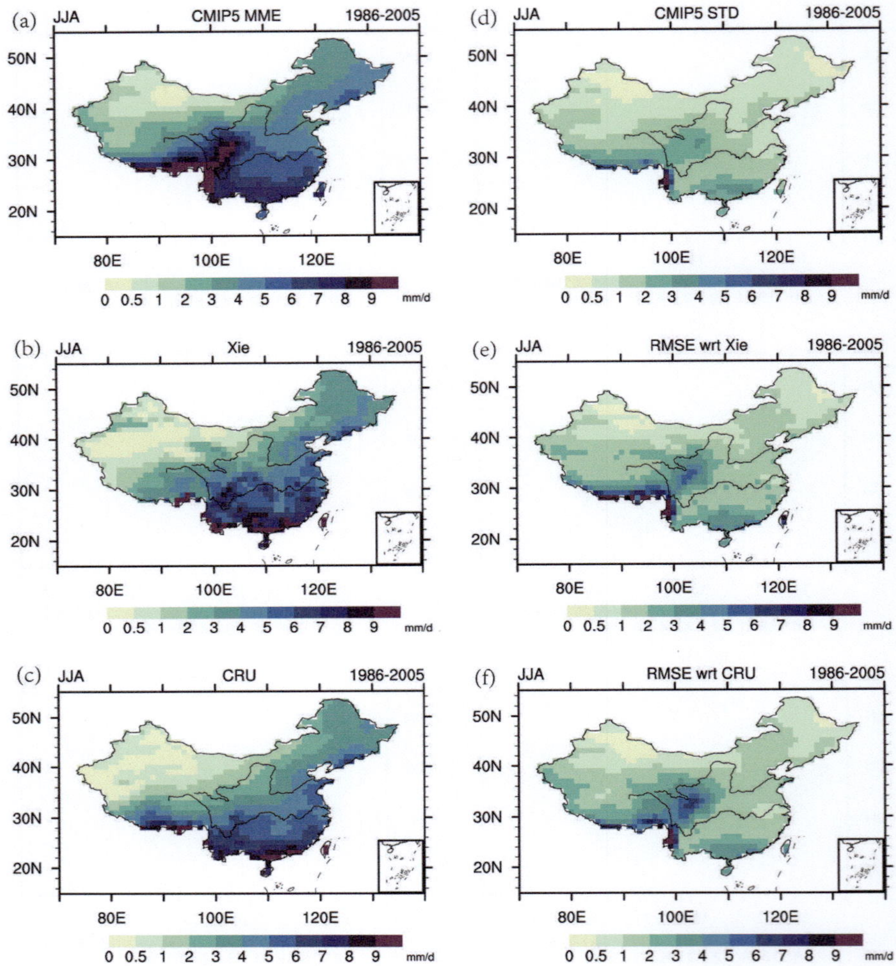

Fig. 2.83 Precipitation in China for June–August averaged over 1986–2005: (**a**) multi-model ensemble mean, (**b**) observation (Xie), (**c**) observation (CRU). (**d**) Inter-model standard deviation among 22 models. RMSE between model simulations and observation (**e**) Xie, (**f**) CRU (unit: mm/day)

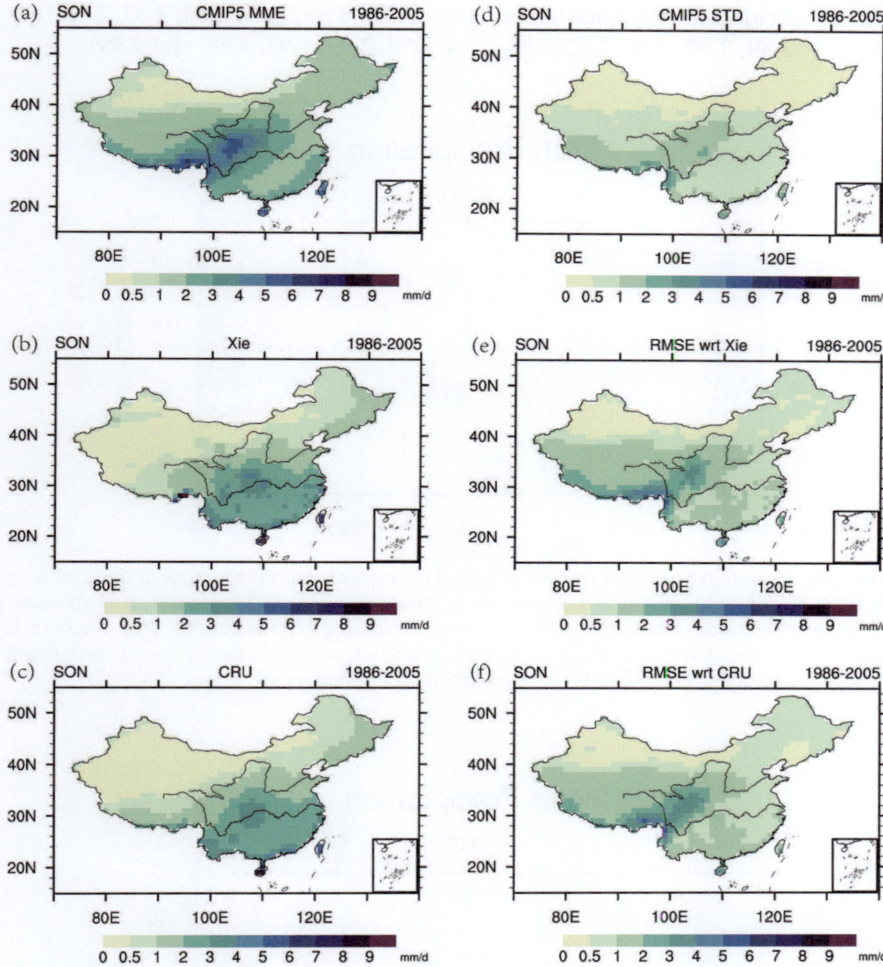

Fig. 2.84 Precipitation in China for September–November averaged over 1986–2005: (**a**) multi-model ensemble mean, (**b**) observation (Xie), (**c**) observation (CRU). (**d**) Inter-model standard deviation among 22 models. RMSE between model simulations and observation (**e**) Xie, (**f**) CRU (unit: mm/day)

2.2.2.2 Projected Precipitation in China (CP) (Figs. 2.85, 2.86, 2.87, 2.88, 2.89, 2.90, 2.91, 2.92, 2.93, 2.94, 2.95, 2.96, 2.97, 2.98, and 2.99)

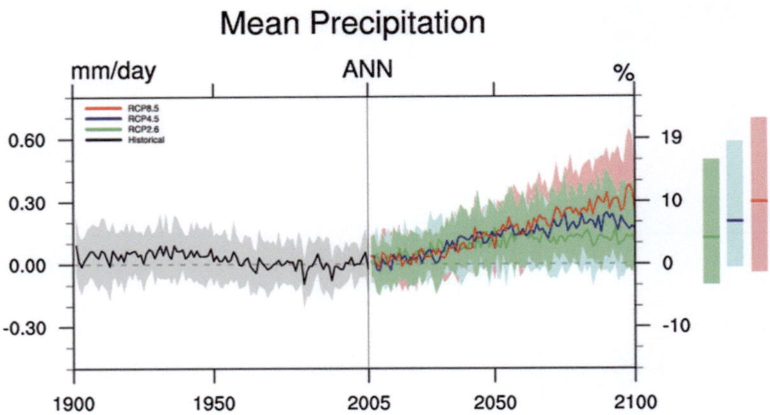

Fig. 2.85 Evolution from 1901 to 2100 of annual mean precipitation anomalies in China relative to 1986–2005 from multi-model ensemble mean. *Shadow* represents inter-model standard deviation. The band in the box (*right side*) represents the mean precipitation over 2081–2100 from multi-model ensemble mean. The *bottom* (*top*) of the box represents the minimum (maximum) mean precipitation over 2081–2100 among multi-model (mm/day; %)

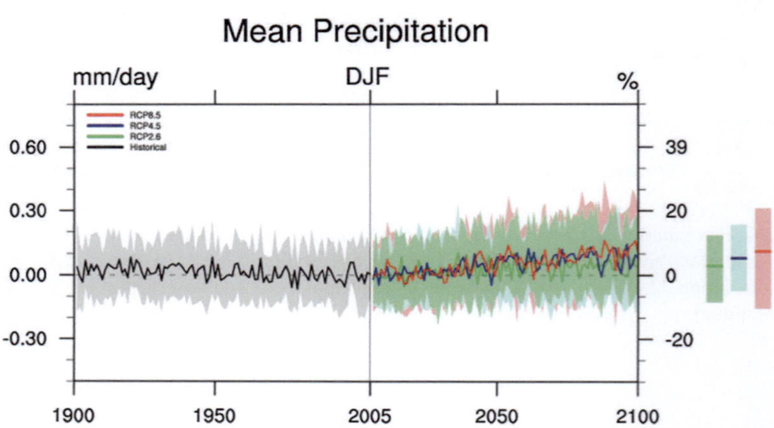

Fig. 2.86 Evolution from 1901 to 2100 of winter (December–February) precipitation anomalies in China relative to 1986–2005 from multi-model ensemble mean. *Shadow* represents inter-model standard deviation. The band in the box (*right side*) represents the mean precipitation over 2081–2100 from multi-model ensemble mean. The *bottom* (*top*) of the box represents the minimum (maximum) mean precipitation over 2081–2100 among multi-model (units: mm/day; %)

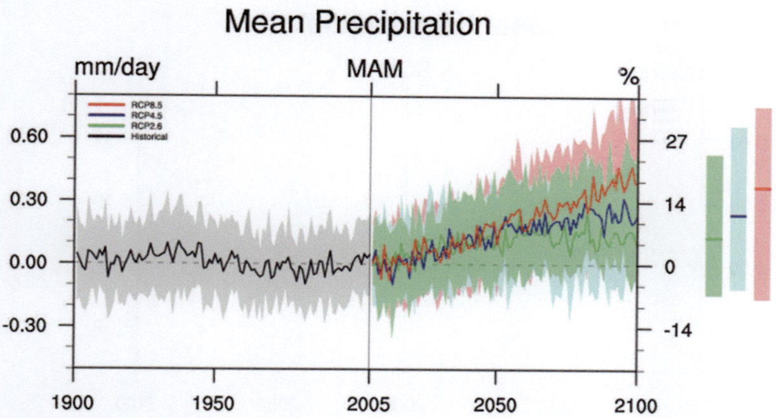

Fig. 2.87 Evolution from 1901 to 2100 of spring (March–May) precipitation anomalies in China relative to 1986–2005 from multi-model ensemble mean. *Shadow* represents inter-model standard deviation. The band in the box (*right side*) represents the mean precipitation over 2081–2100 from multi-model ensemble mean. The *bottom* (*top*) of the box represents the minimum (maximum) mean precipitation over 2081–2100 among multi-model (units: mm/day; %)

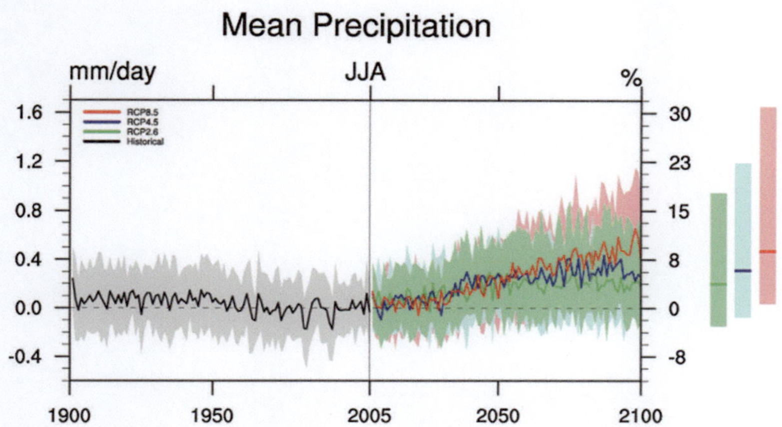

Fig. 2.88 Evolution from 1901 to 2100 of summer (June–August) precipitation anomalies in China relative to 1986–2005 from multi-model ensemble mean. *Shadow* represents inter-model standard deviation. The band in the box (*right side*) represents the mean precipitation over 2081–2100 from multi-model ensemble mean. The *bottom* (*top*) of the box represents the minimum (maximum) mean precipitation over 2081–2100 among multi-model (units: mm/day; %)

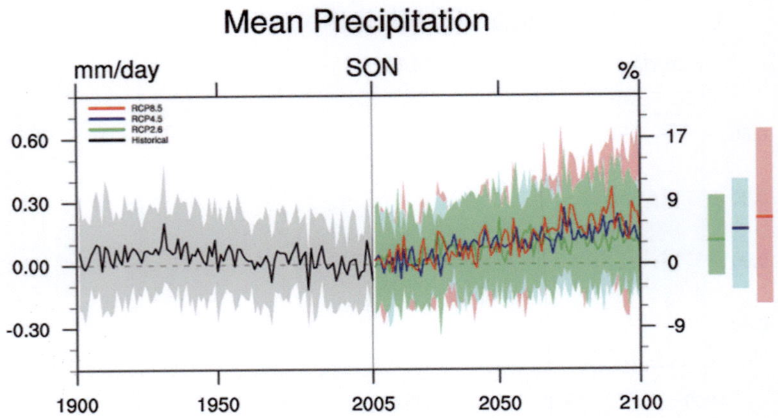

Fig. 2.89 Evolution from 1901 to 2100 of autumn (September–November) precipitation anomalies in China relative to 1986–2005 from multi-model ensemble mean. *Shadow* represents inter-model standard deviation. The band in the box (*right side*) represents the mean precipitation over 2081–2100 from multi-model ensemble mean. The *bottom* (*top*) of the box represents the minimum (maximum) mean precipitation over 2081–2100 among multi-model (units: mm/day; %)

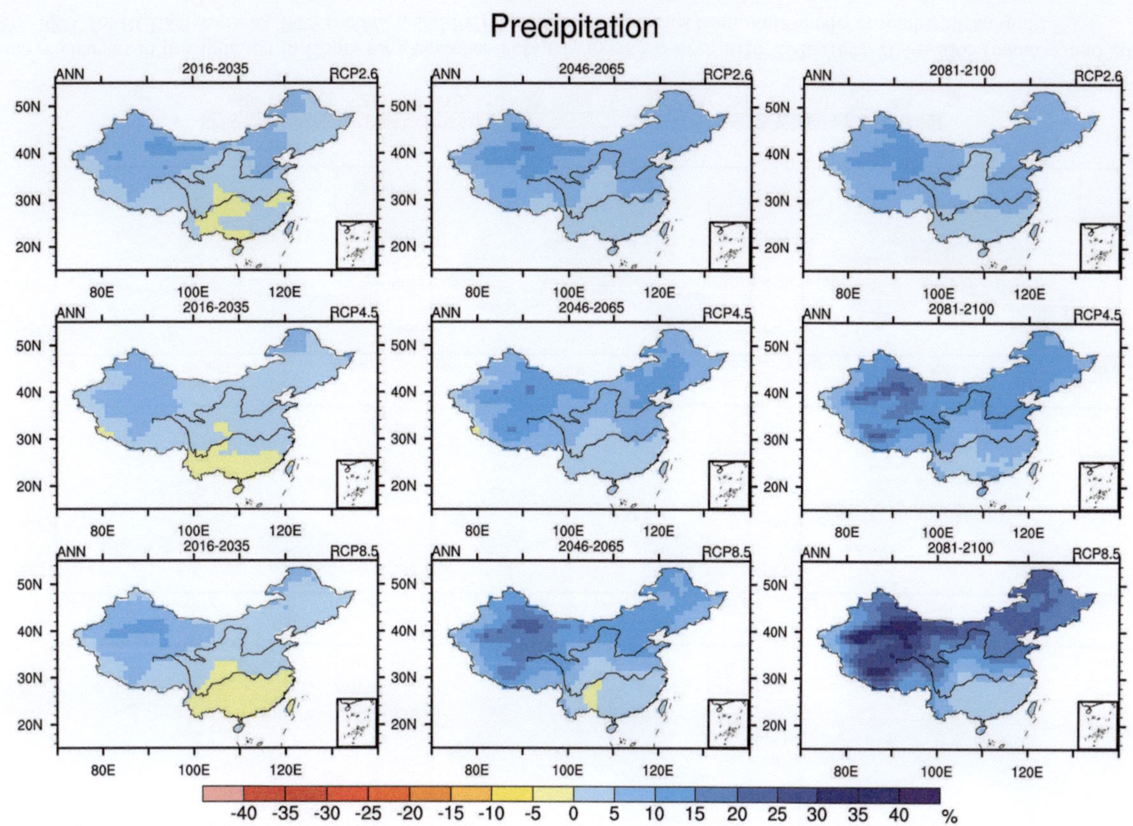

Fig. 2.90 Relative changes in annual mean precipitation in China averaged over 2016–2035 (*left*), 2046–2065 (*middle*), and 2081–2100 (*right*), relative to 1986–2005, for RCP2.6 (*top*), RCP4.5 (*middle*), and RCP8.5 (*bottom*) scenarios from multi-model ensemble mean (unit: %)

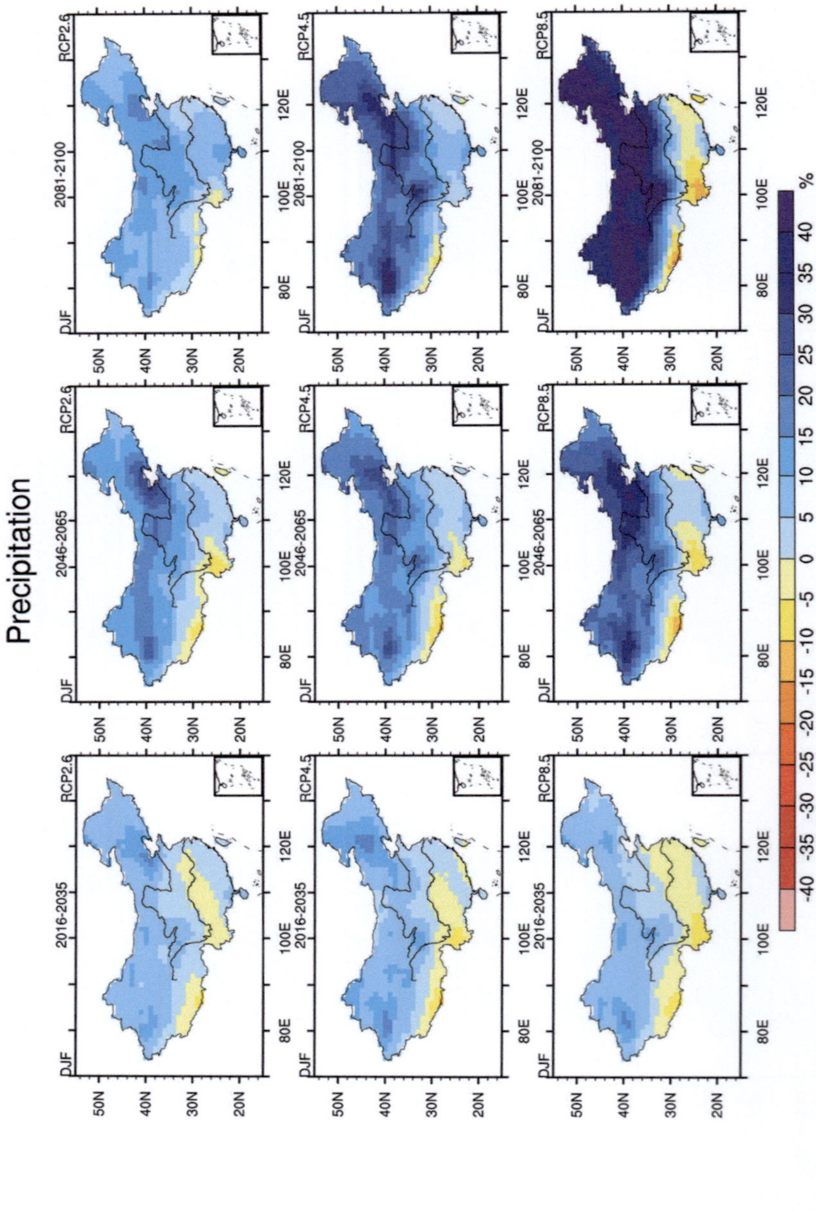

Fig. 2.91 Relative changes in precipitation in China for December–February averaged over 2016–2035 (*left*), 2046–2065 (*middle*), and 2081–2100 (*right*), relative to 1986–2005, for RCP2.6 (*top*), RCP4.5 (*middle*), and RCP8.5 (*bottom*) scenarios from multi-model ensemble mean (unit: %)

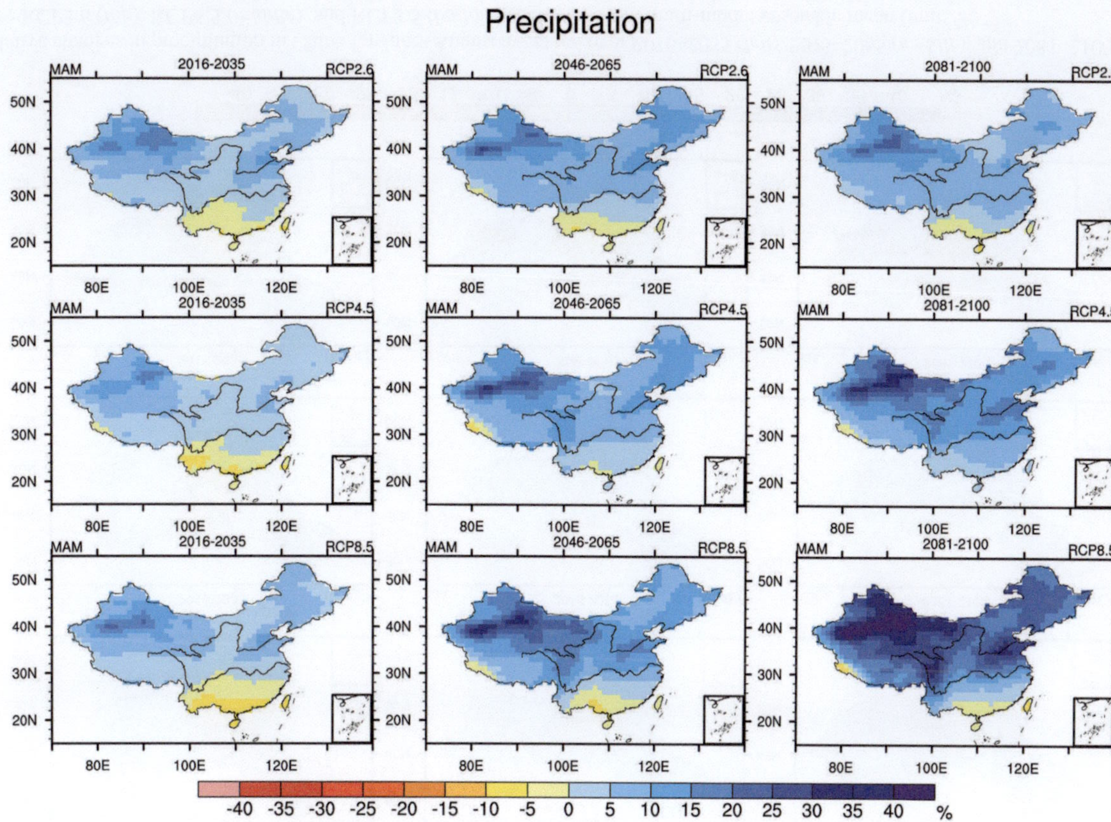

Fig. 2.92 Relative changes in precipitation in China for March–May averaged over 2016–2035 (*left*), 2046–2065 (*middle*), and 2081–2100 (*right*), relative to 1986–2005, for RCP2.6 (*top*), RCP4.5 (*middle*), and RCP8.5 (*bottom*) scenarios from multi-model ensemble mean (unit: %)

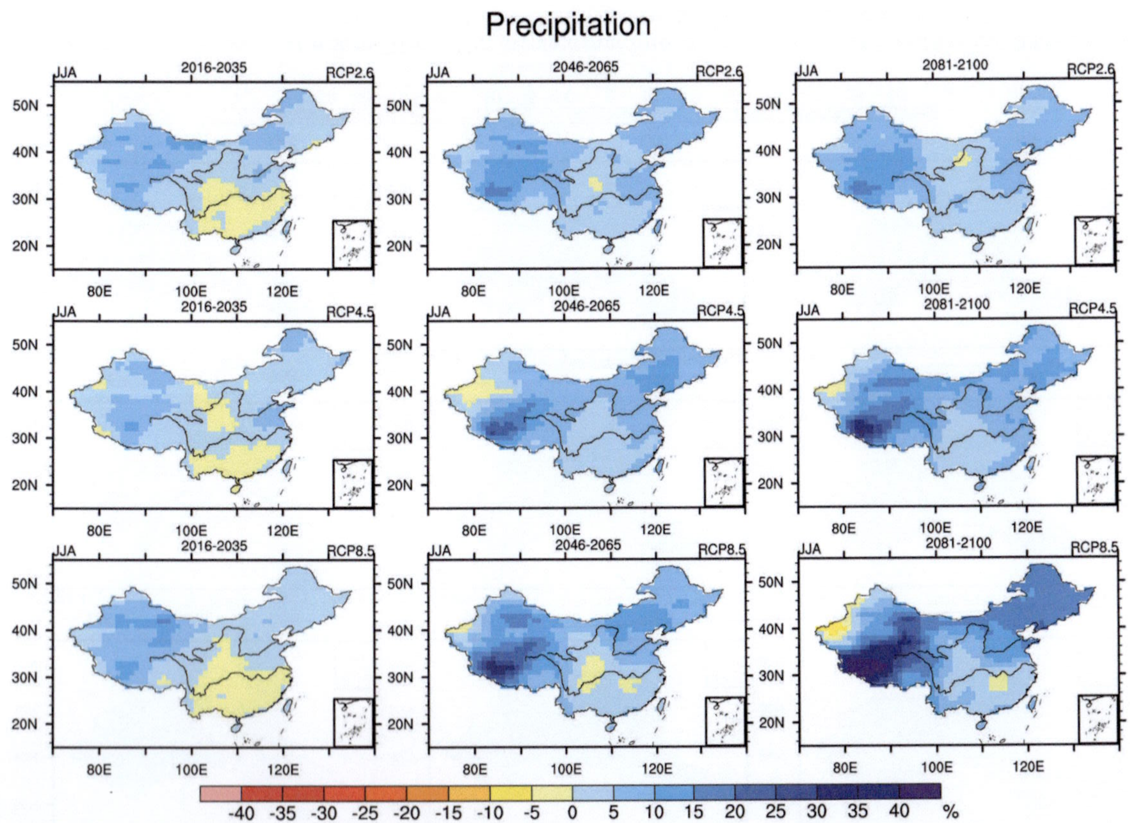

Fig. 2.93 Relative changes in precipitation in China for June–August averaged over 2016–2035 (*left*), 2046–2065 (*middle*), and 2081–2100 (*right*), relative to 1986–2005, for RCP2.6 (*top*), RCP4.5 (*middle*), and RCP8.5 (*bottom*) scenarios from multi-model ensemble mean (unit: %)

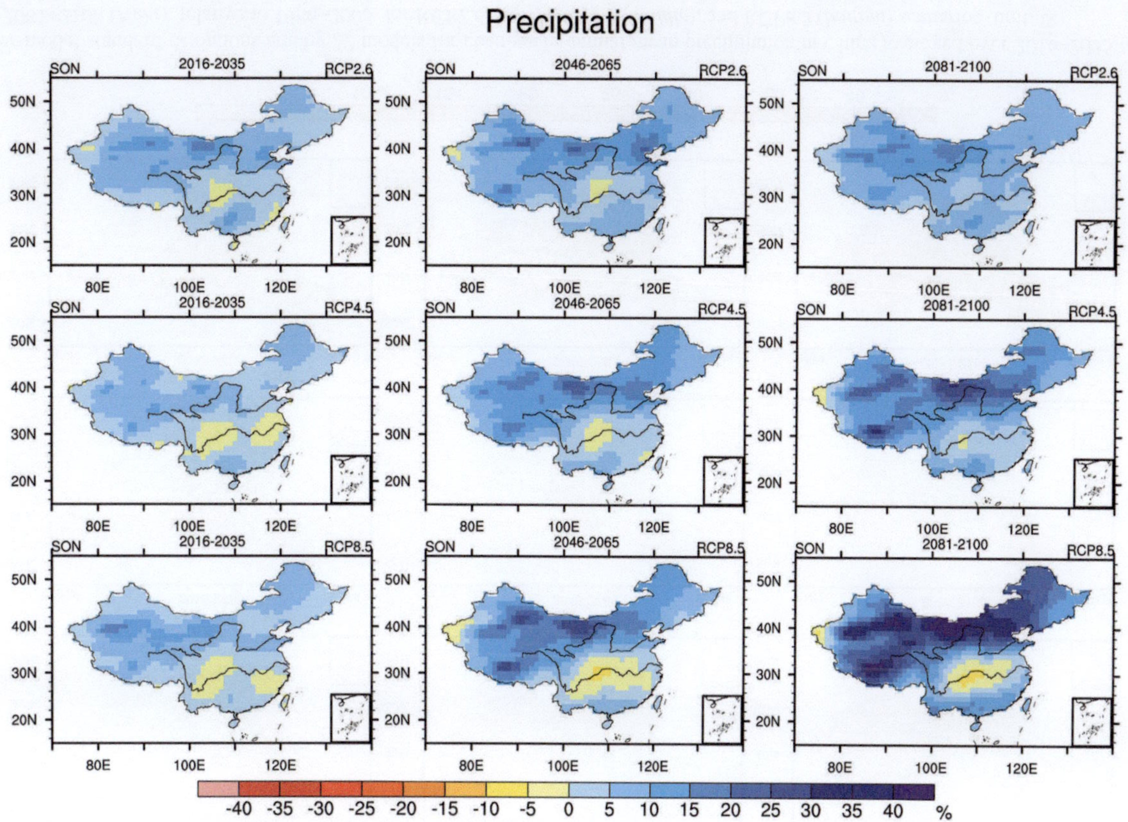

Fig. 2.94 Relative changes in precipitation in China for September–November averaged over 2016–2035 (*left*), 2046–2065 (*middle*), and 2081–2100 (*right*), relative to 1986–2005, for RCP2.6 (*top*), RCP4.5 (*middle*), and RCP8.5 (*bottom*) scenarios from multi-model ensemble mean (unit: %)

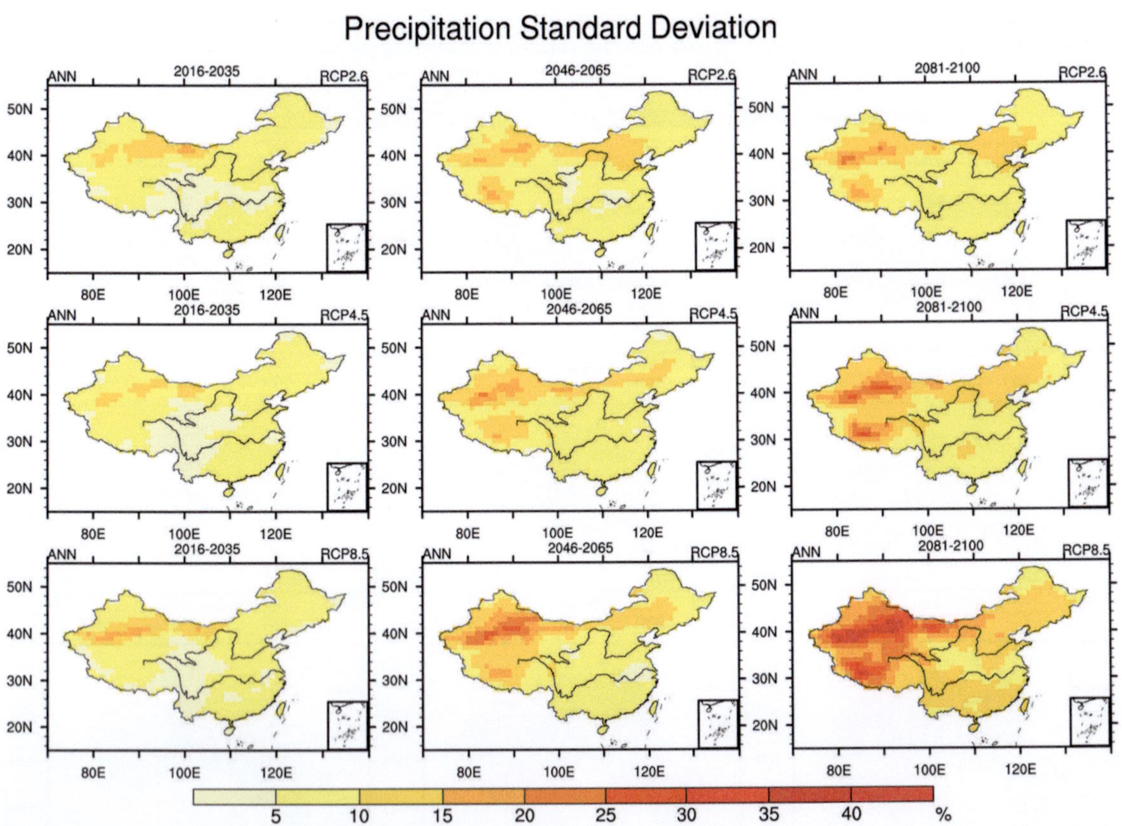

Fig. 2.95 Inter-model standard deviations among 22 models for changes in annual mean precipitation in China averaged over 2016–2035 (*left*), 2046–2065 (*middle*), and 2081–2100 (*right*), relative to 1986–2005, for RCP2.6 (*top*), RCP4.5 (*middle*), and RCP8.5 (*bottom*) scenarios (unit: %)

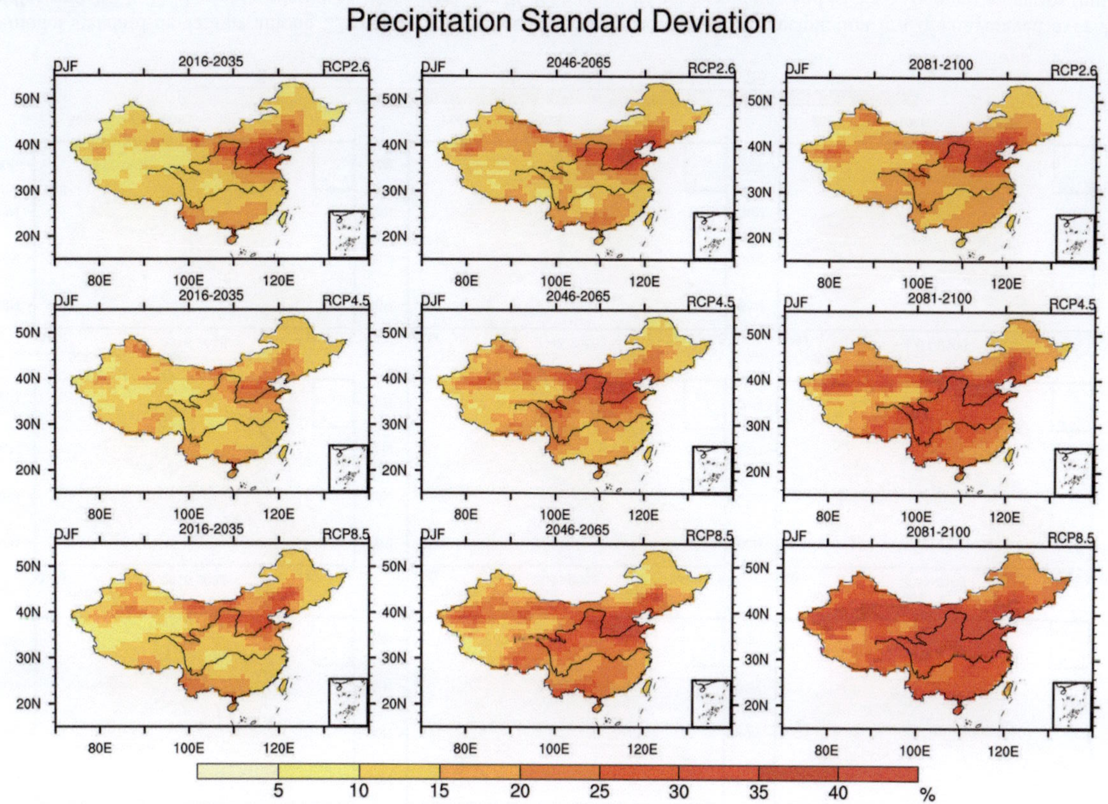

Fig. 2.96 Inter-model standard deviations among 22 models for changes in winter (December–February) precipitation in China averaged over 2016–2035 (*left*), 2046–2065 (*middle*), and 2081–2100 (*right*), relative to 1986–2005, for RCP2.6 (*top*), RCP4.5 (*middle*), and RCP8.5 (*bottom*) scenarios (unit: %)

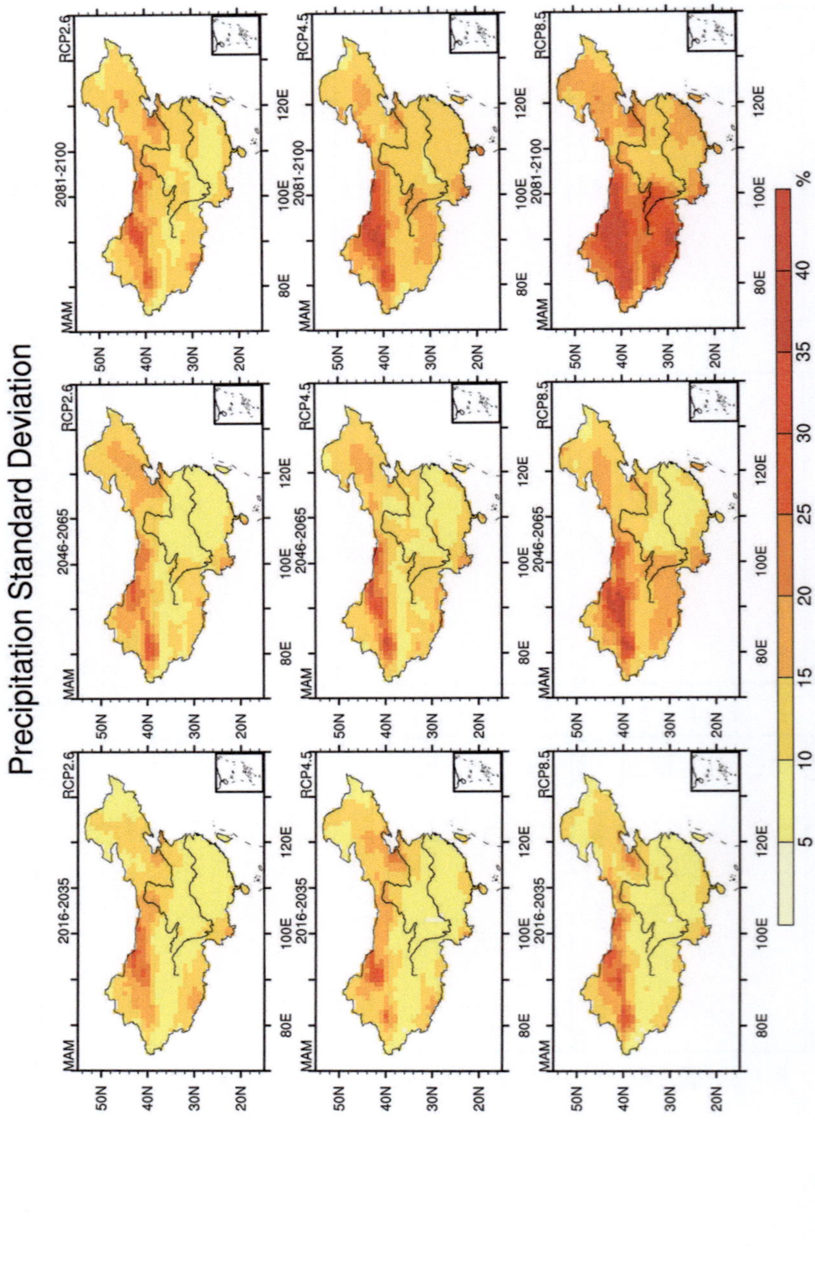

Fig. 2.97 Inter-model standard deviations among 22 models for changes in spring (March–May) precipitation in China averaged over 2016–2035 (*left*), 2046–2065 (*middle*), and 2081–2100 (*right*), relative to 1986–2005, for RCP2.6 (*top*), RCP4.5 (*middle*), and RCP8.5 (*bottom*) scenarios (unit: %)

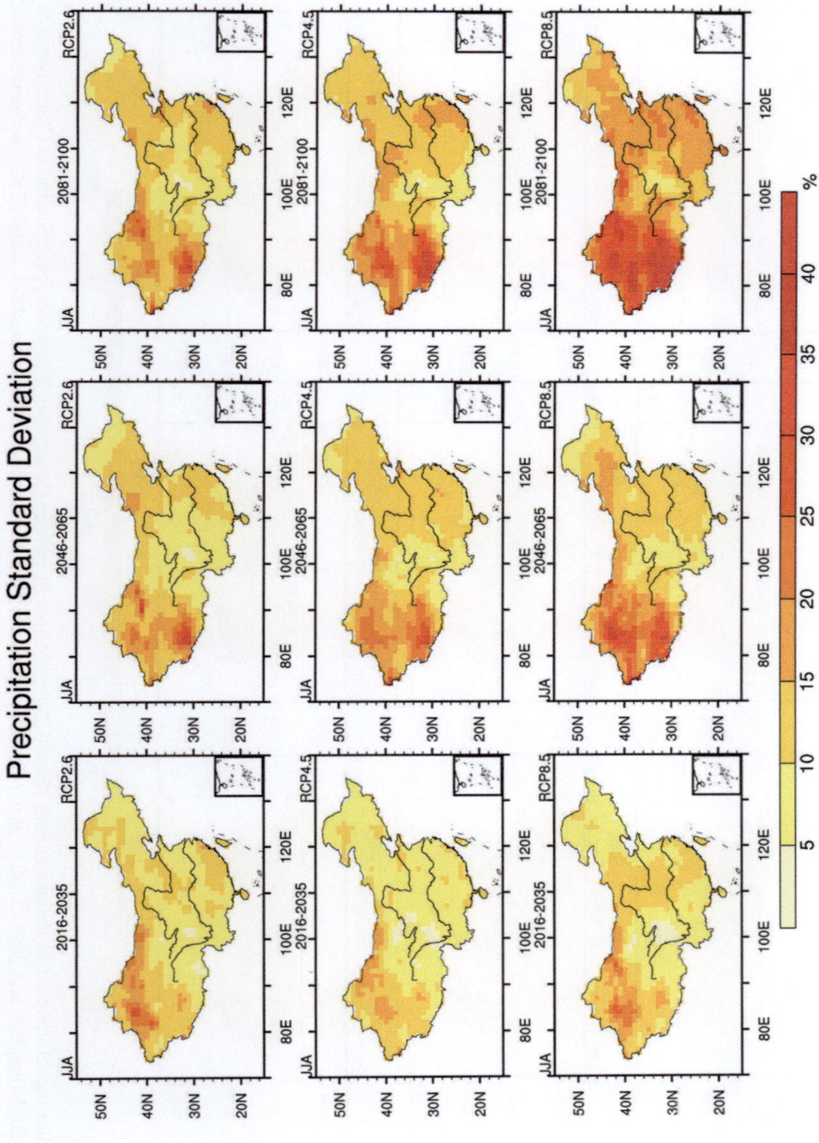

Fig. 2.98 Inter-model standard deviations among 22 models for changes in summer (June–August) precipitation in China averaged over 2016–2035 (*left*), 2046–2065 (*middle*), and 2081–2100 (*right*), relative to 1986–2005, for RCP2.6 (*top*), RCP4.5 (*middle*), and RCP8.5 (*bottom*) scenarios (unit: %)

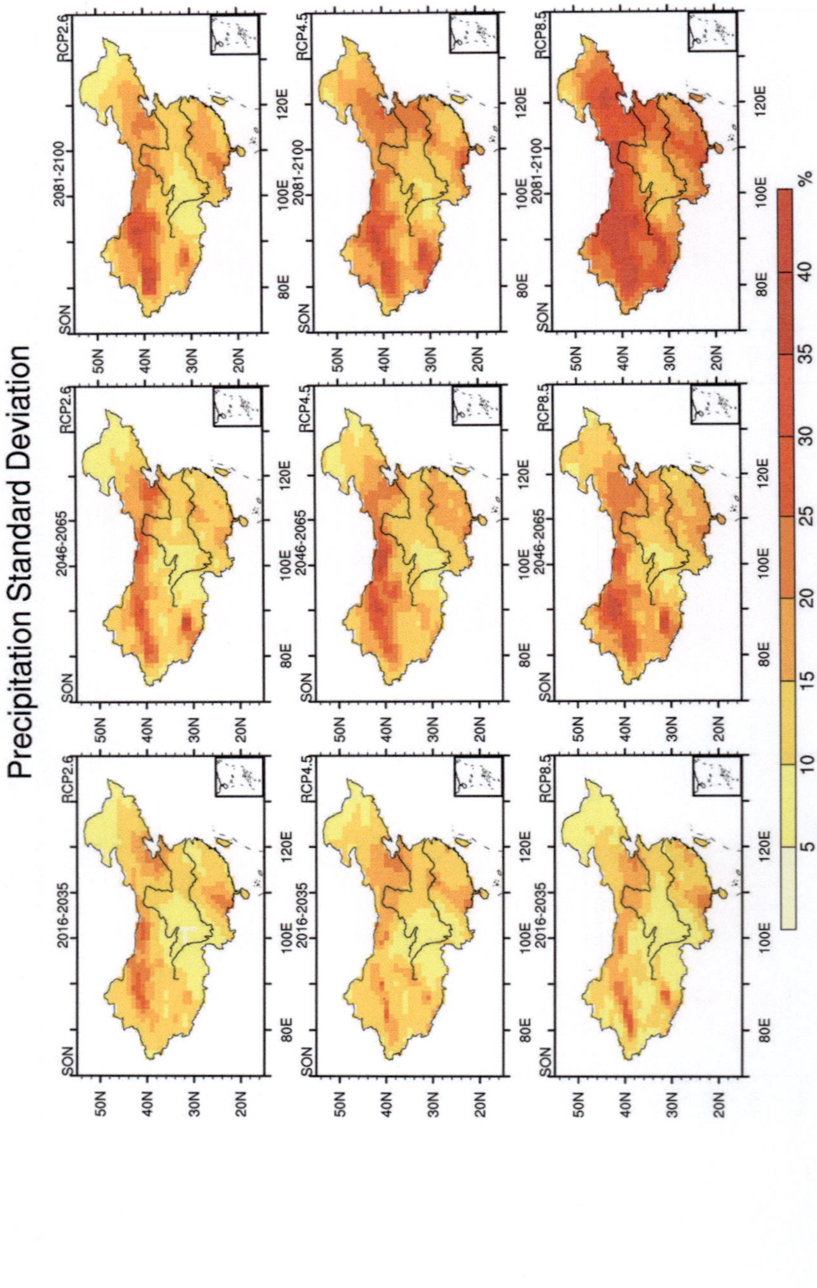

Fig. 2.99 Inter-model standard deviations among 22 models for changes in autumn (September–November) precipitation in China averaged over 2016–2035 (*left*), 2046–2065 (*middle*), and 2081–2100 (*right*), relative to 1986–2005, for RCP2.6 (*top*), RCP4.5 (*middle*), and RCP8.5 (*bottom*) scenarios (unit: %)

2.3 Atmospheric Circulation in the Asia-Pacific

2.3.1 Historical Atmospheric Circulation (Figs. 2.100, 2.101, 2.102, and 2.103)

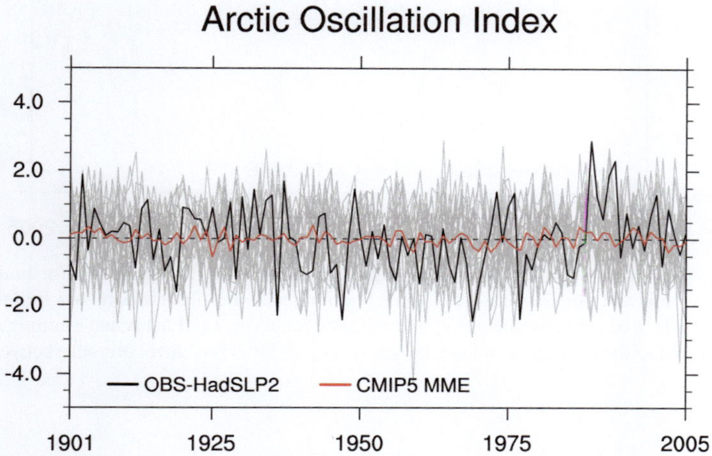

Fig. 2.100 Arctic Oscillation index anomalies relative to 1986–2005 for December–February from observation and simulations. The *black line* indicates the observation, *red line* the multi-model ensemble, and *gray line* the individual model simulation

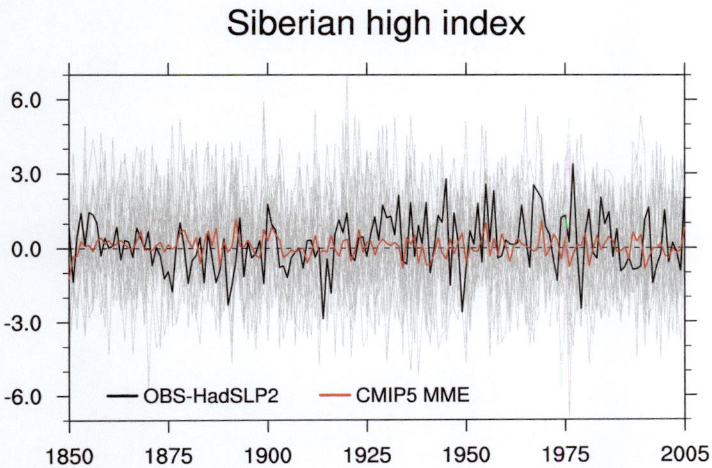

Fig. 2.101 Siberian High index anomalies relative to 1986–2005 for December–February from observation and simulations. The *black line* indicates the observation, *red line* the multi-model ensemble, and *gray line* the individual model simulation. The Siberian High index is defined as the regional-averaged sea level pressure over (40–60°N, 80–120°E)

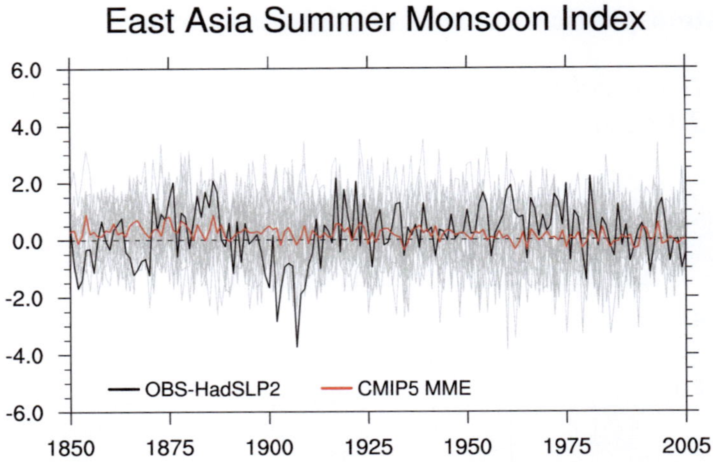

Fig. 2.102 East Asian Summer Monsoon index anomalies relative to 1986–2005 for June–August from observation and simulations. The *black line* indicates the observation, *red line* the multi-model ensemble, and *gray line* the individual model simulation. The East Asian Summer Monsoon index is defined as the difference of meridional-mean (10–50°N) sea level pressure between 160°E and 110°E

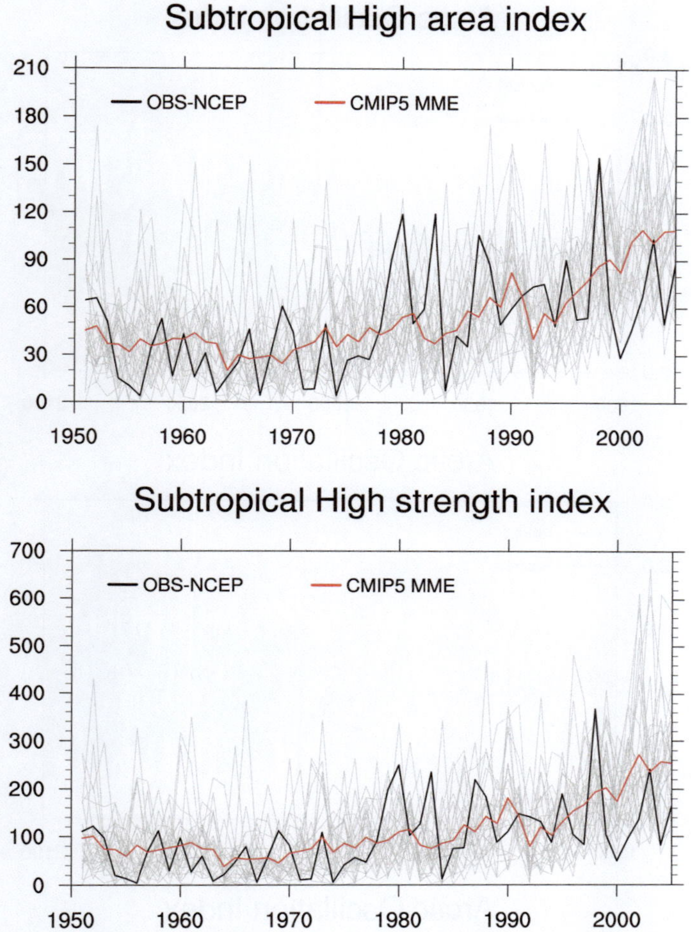

Fig. 2.103 The Western Pacific Subtropical High area (*upper*) and strength (*lower*) indices for June–August from observation and simulations. The *black line* indicates the observation, *red line* the multi-model ensemble, and *gray line* the individual model simulation

2.3.2 Projected Atmospheric Circulation (Figs. 2.104, 2.105, 2.106, and 2.107)

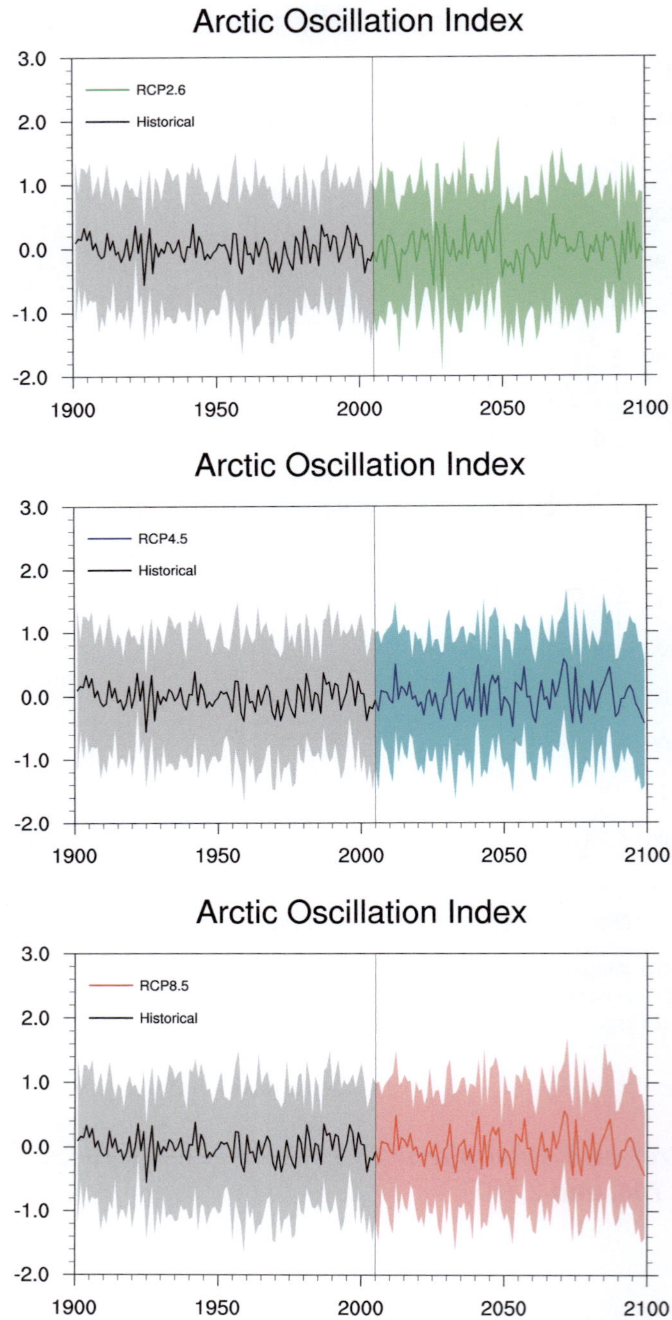

Fig. 2.104 Arctic Oscillation index anomalies relative to 1986–2005 for December–February from historical simulation and future projection under the RCP scenarios

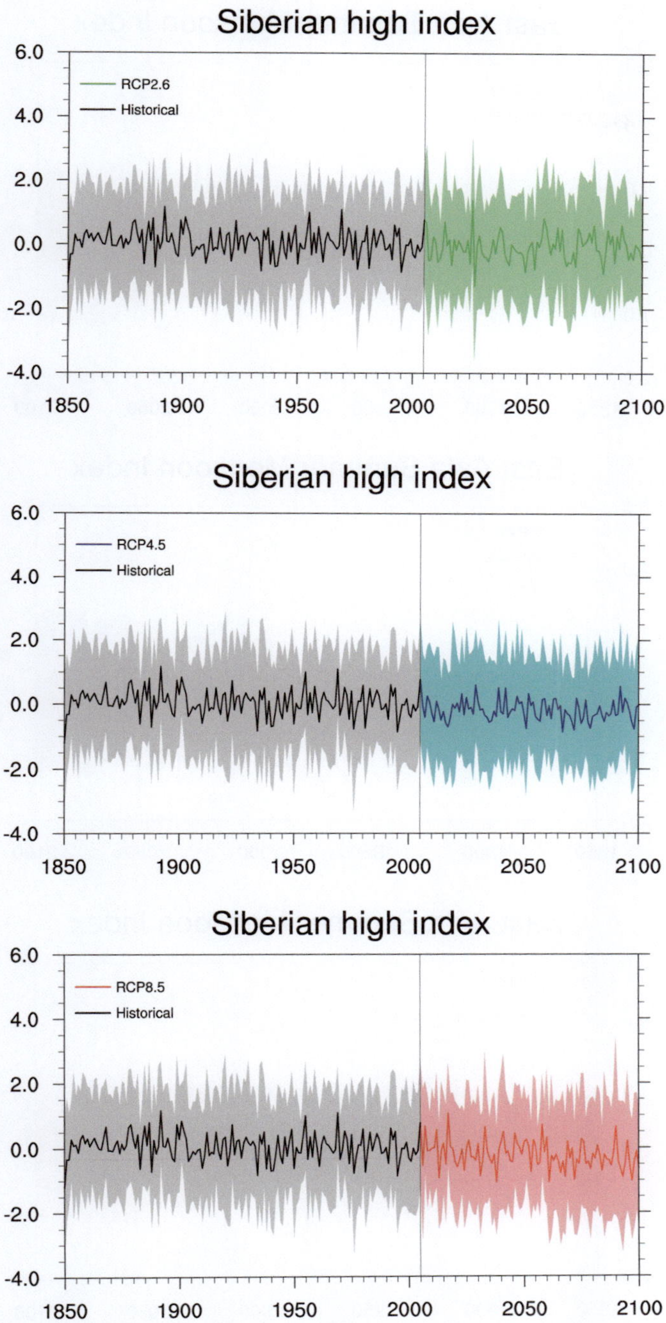

Fig. 2.105 Siberian High index anomalies relative to 1986–2005 for December–February from historical simulation and future projection under the RCP scenarios

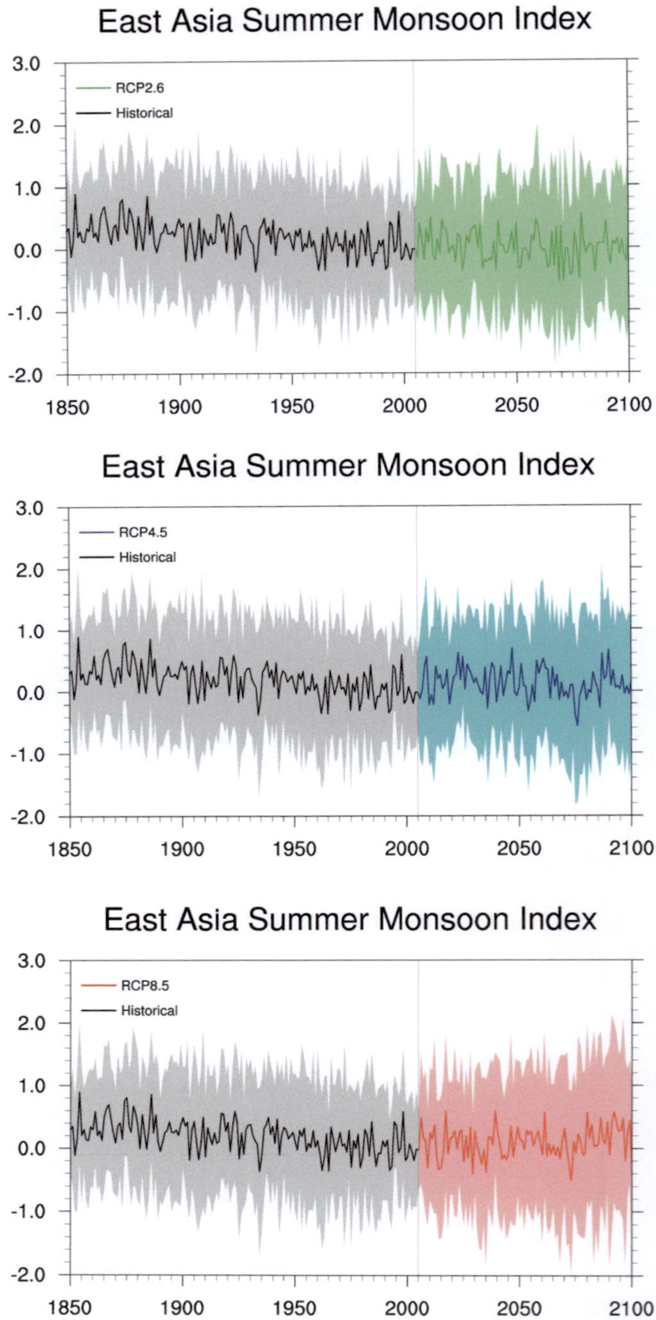

Fig. 2.106 East Asian Summer Monsoon index anomalies relative to 1986–2005 for June–August from historical simulation and future projection under the RCP scenarios

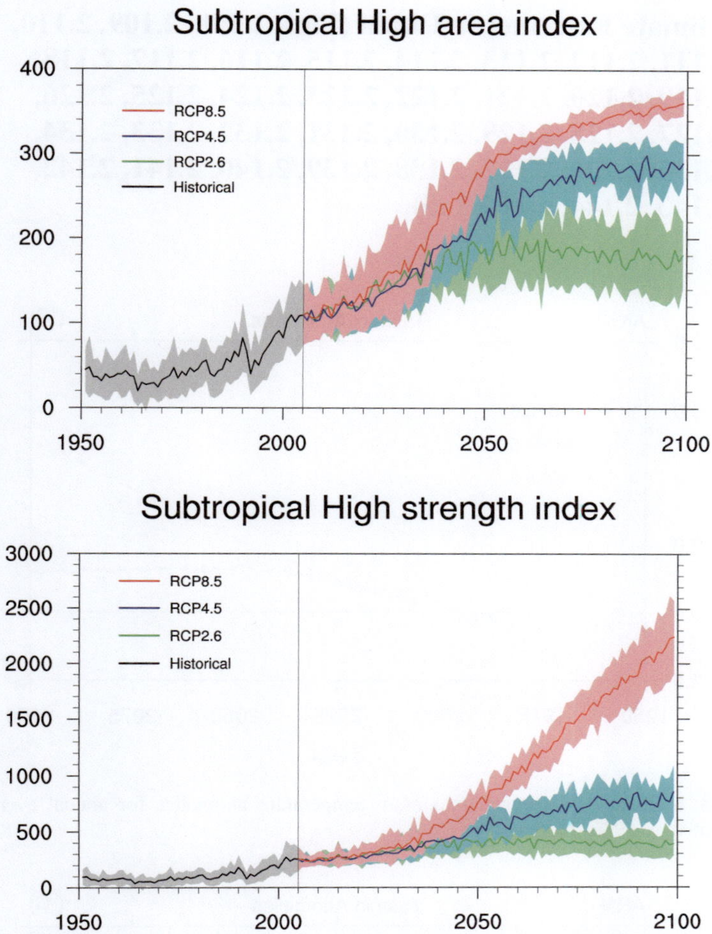

Fig. 2.107 The Western Pacific Subtropical High area and strength indices for June–August from historical simulation and future projection under the RCP scenarios

2.4 Climate Extremes in China (Figs. 2.108, 2.109, 2.110, 2.111, 2.112, 2.113, 2.114, 2.115, 2.116, 2.117, 2.118, 2.119, 2.120, 2.121, 2.122, 2.123, 2.124, 2.125, 2.126, 2.127, 2.128, 2.129, 2.130, 2.131, 2.132, 2.133, 2.134, 2.135, 2.136, 2.137, 2.138, 2.139, 2.140, 2.141, 2.142, 2.143, 2.144, and 2.145)

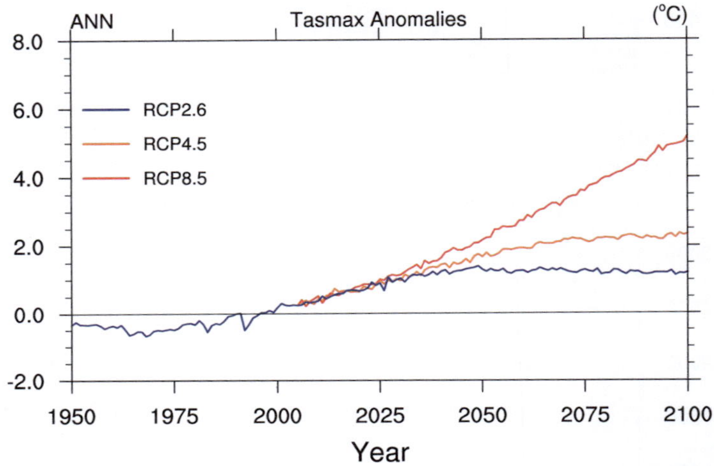

Fig. 2.108 Multi-model averaged maximum temperature anomalies for annual average over China, relative to 1986–2005

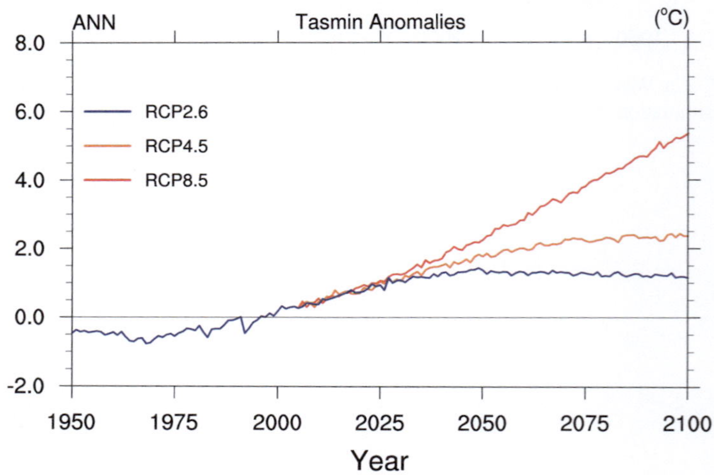

Fig. 2.109 Multi-model averaged minimum temperature anomalies for annual average over China, relative to 1986–2005

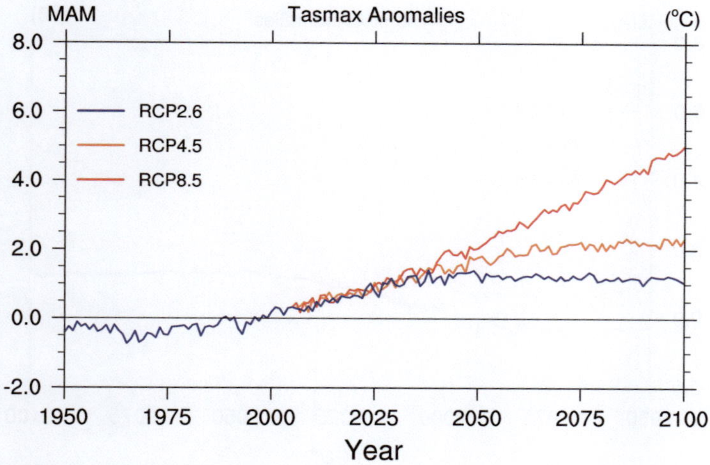

Fig. 2.110 Multi-model averaged maximum temperature anomalies for March–May over China, relative to 1986–2005

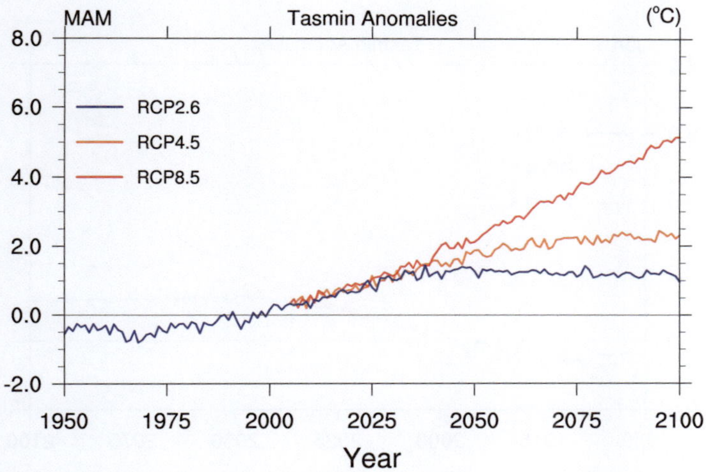

Fig. 2.111 Multi-model averaged minimum temperature anomalies for March–May over China, relative to 1986–2005

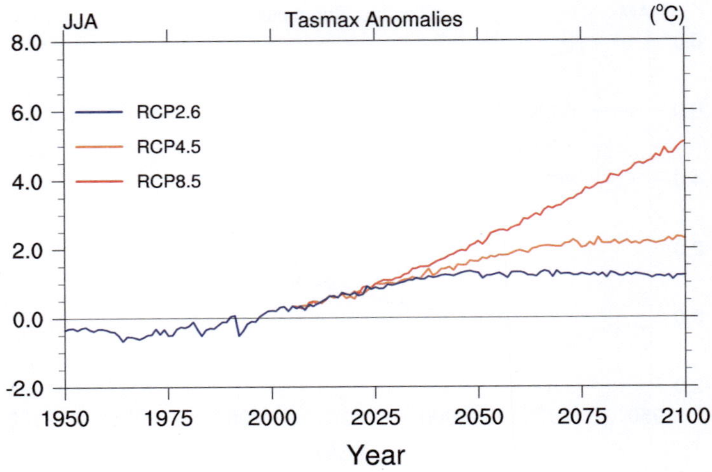

Fig. 2.112 Multi-model averaged maximum temperature anomalies for June–August over China, relative to 1986–2005

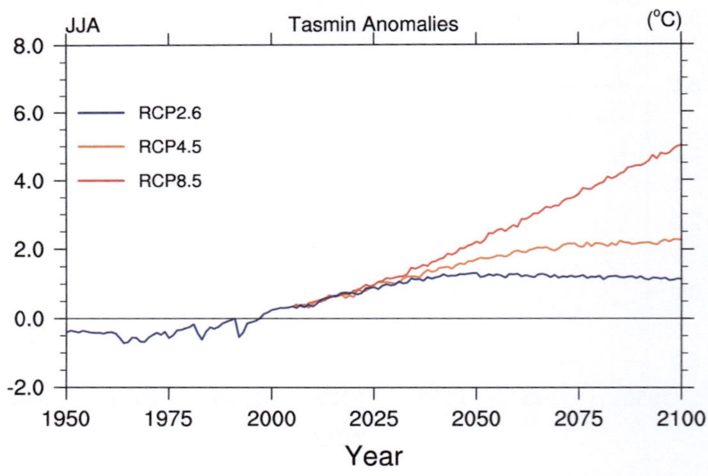

Fig. 2.113 Multi-model averaged minimum temperature anomalies for June–August over China, relative to 1986–2005

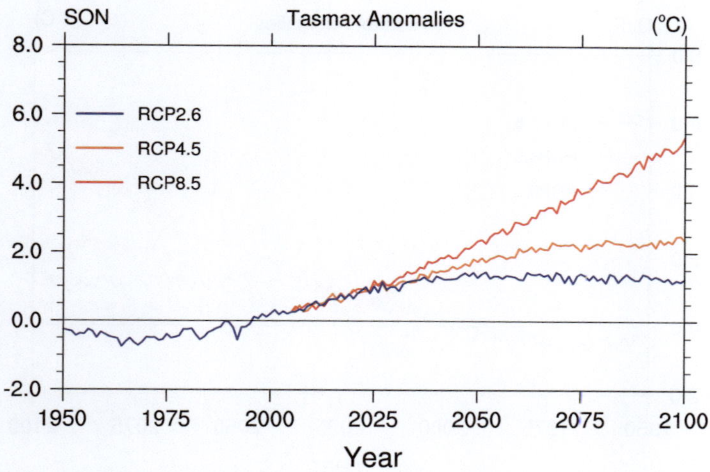

Fig. 2.114 Multi-model averaged maximum temperature anomalies for September–November over China, relative to 1986–2005

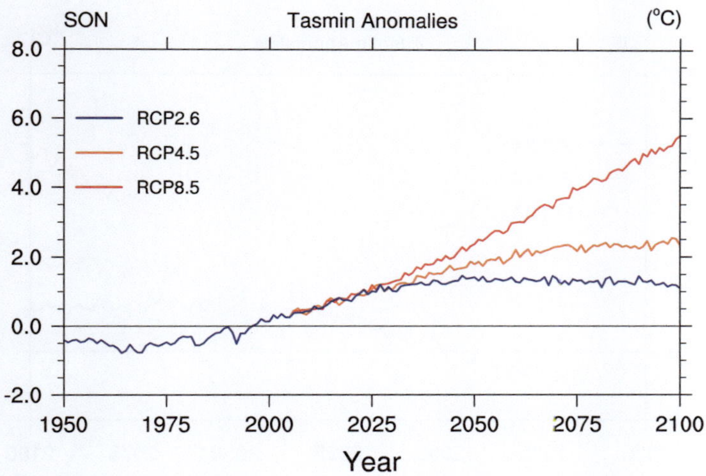

Fig. 2.115 Multi-model averaged minimum temperature anomalies for September–November over China, relative to 1986–2005

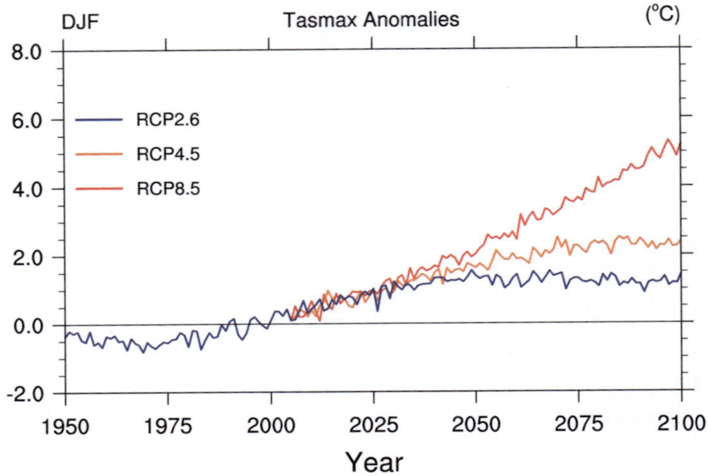

Fig. 2.116 Multi-model averaged maximum temperature anomalies for December–February over China, relative to 1986–2005

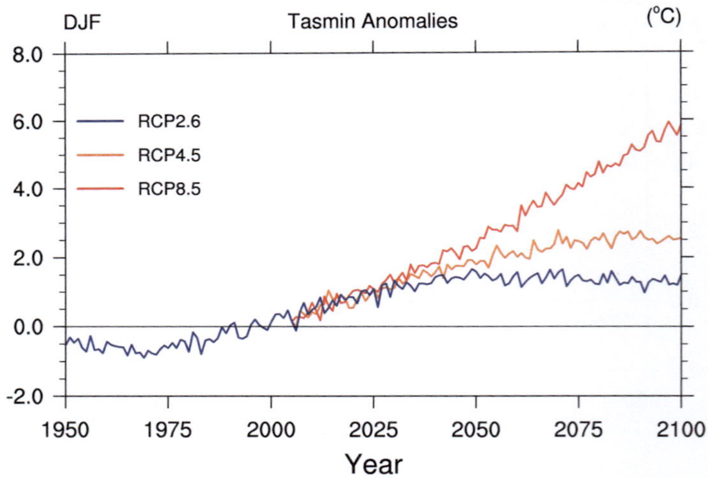

Fig. 2.117 Multi-model averaged minimum temperature anomalies for December–February over China, relative to 1986–2005

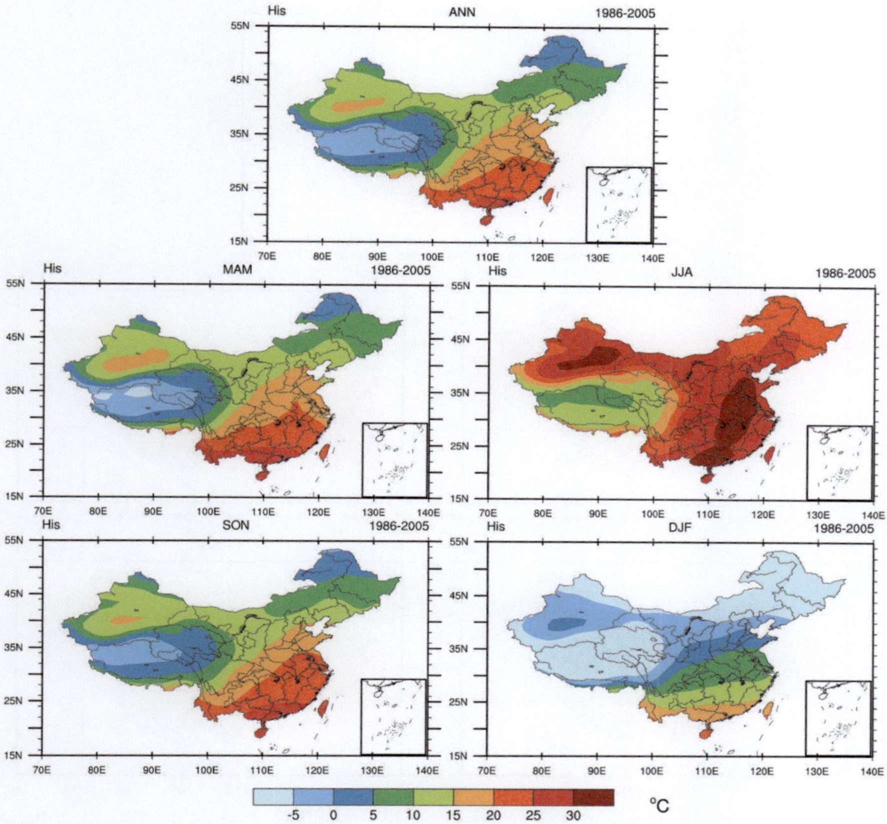

Fig. 2.118 Maximum temperature in China for annual and season averaged over 1986–2005 from multi-model average

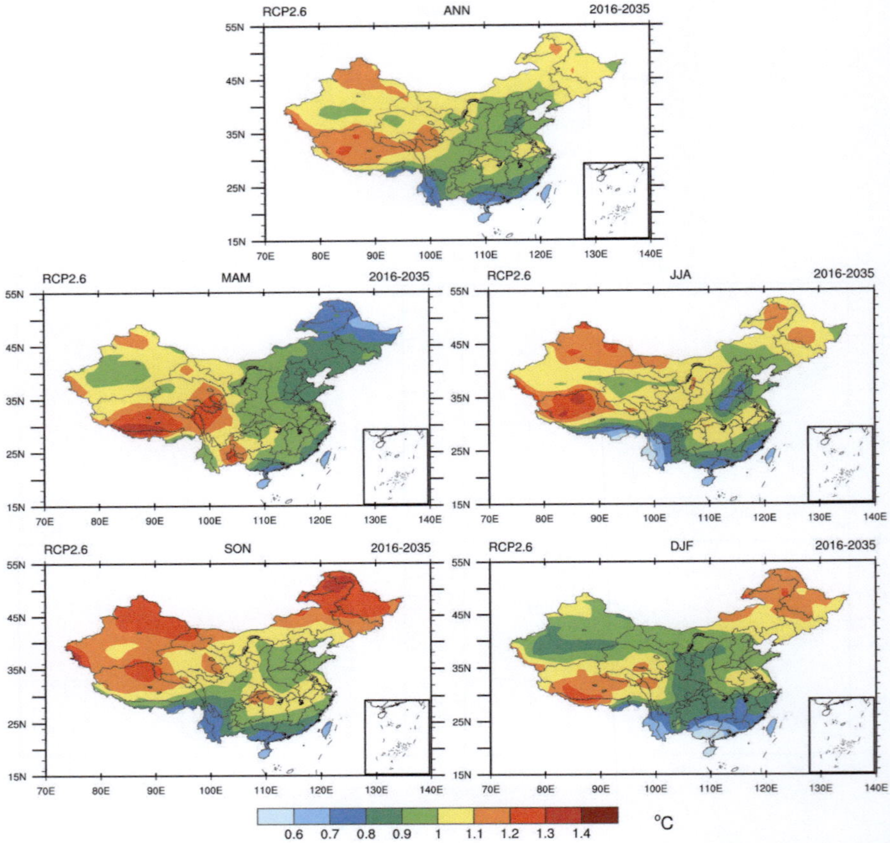

Fig. 2.119 Maximum temperature change in China for annual and season averaged over 2016–2035 from multi-model average under RCP2.6, relative to 1986–2005

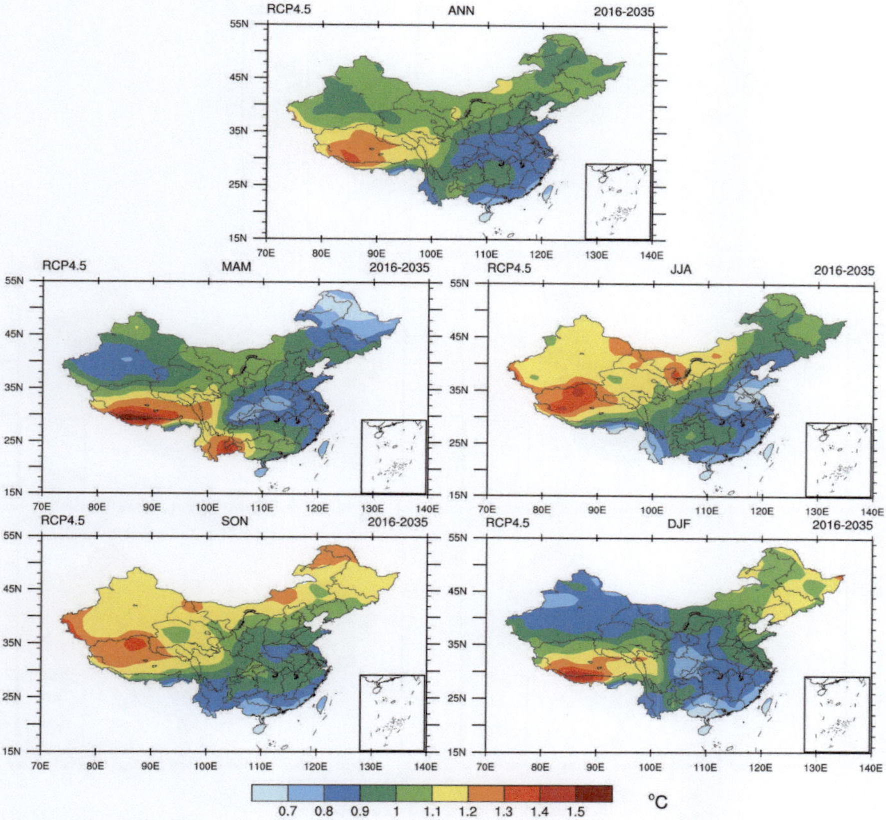

Fig. 2.120 Maximum temperature change in China for annual and season averaged over 2016–2035 from multi-model average under RCP4.5, relative to 1986–2005

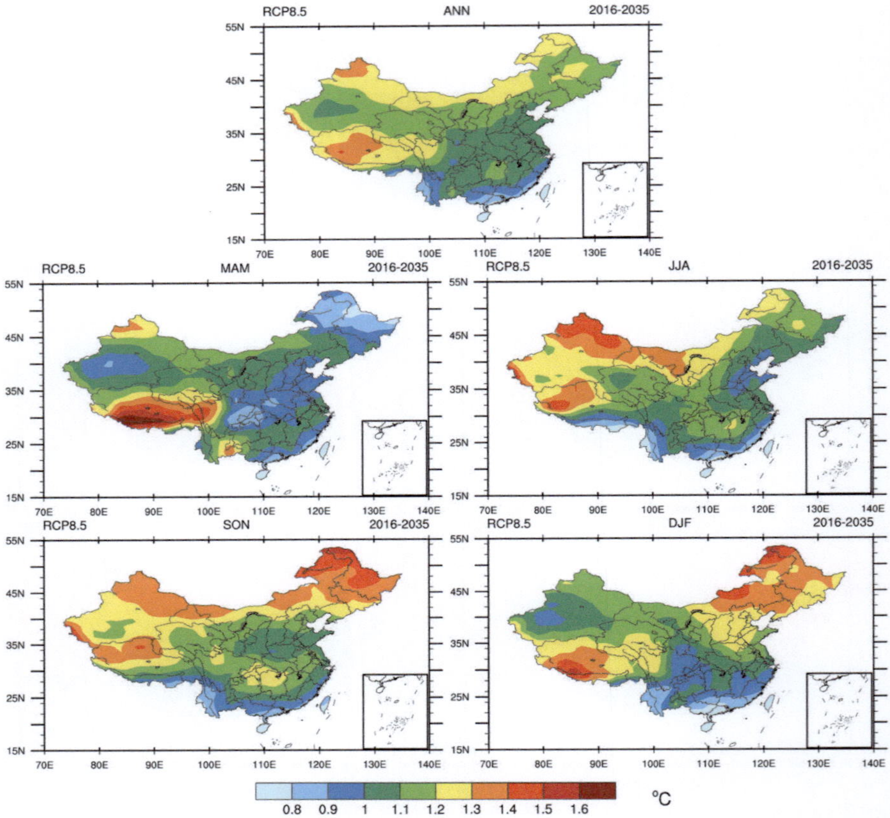

Fig. 2.121 Maximum temperature change in China for annual and season averaged over 2016–2035 from multi-model average under RCP8.5, relative to 1986–2005

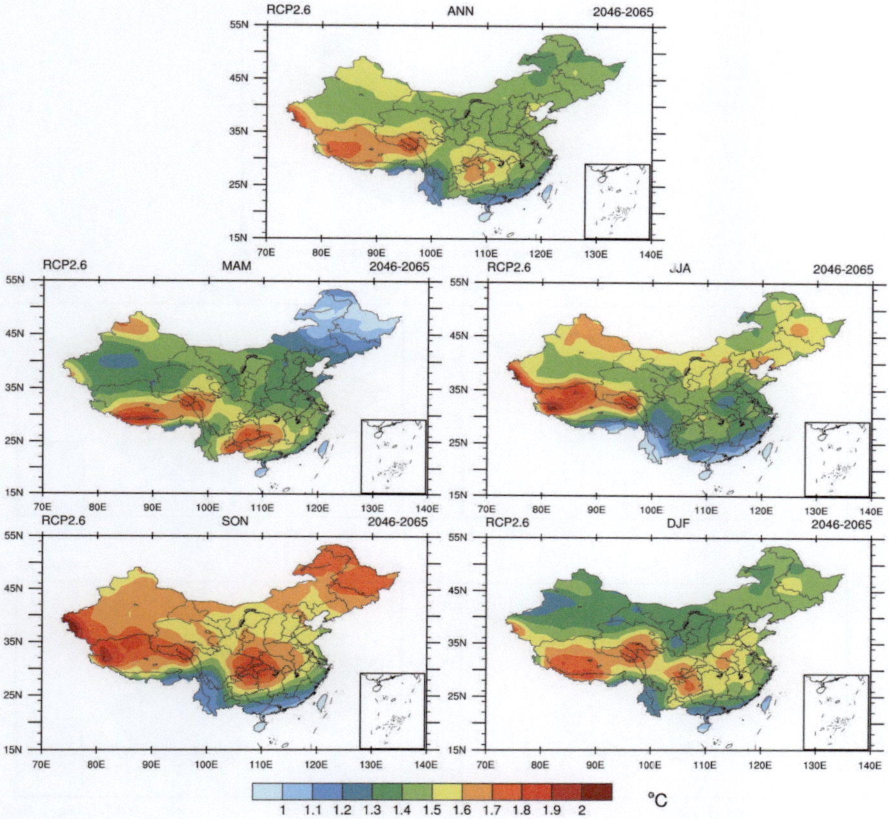

Fig. 2.122 Maximum temperature change in China for annual and season averaged over 2046–2065 from multi-model average under RCP2.6, relative to 1986–2005

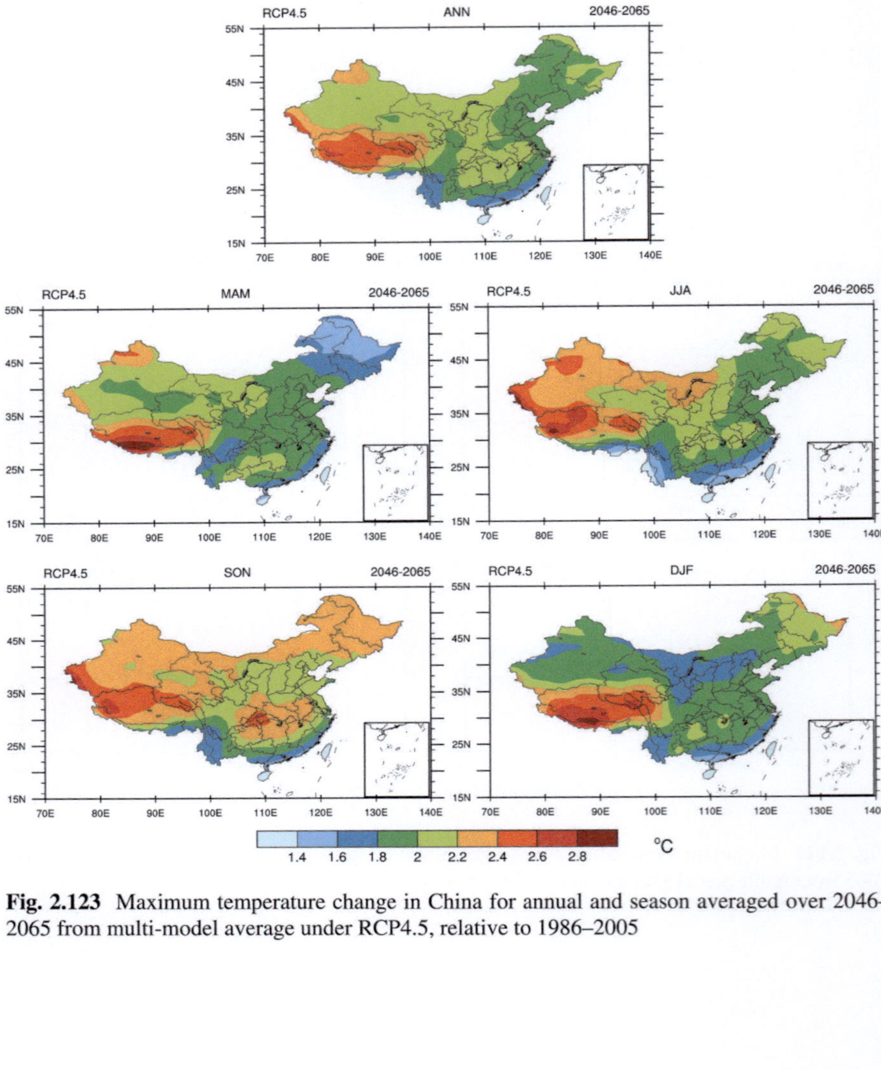

Fig. 2.123 Maximum temperature change in China for annual and season averaged over 2046–2065 from multi-model average under RCP4.5, relative to 1986–2005

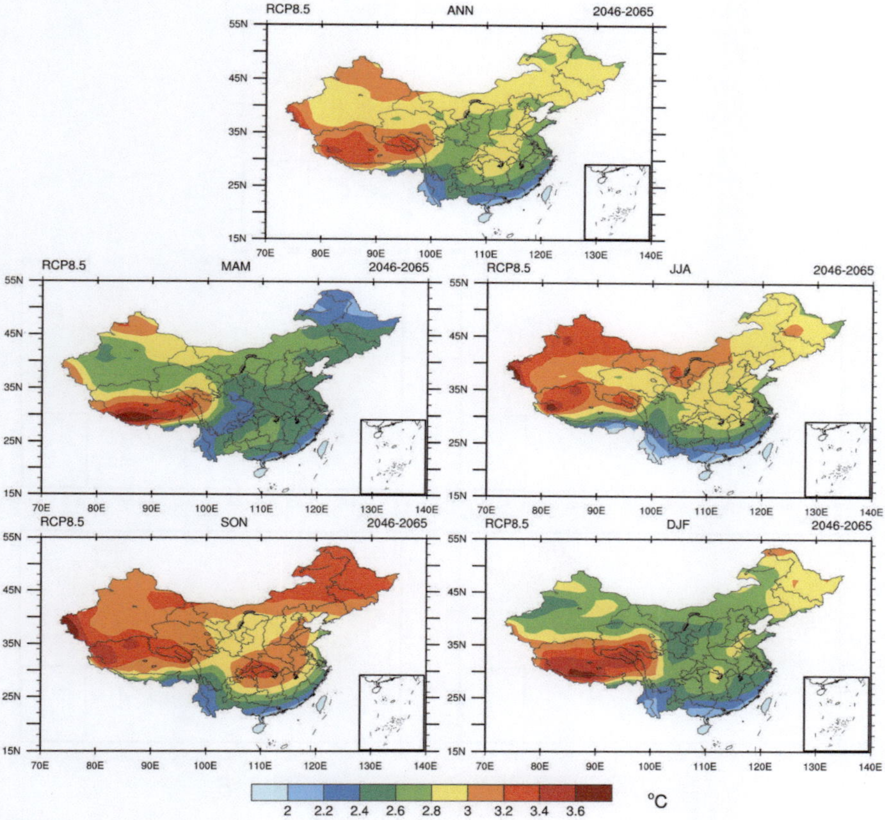

Fig. 2.124 Maximum temperature change in China for annual and season averaged over 2046–2065 from multi-model average under RCP8.5, relative to 1986–2005

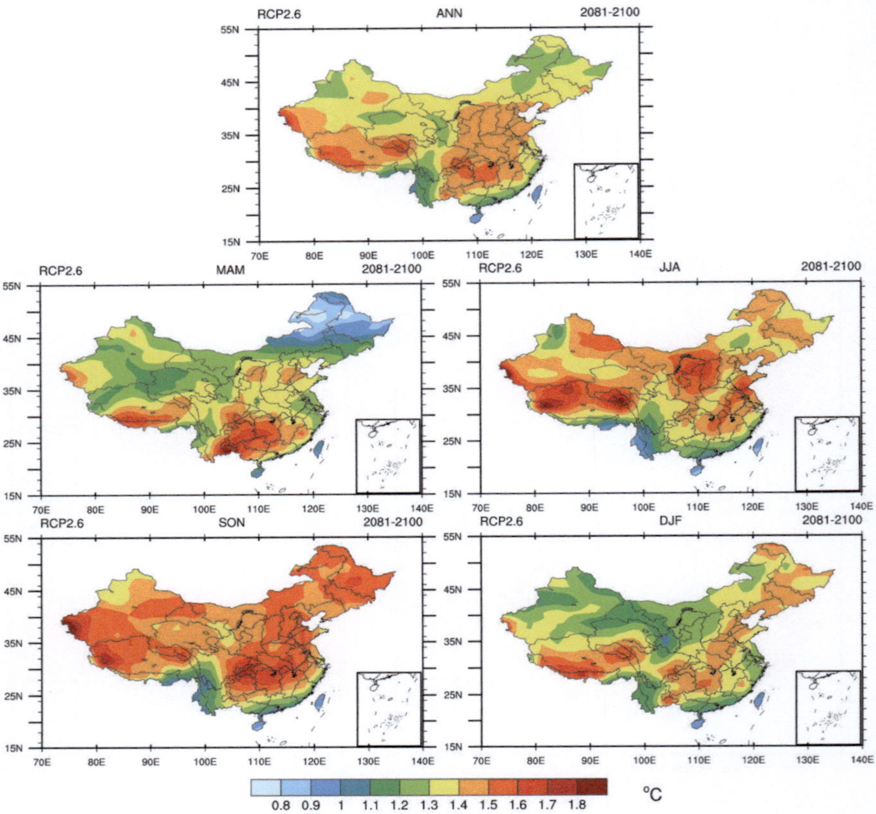

Fig. 2.125 Maximum temperature change in China for annual and season averaged over 2081–2100 from multi-model average under RCP2.6, relative to 1986–2005

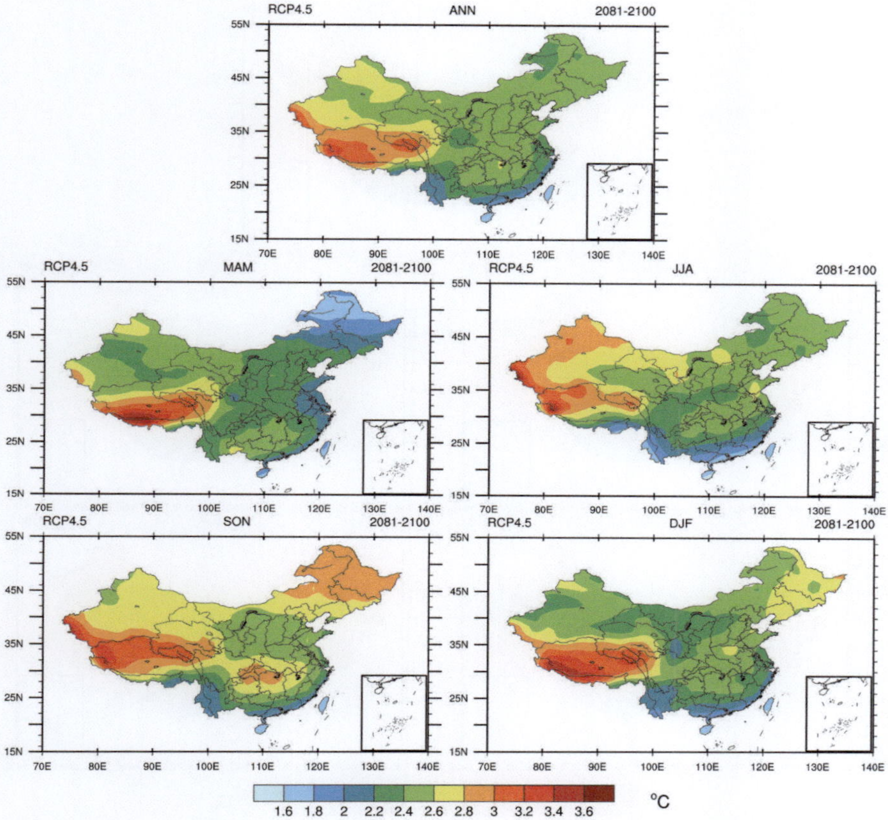

Fig. 2.126 Maximum temperature change in China for annual and season averaged over 2081–2100 from multi-model average under RCP4.5, relative to 1986–2005

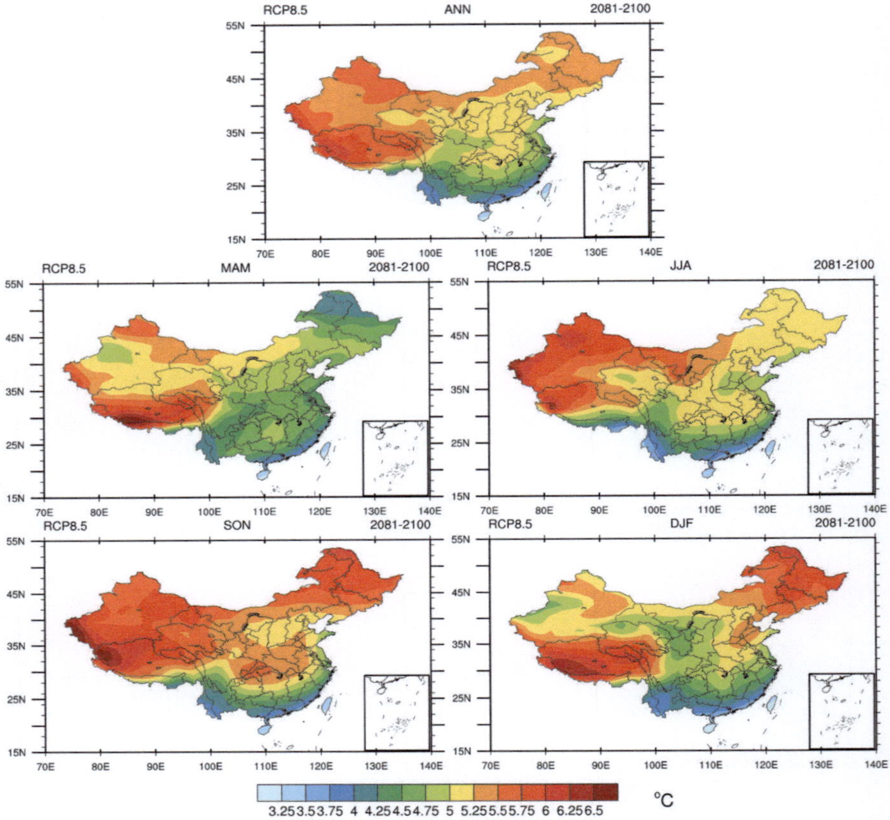

Fig. 2.127 Maximum temperature change in China for annual and season averaged over 2081–2100 from multi-model average under RCP8.5, relative to 1986–2005

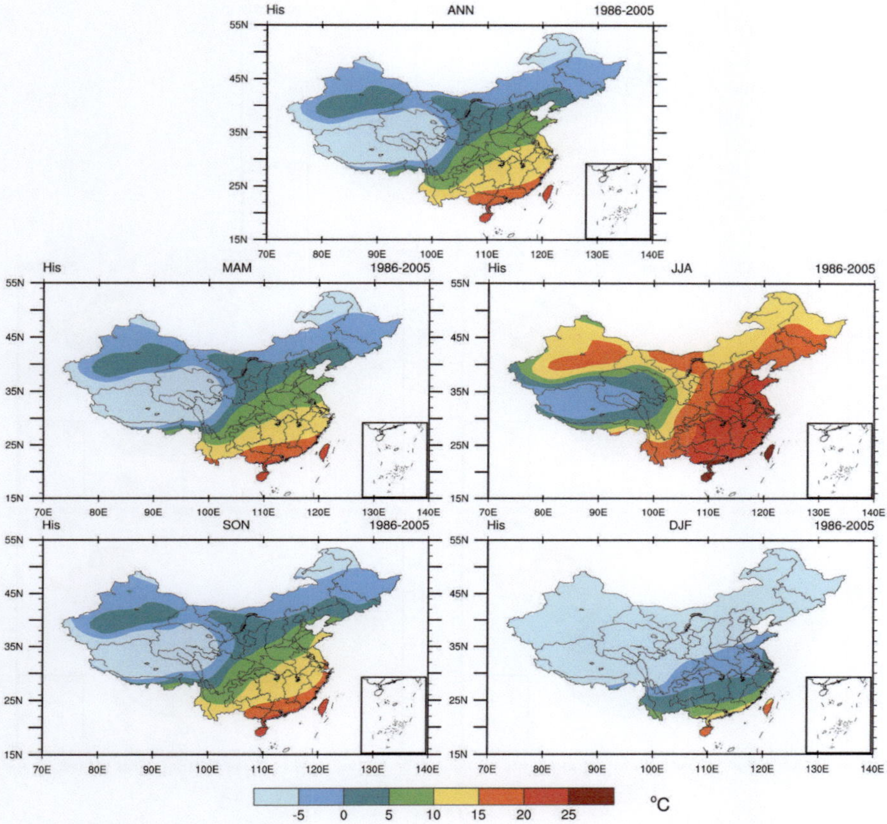

Fig. 2.128 Minimum temperature in China for annual and season averaged over 1986–2005 from multi-model average

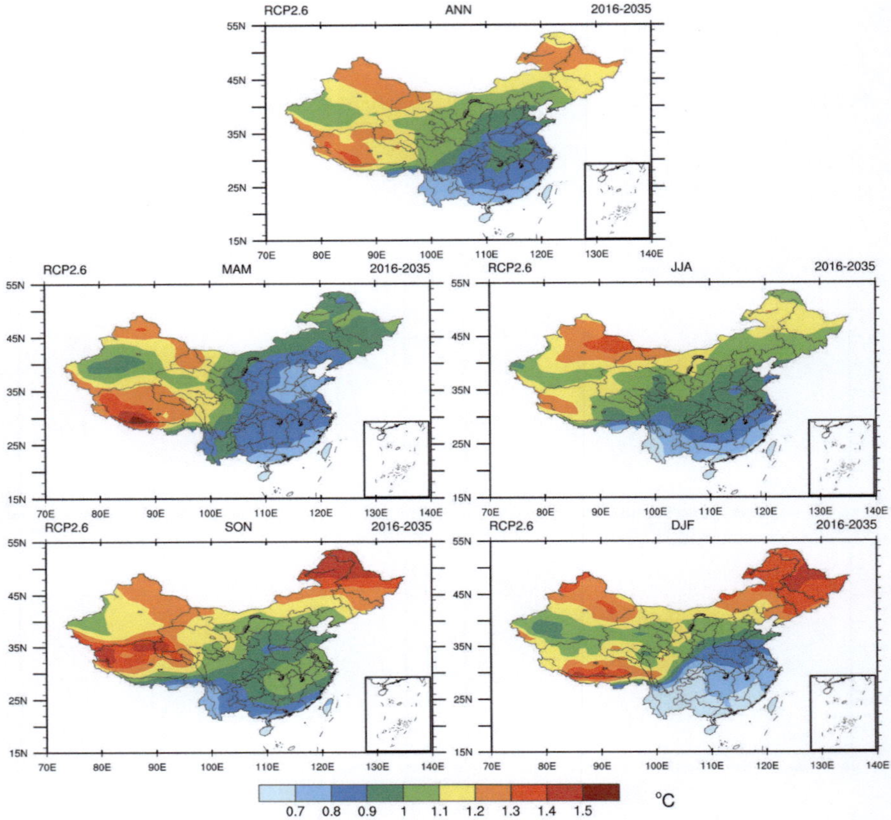

Fig. 2.129 Minimum temperature change in China for annual and season averaged over 2016–2035 from multi-model average under RCP2.6, relative to 1986–2005

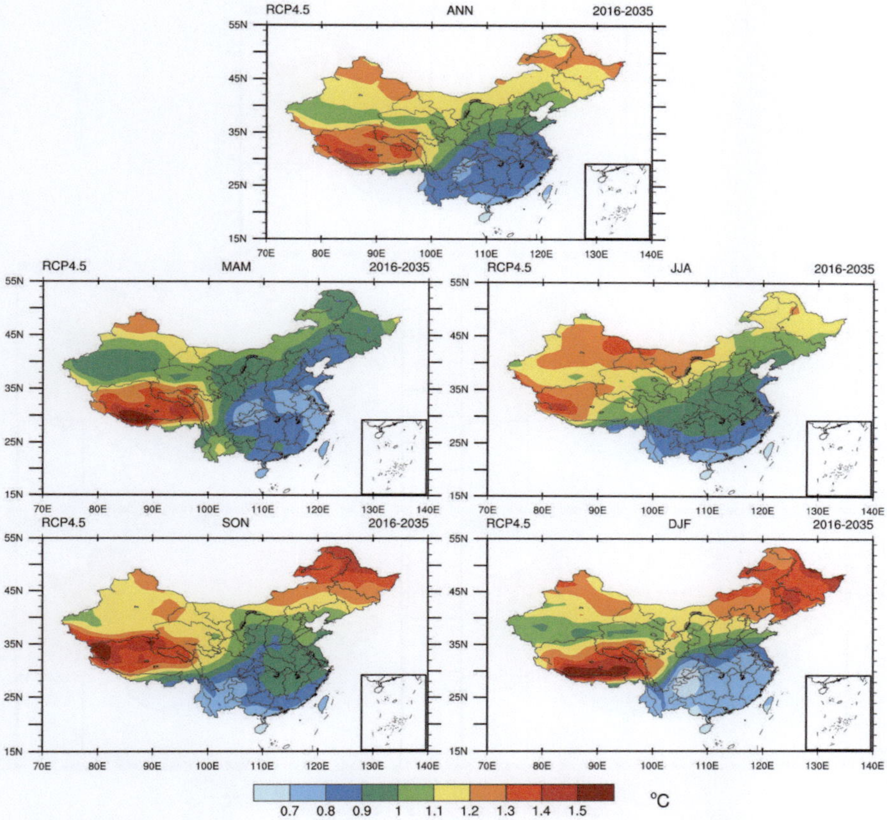

Fig. 2.130 Minimum temperature change in China for annual and season averaged over 2016–2035 from multi-model average under RCP4.5, relative to 1986–2005

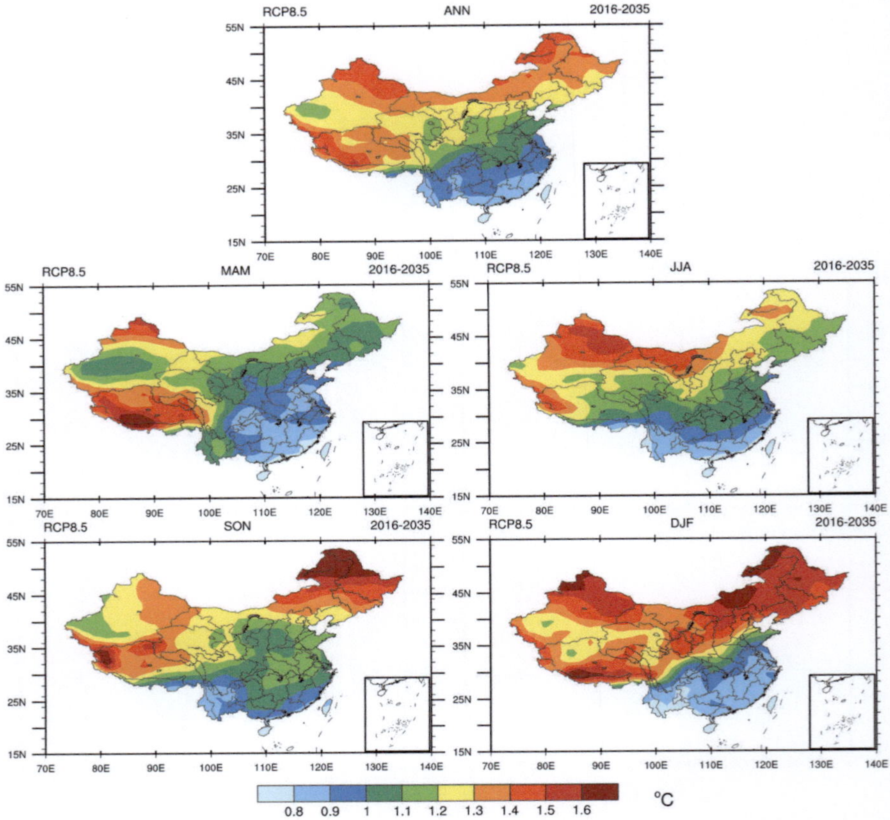

Fig. 2.131 Minimum temperature change in China for annual and season averaged over 2016–2035 from multi-model average under RCP8.5, relative to 1986–2005

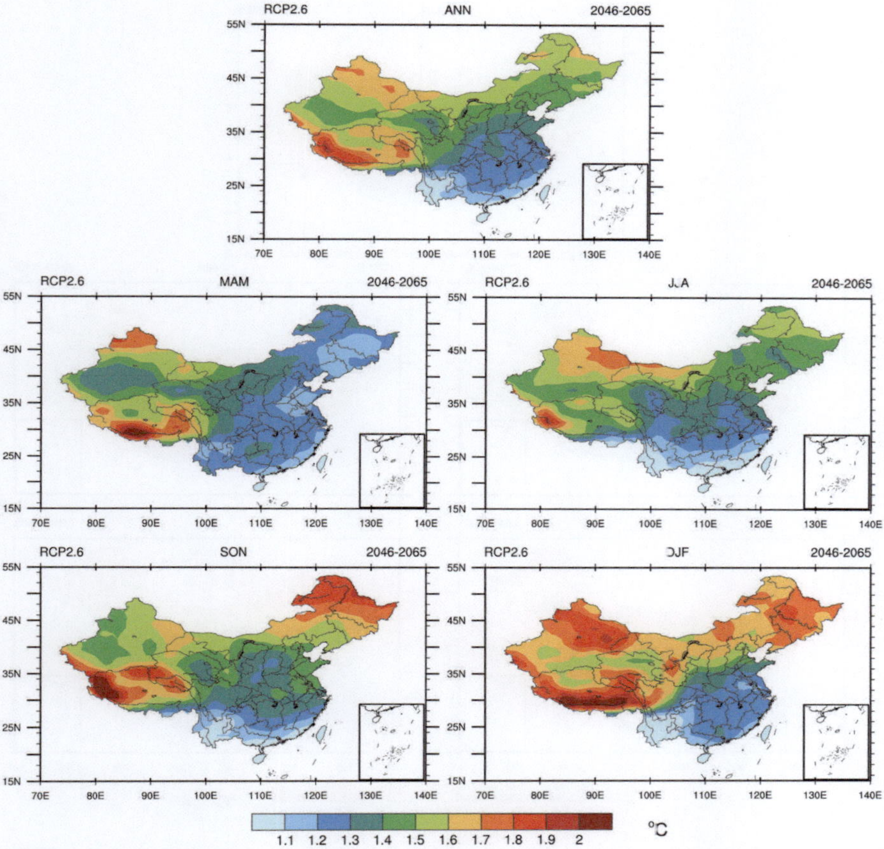

Fig. 2.132 Minimum temperature change in China for annual and season averaged over 2046–2065 from multi-model average under RCP2.6, relative to 1986–2005

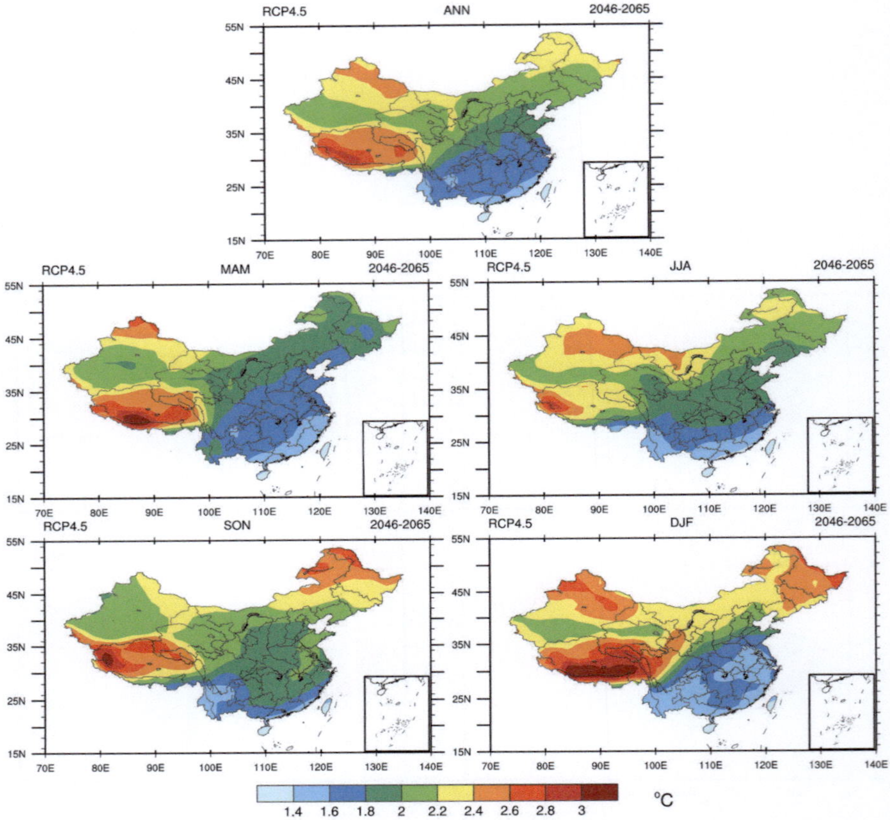

Fig. 2.133 Minimum temperature change in China for annual and season averaged over 2046–2065 from multi-model average under RCP4.5, relative to 1986–2005

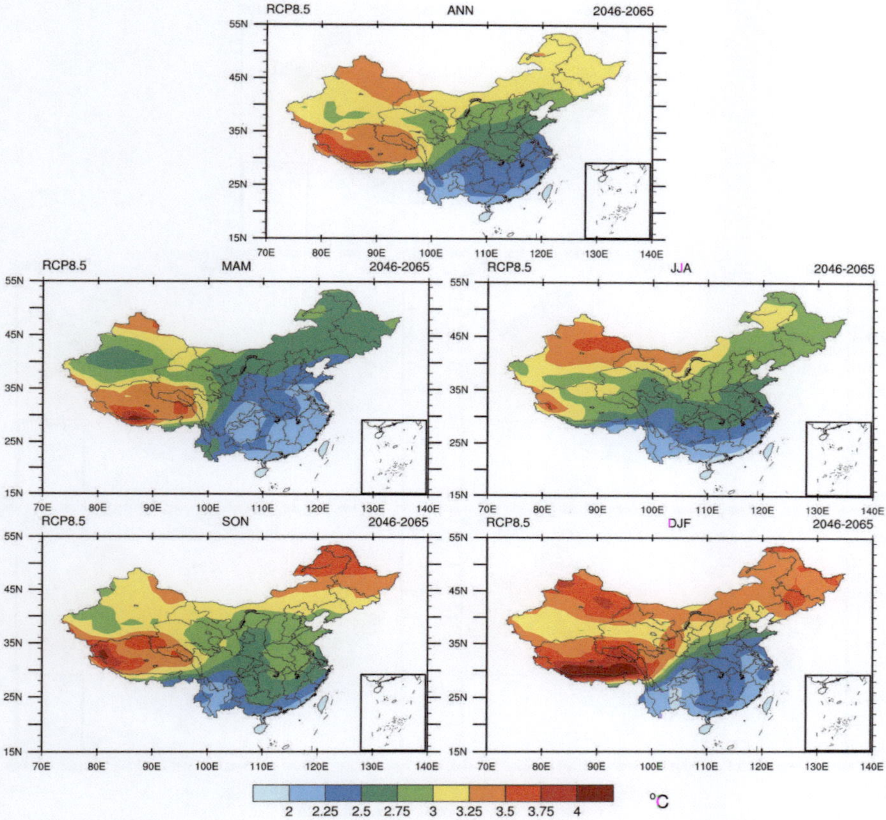

Fig. 2.134 Minimum temperature change in China for annual and season averaged over 2046–2065 from multi-model average under RCP8.5, relative to 1986–2005

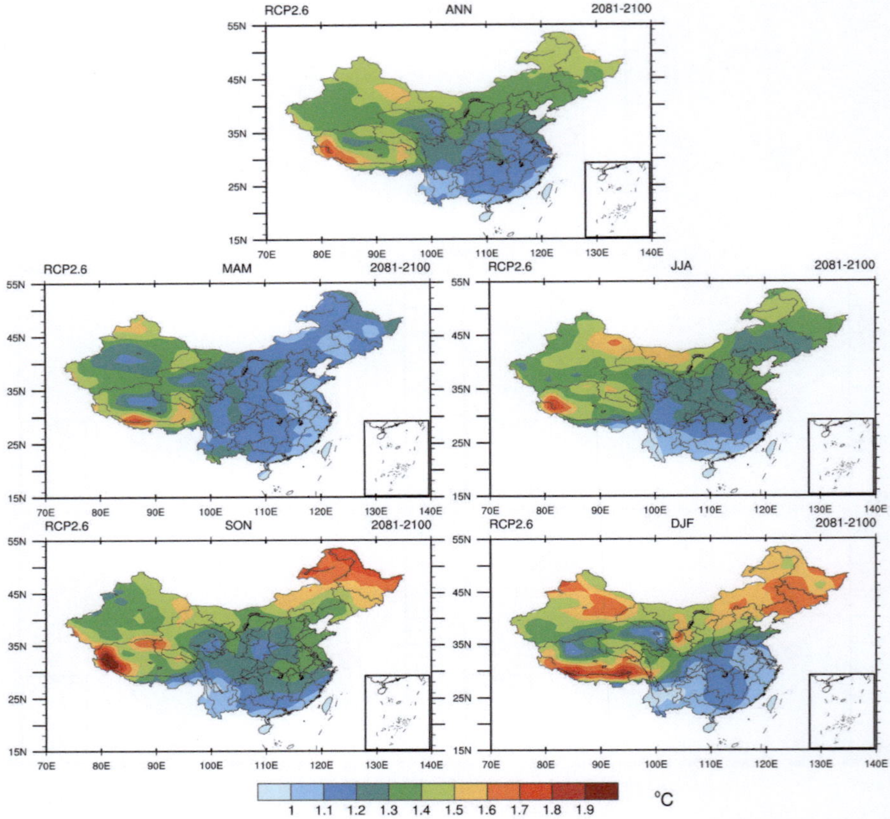

Fig. 2.135 Minimum temperature change in China for annual and season averaged over 2081–2100 from multi-model average under RCP2.6, relative to 1986–2005

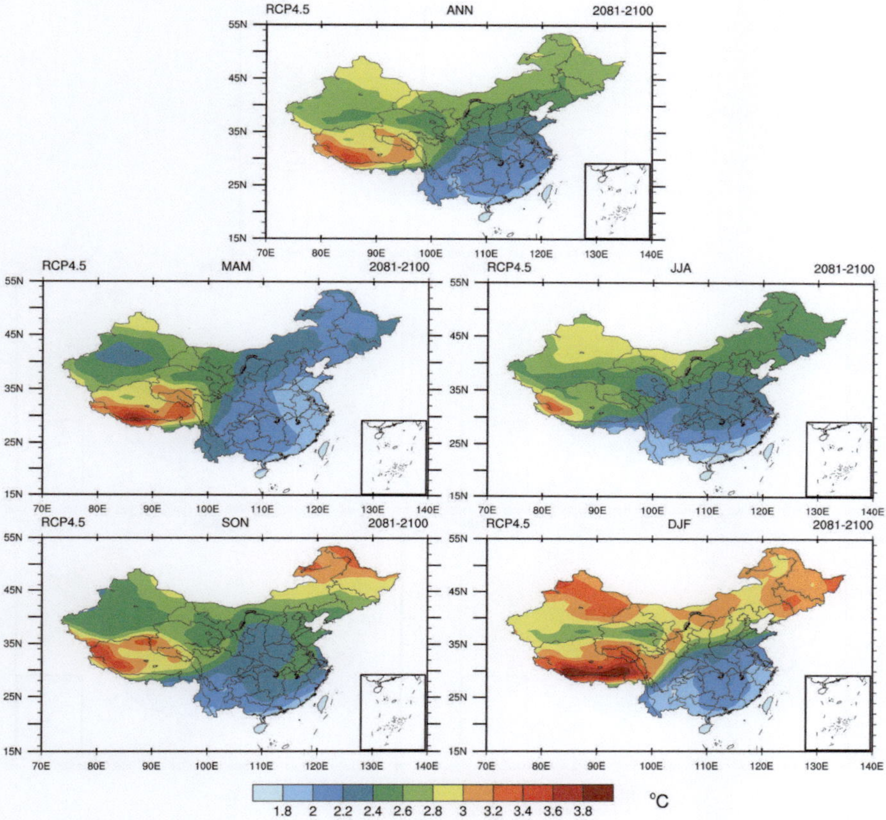

Fig. 2.136 Minimum temperature change in China for annual and season averaged over 2081–2100 from multi-model average under RCP4.5, relative to 1986–2005

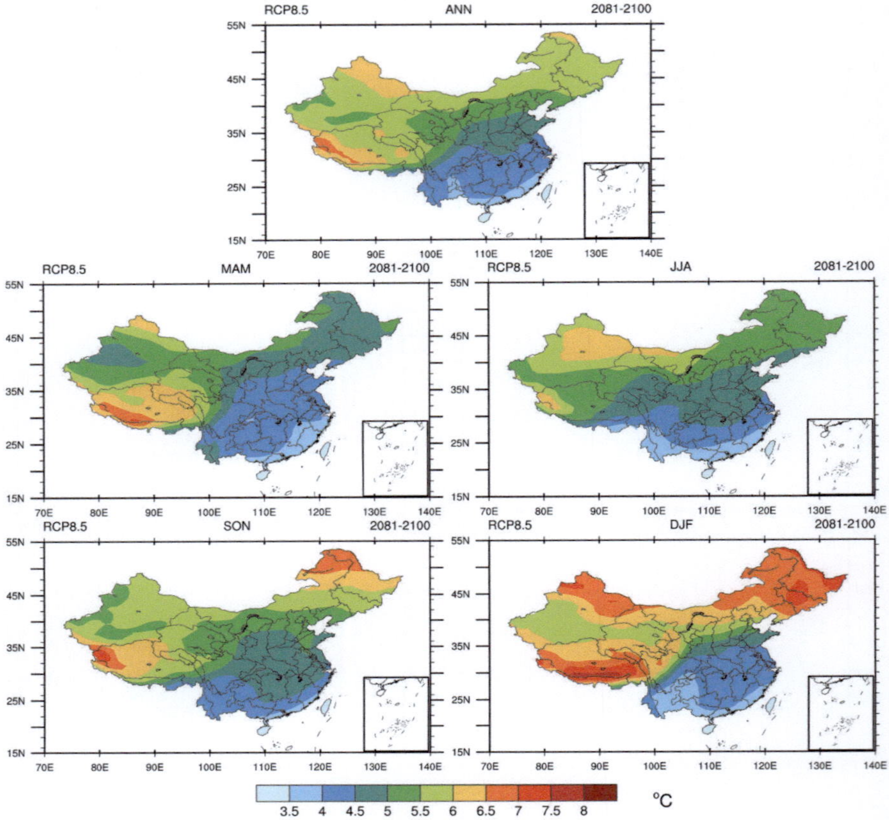

Fig. 2.137 Minimum temperature change in China for annual and season averaged over 2081–2100 from multi-model average under RCP8.5, relative to 1986–2005

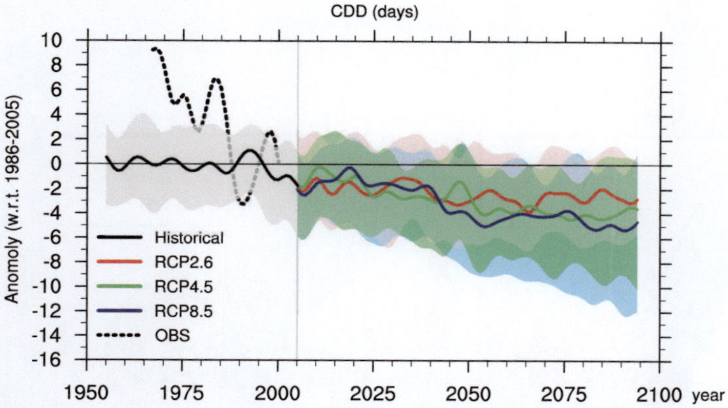

Fig. 2.138 Multi-model consecutive dry days (CDD) anomalies for annual average over China, relative to 1986–2005

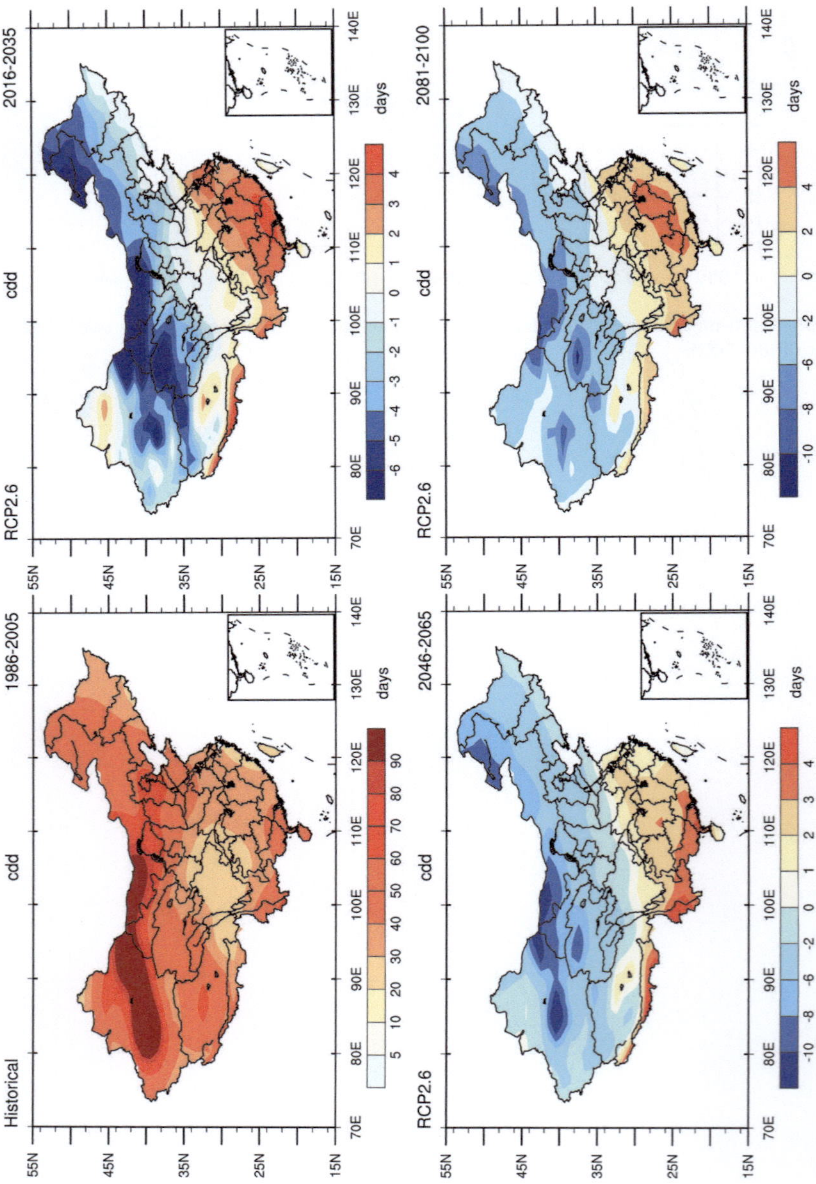

Fig. 2.139 Historical annual averaged consecutive dry days (CDD) and changes of future in China from multi-model average under RCP2.6, relative to 1986–2005

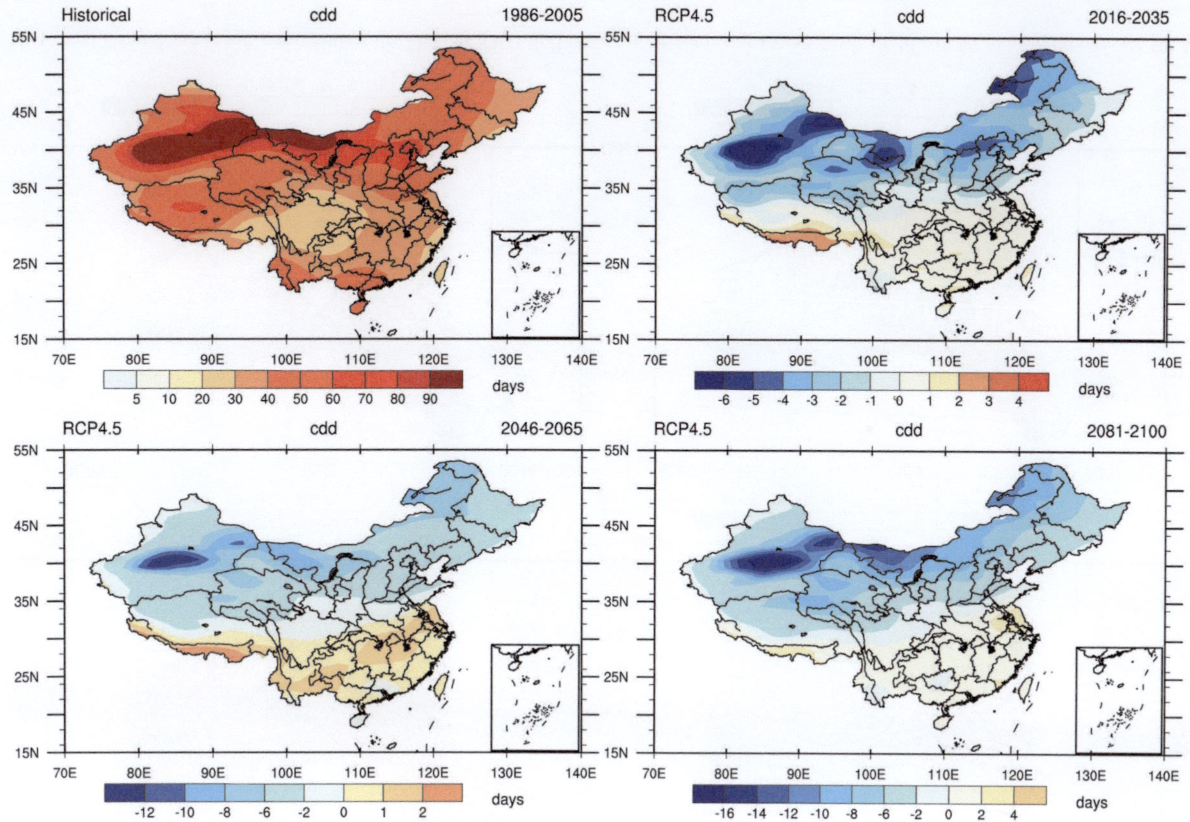

Fig. 2.140 Historical annual averaged consecutive dry days (CDD) and changes of future in China from multi-model average under RCP4.5, relative to 1986–2005

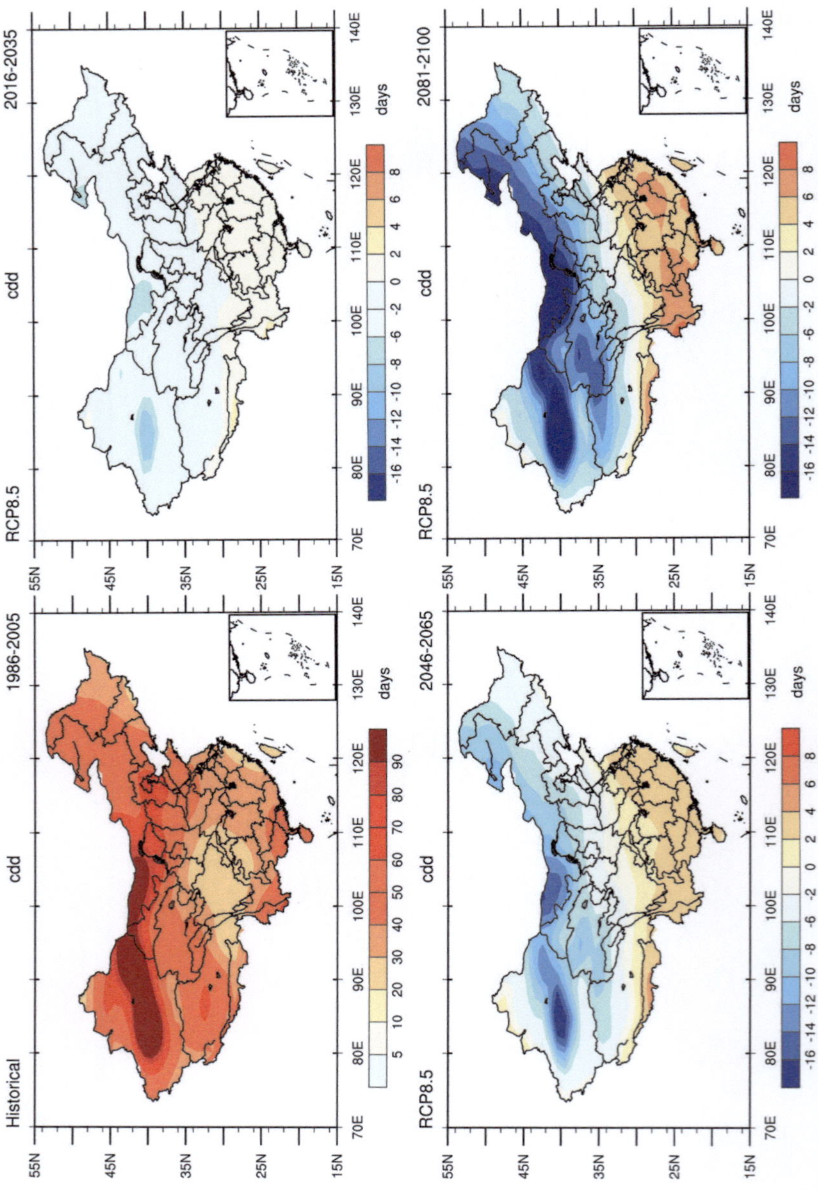

Fig. 2.141 Historical annual averaged consecutive dry days (CDD) and changes of future in China from multi-model average under RCP8.5, relative to 1986–2005

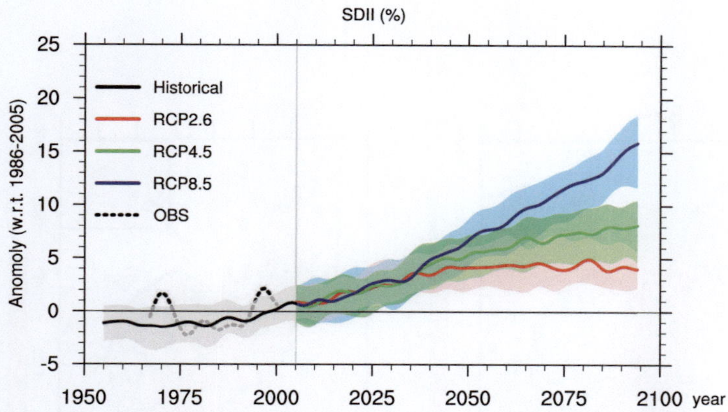

Fig. 2.142 Multi-model simple daily precipitation intensity (SDII) anomalies for annual average over China, relative to 1986–2005

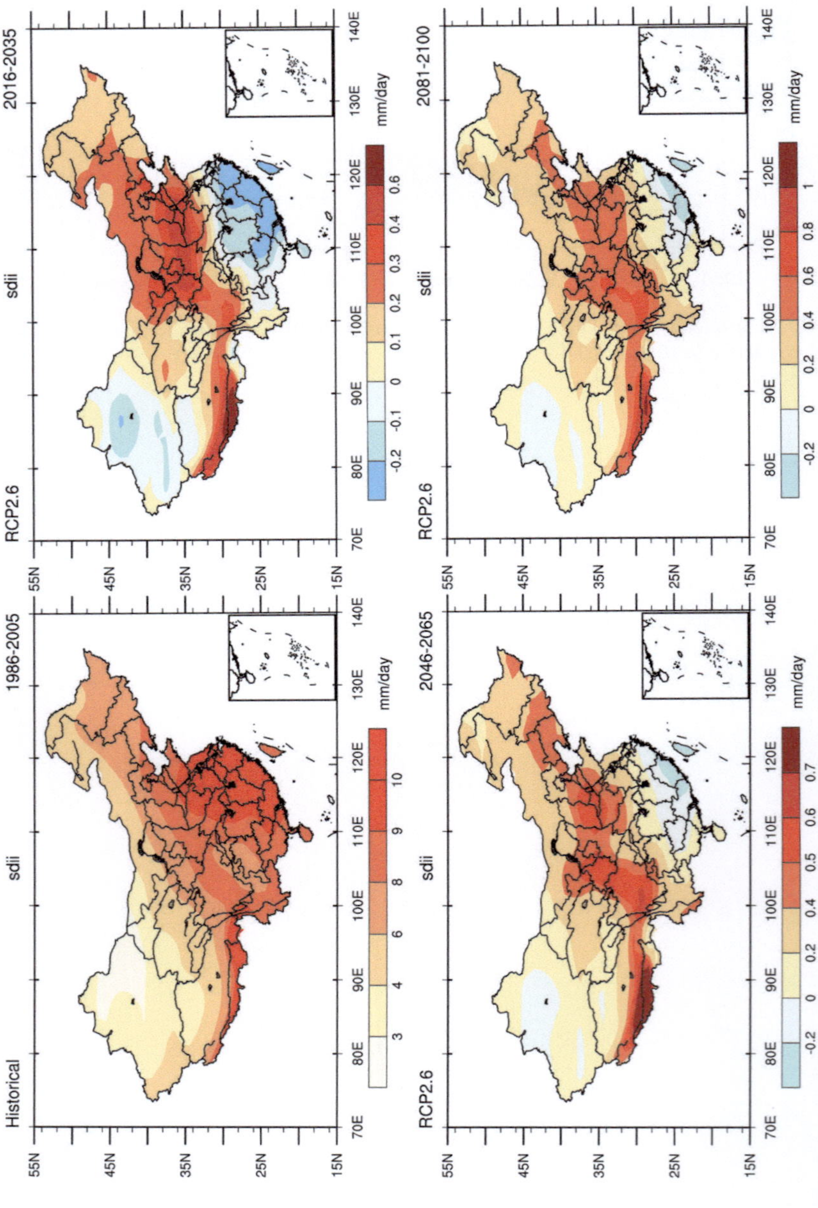

Fig. 2.143 Historical annual averaged simple daily precipitation intensity (SDII) and changes of future in China from multi-model average under RCP2.6, relative to 1986–2005

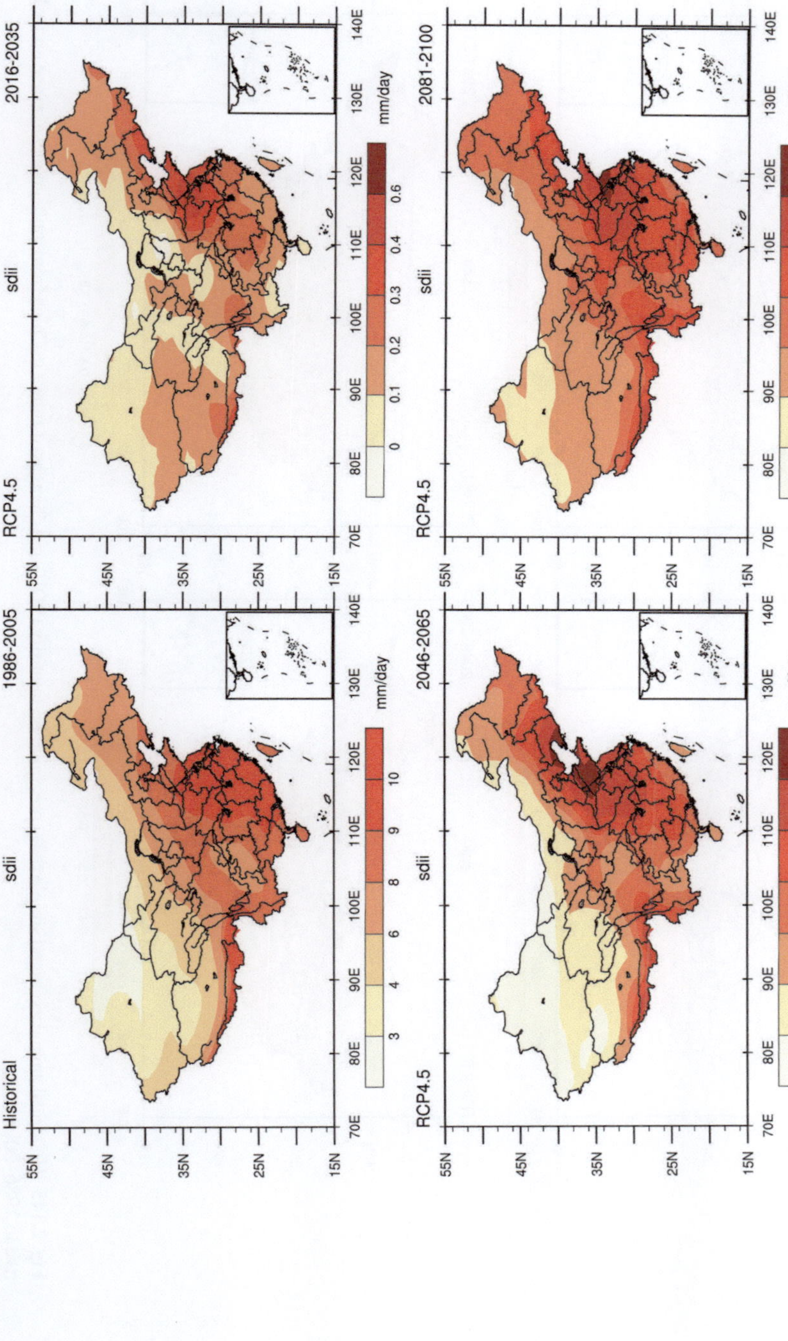

Fig. 2.144 Historical annual averaged simple daily precipitation intensity (SDII) and changes of future in China from multi-model average under RCP4.5, relative to 1986–2005

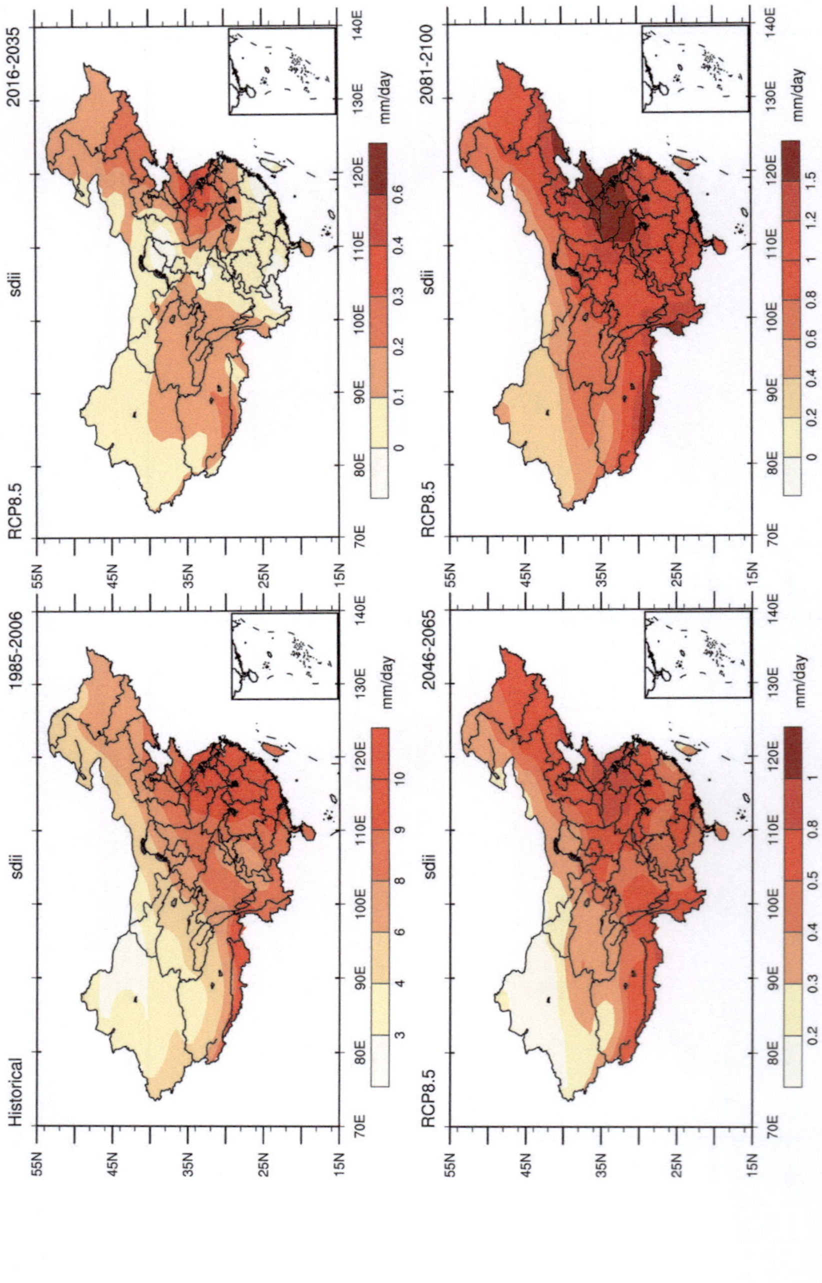

Fig. 2.145 Historical annual averaged simple daily precipitation intensity (SDII) and changes of future in China from multi-model average under RCP8.5, relative to 1986–2005

2.5 SST in the South China Sea (Figs. 2.146, 2.147, 2.148, 2.149, 2.150, 2.151, 2.152, 2.153, 2.154, 2.155, 2.156, 2.157, 2.158, 2.159, and 2.160)

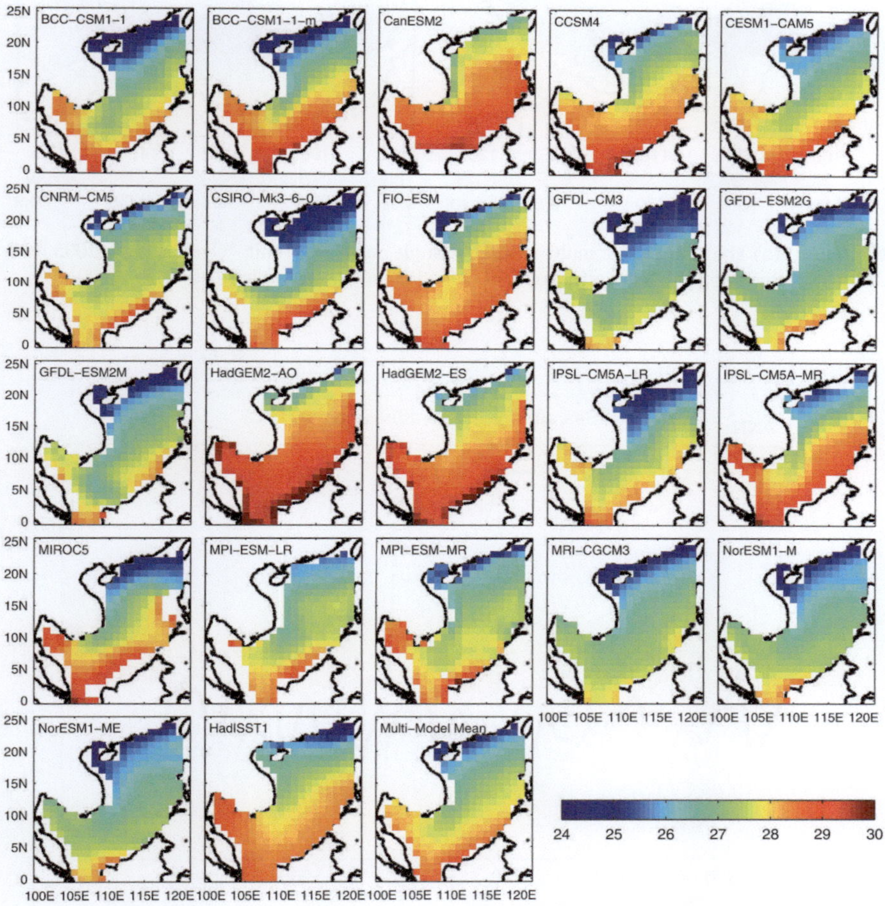

Fig. 2.146 The spatial distribution of annual mean SST (unit: °C) in the SCS averaged over 1986–2005 from 21 individual models, observation (HadISST1), and multi-model ensemble mean

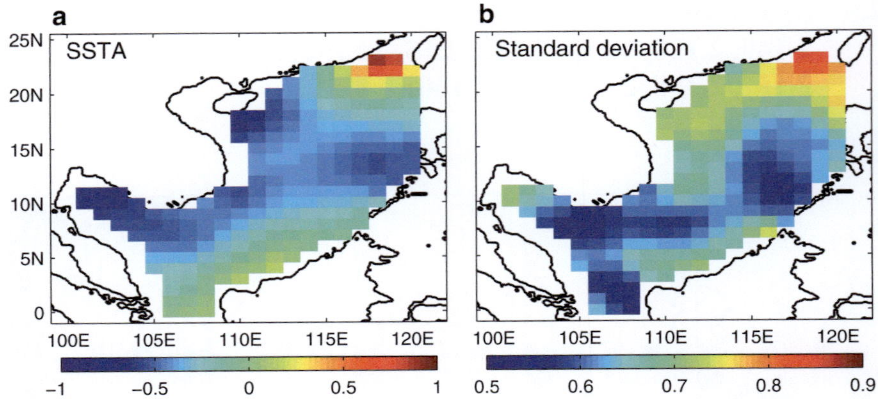

Fig. 2.147 (**a**) The deviation of multi-model ensemble mean SST (unit: °C) over 1986–2005 from the observation and (**b**) inter-model standard deviation among 21 models

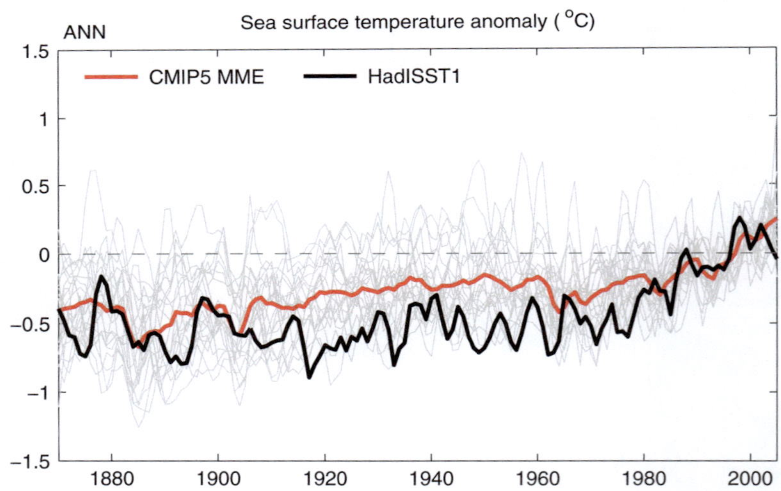

Fig. 2.148 Evolution of annual mean SST anomalies (unit: °C) in the SCS, relative to 1986–2005, from 21 individual models (*gray lines*), observation (HadISST1, *black line*), and multi-model ensemble mean (*red line*). A three-year running mean is applied to the time series

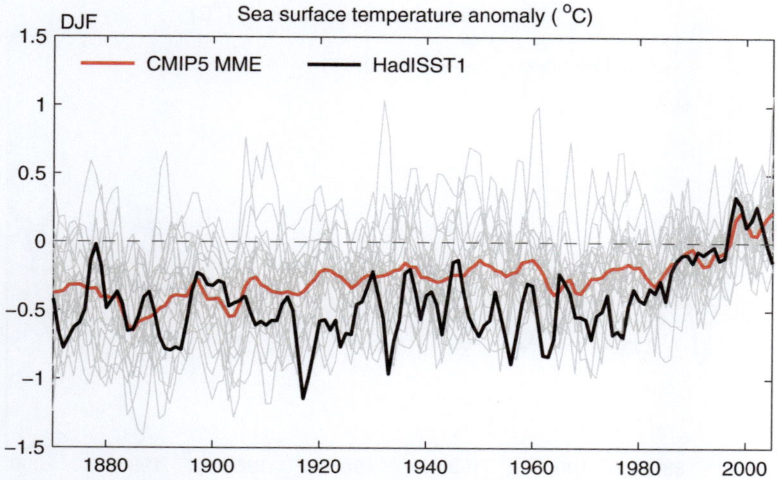

Fig. 2.149 Similar as Fig. 2.148, except for winter (December–February)

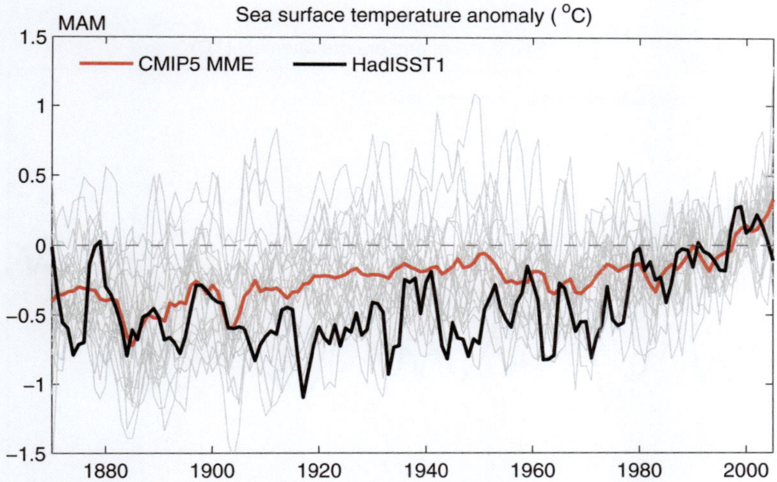

Fig. 2.150 Similar as Fig. 2.148, except for spring (March–May)

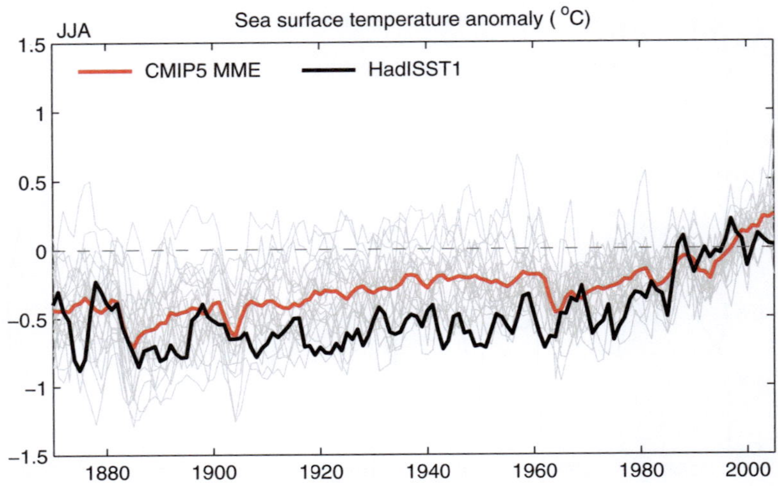

Fig. 2.151 Similar as Fig. 2.148, except for summer (June–August)

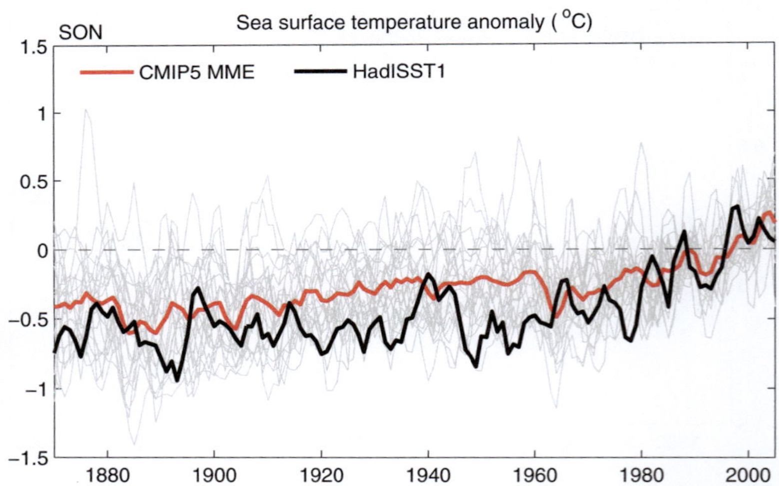

Fig. 2.152 Similar as Fig. 2.148, except for autumn (September–November)

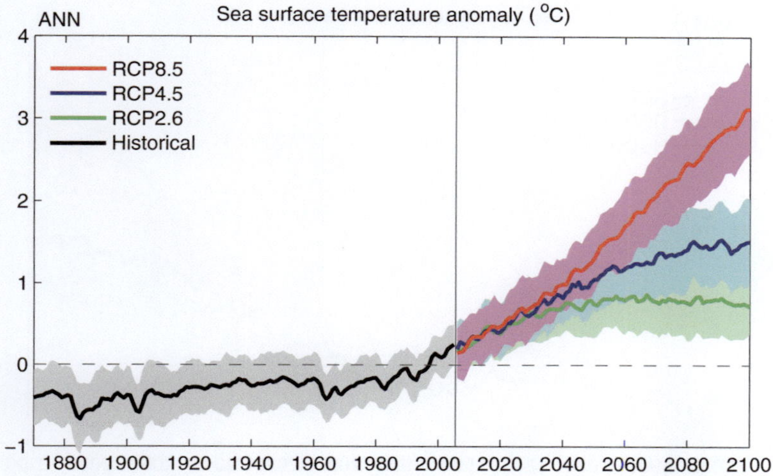

Fig. 2.153 Evolution of annual mean SST anomalies in the SCS (relative to 1986–2005) from multi-model ensemble mean of 21 models. *Shadow* represents inter-model standard deviation (unit: °C). A three-year running mean is applied to the time series

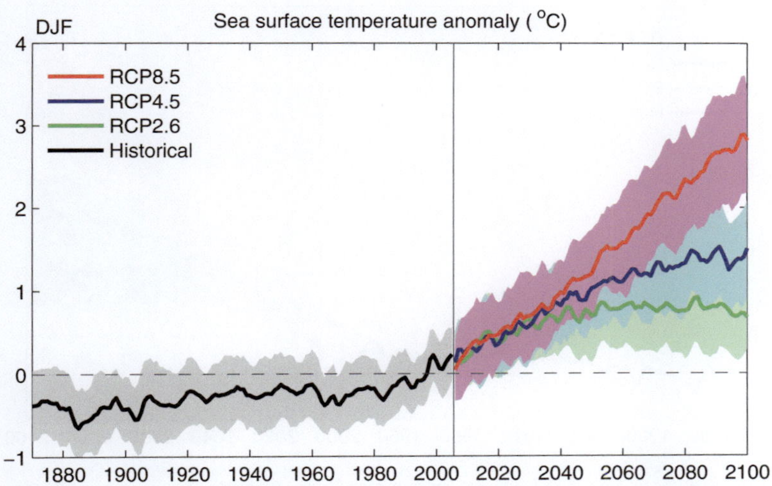

Fig. 2.154 Similar as Fig. 2.153, except for winter (December–February)

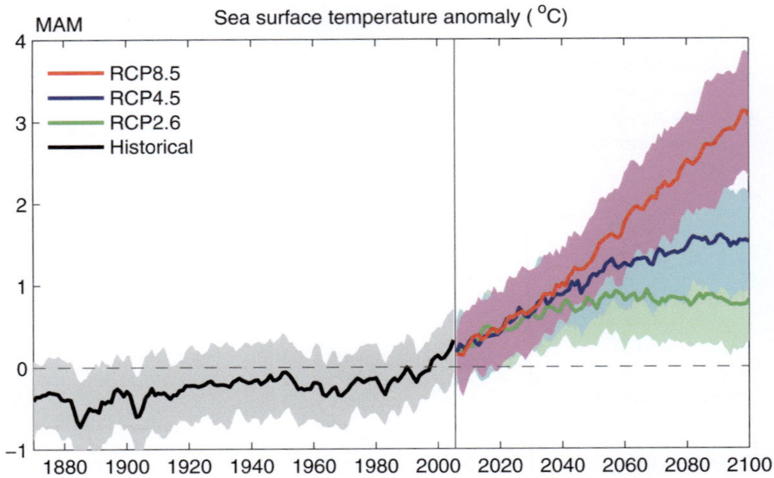

Fig. 2.155 Similar as Fig. 2.153, except for spring (March–May)

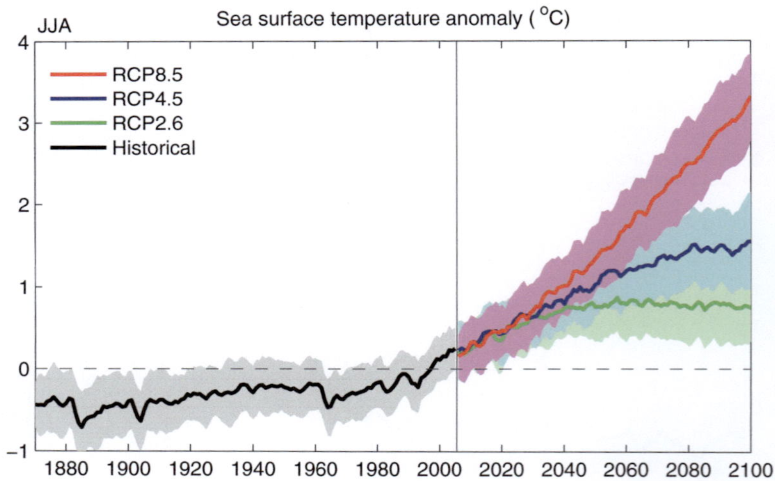

Fig. 2.156 Similar as Fig. 2.153, except for summer (June–August)

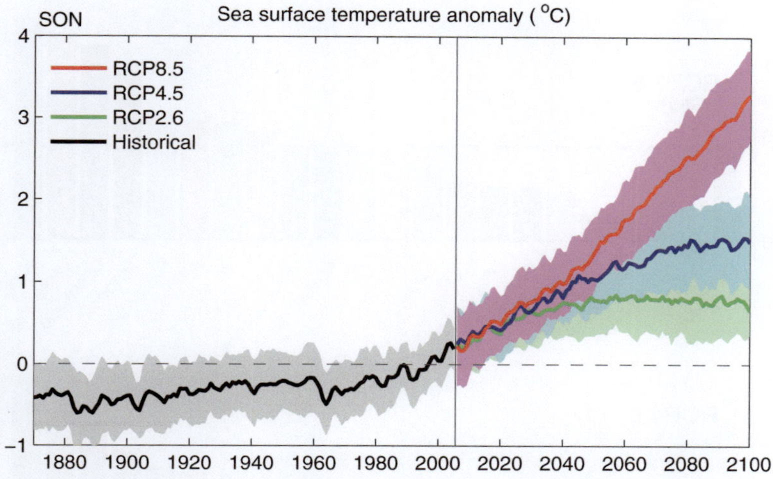

Fig. 2.157 Similar as Fig. 2.153, except for autumn (September–November)

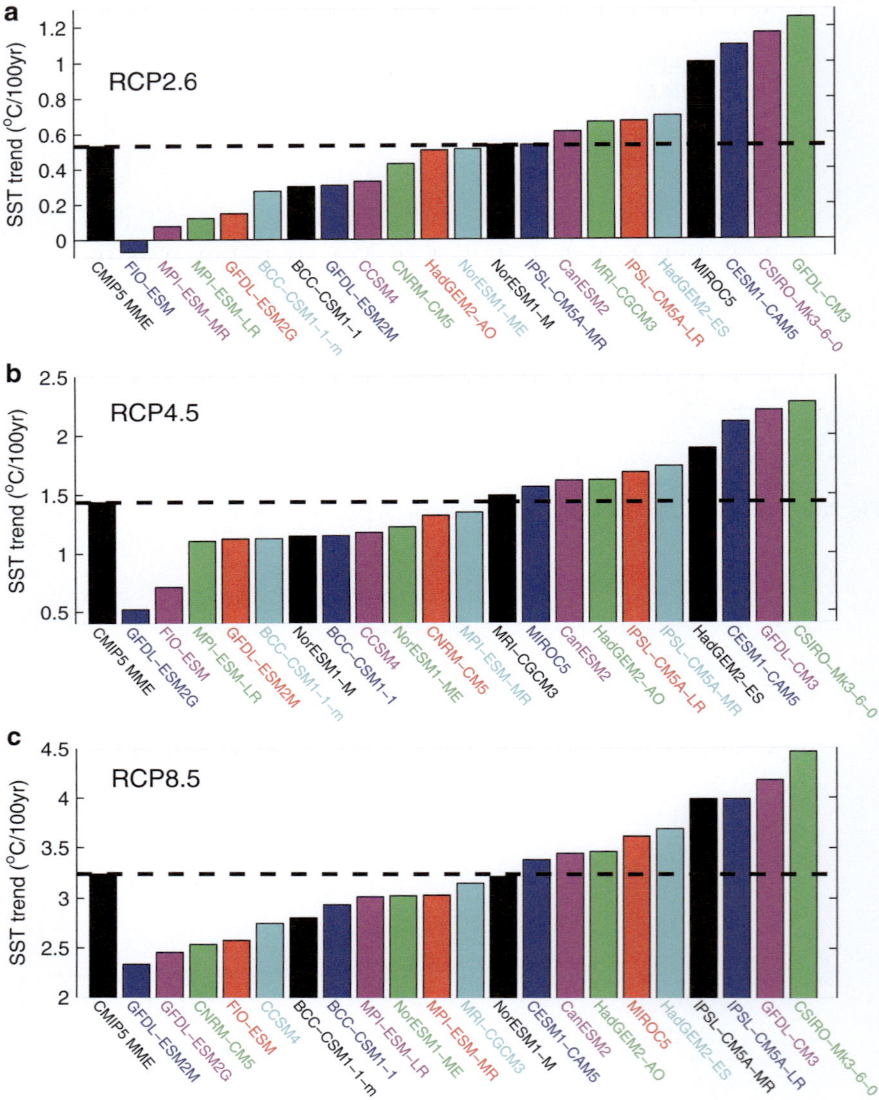

Fig. 2.158 Linear trend (°C/100 year) of SST warming over the SCS for the period 2006–2100 under (**a**) RCP2.6, (**b**) RCP4.5, and (**c**) RCP8.5 scenarios

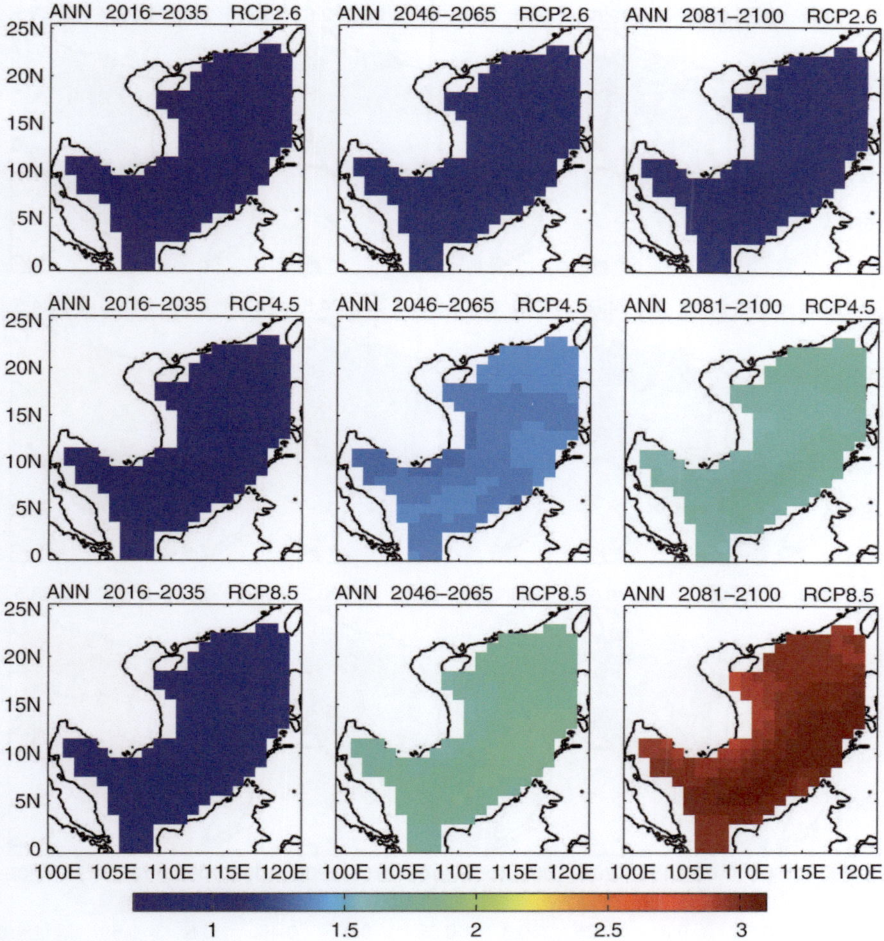

Fig. 2.159 Changes in annual mean SST in the SCS averaged over 2016–2035 (*left*), 2046–2065 (*middle*), and 2081–2100 (*right*), relative to 1986–2005, for RCP2.6 (*upper*), RCP4.5 (*middle*), and RCP8.5 (*low*) scenarios from multi-model ensemble mean (unit: °C)

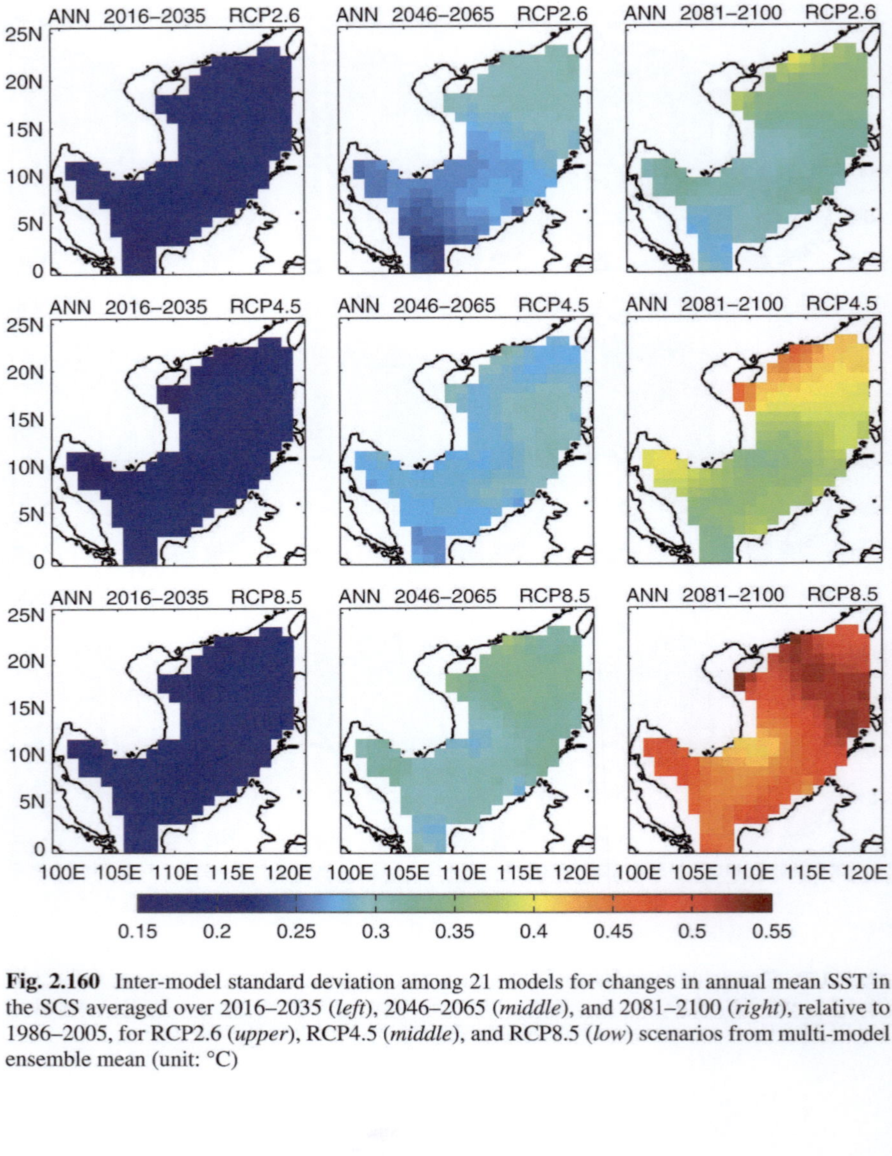

Fig. 2.160 Inter-model standard deviation among 21 models for changes in annual mean SST in the SCS averaged over 2016–2035 (*left*), 2046–2065 (*middle*), and 2081–2100 (*right*), relative to 1986–2005, for RCP2.6 (*upper*), RCP4.5 (*middle*), and RCP8.5 (*low*) scenarios from multi-model ensemble mean (unit: °C)

2.6 Arctic Sea Ice (Figs. 2.161, 2.162, 2.163, 2.164, 2.165, 2.166, 2.167, 2.168, 2.169, 2.170, 2.171, 2.172, 2.173, 2.174, 2.175, 2.176, 2.177, 2.178, 2.179, 2.180, 2.181, 2.182, 2.183, and 2.184)

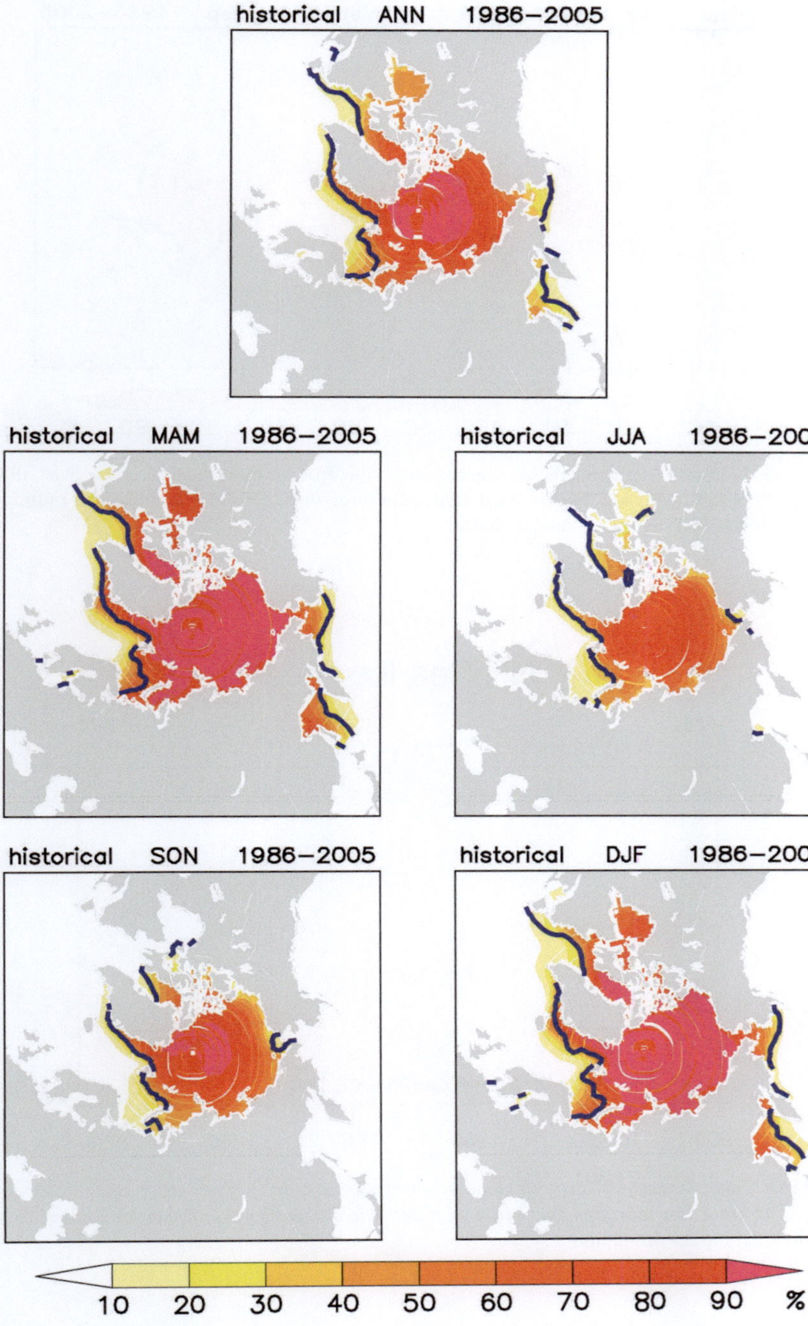

Fig. 2.161 Annual and seasonal mean of Northern Hemisphere sea ice concentration from multi-model average over 1986–2005. The *blue line* indicates the observed outline of 20-year (1986–2005) averaged sea ice extent

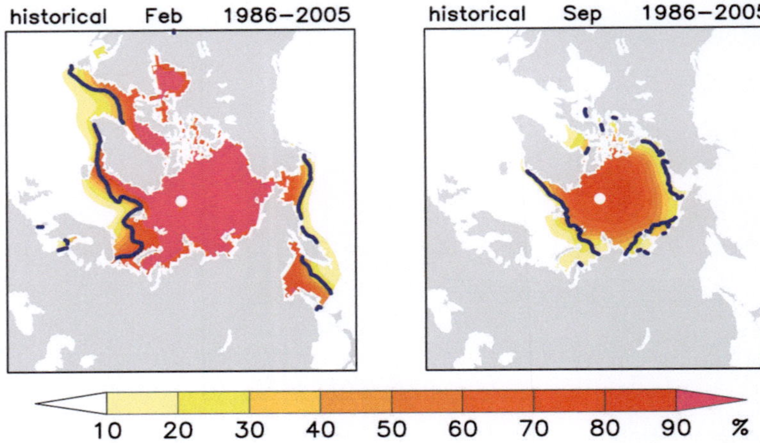

Fig. 2.162 Northern Hemisphere sea ice concentration from multi-model average over 1986–2005 for February (*left*) and September (*right*). The *blue line* indicates the observed outline of 20-year (1986–2005) averaged sea ice extent

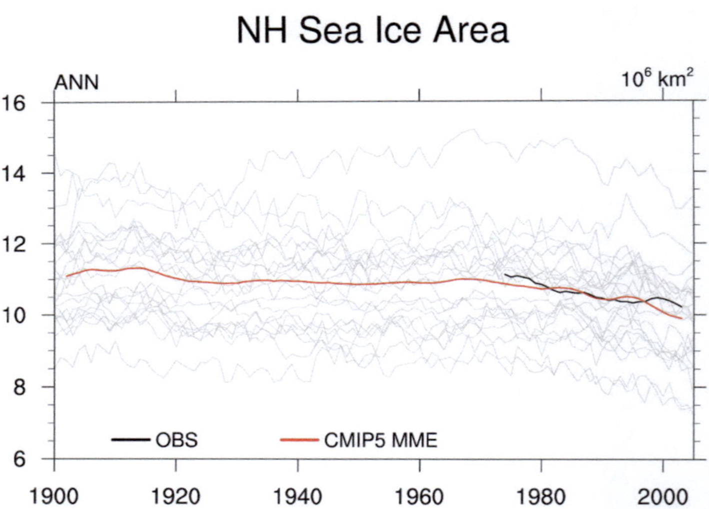

Fig. 2.163 Annual mean of Northern Hemisphere sea ice area from observation and model simulations. The *black line* indicates the observation, *red line* the multi-model ensemble, and *gray line* the individual model simulation

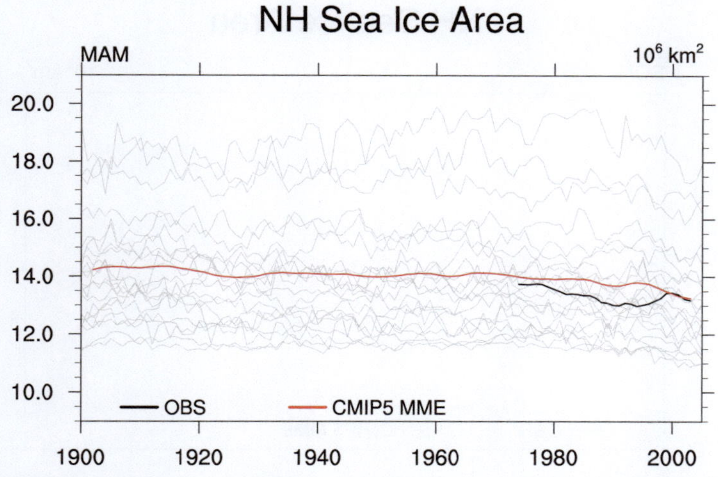

Fig. 2.164 Spring mean (March–May) of Northern Hemisphere sea ice area from observation and model simulations. The *black line* indicates the observation, *red line* the multi-model ensemble, and *gray line* the individual model simulation

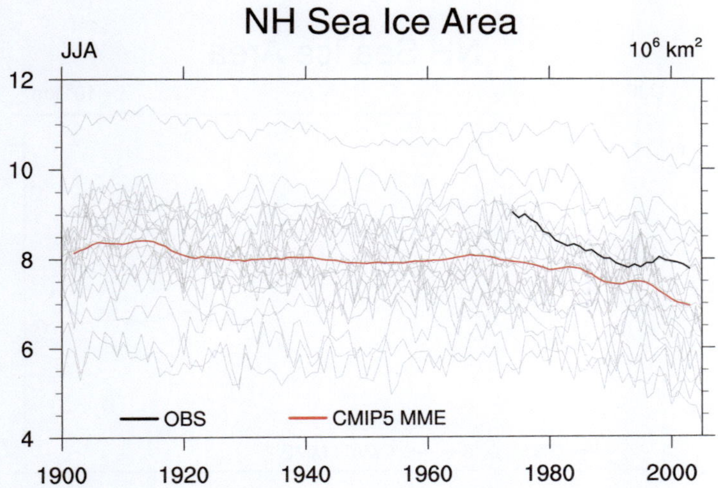

Fig. 2.165 Summer mean (June–August) of Northern Hemisphere sea ice area from observation and model simulations. The *black line* indicates the observation, *red line* the multi-model ensemble, and *gray line* the individual model simulation

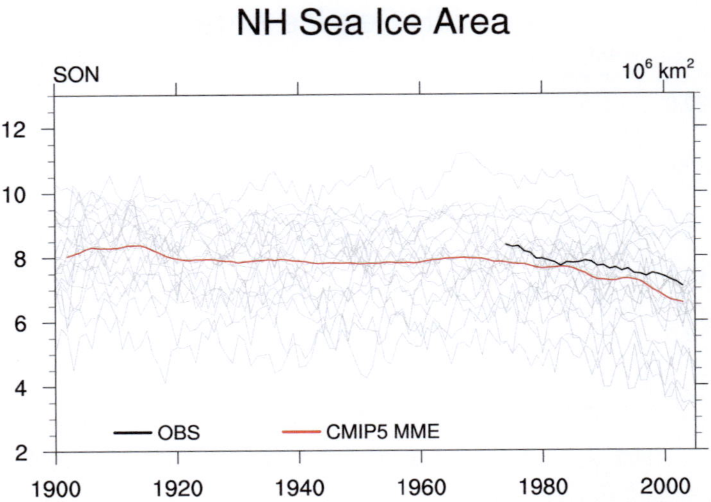

Fig. 2.166 Autumn mean (September–November) of Northern Hemisphere sea ice area from observation and model simulations. The *black line* indicates the observation, *red line* the multi-model ensemble, and *gray line* the individual model simulation

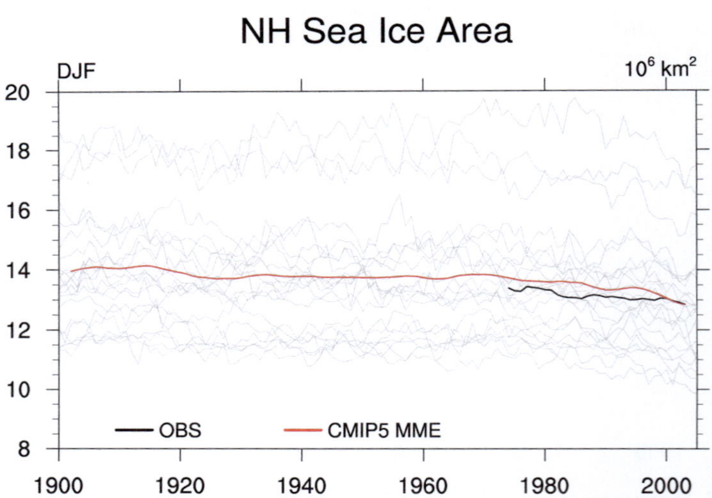

Fig. 2.167 Winter mean (December–February) of Northern Hemisphere sea ice area from observation and model simulations. The *black line* indicates the observation, *red line* the multi-model ensemble, and *gray line* the individual model simulation

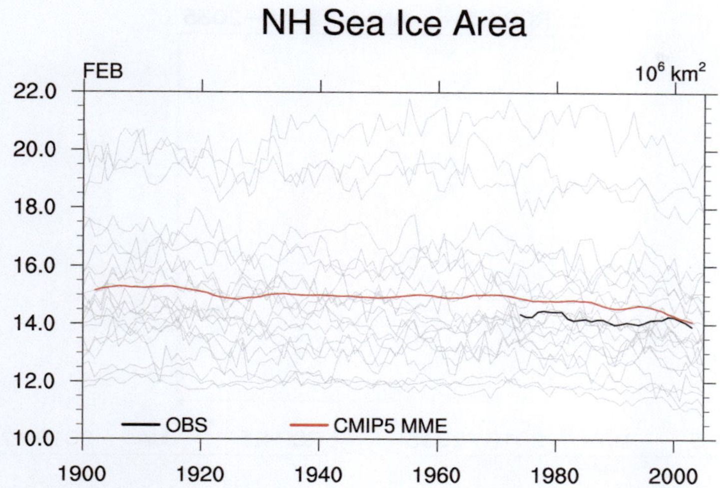

Fig. 2.168 Northern Hemisphere sea ice area from observation and model simulations for February. The *black line* indicates the observation, *red line* the multi-model ensemble, and *gray line* the individual model simulation

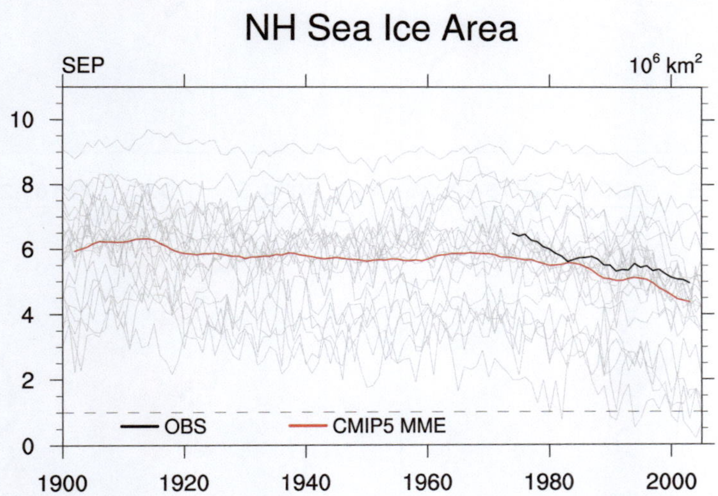

Fig. 2.169 Northern Hemisphere sea ice area from observation and model simulations for September. The *black line* indicates the observation, *red line* the multi-model ensemble, and *gray line* the individual model simulation

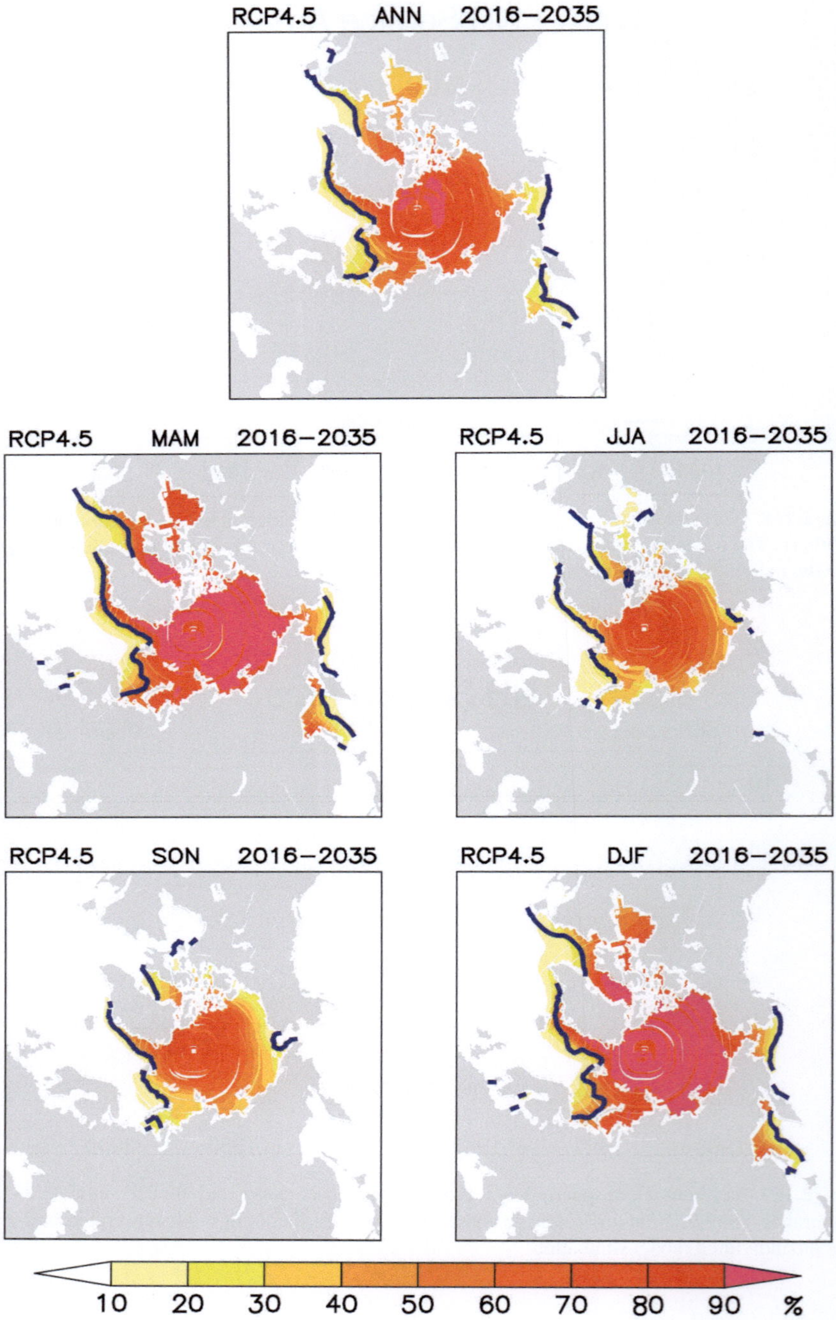

Fig. 2.170 Annual and seasonal mean of Northern Hemisphere sea ice concentration from multi-model average over 2016–2035 under the RCP4.5 scenario. The *blue line* indicates the observed outline of 20-year (1986–2005) averaged sea ice extent

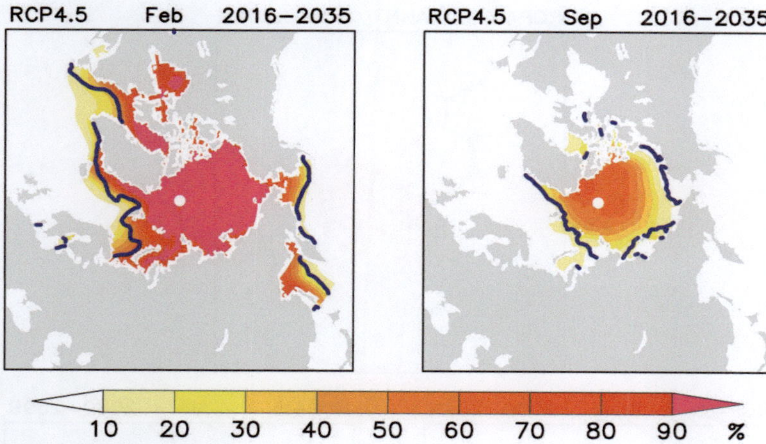

Fig. 2.171 Northern Hemisphere sea ice concentration from multi-model average over 2016–2035 for February (*left*) and September (*right*) under the RCP4.5 scenario. The *blue line* indicates the observed outline of 20-year (1986–2005) averaged sea ice extent

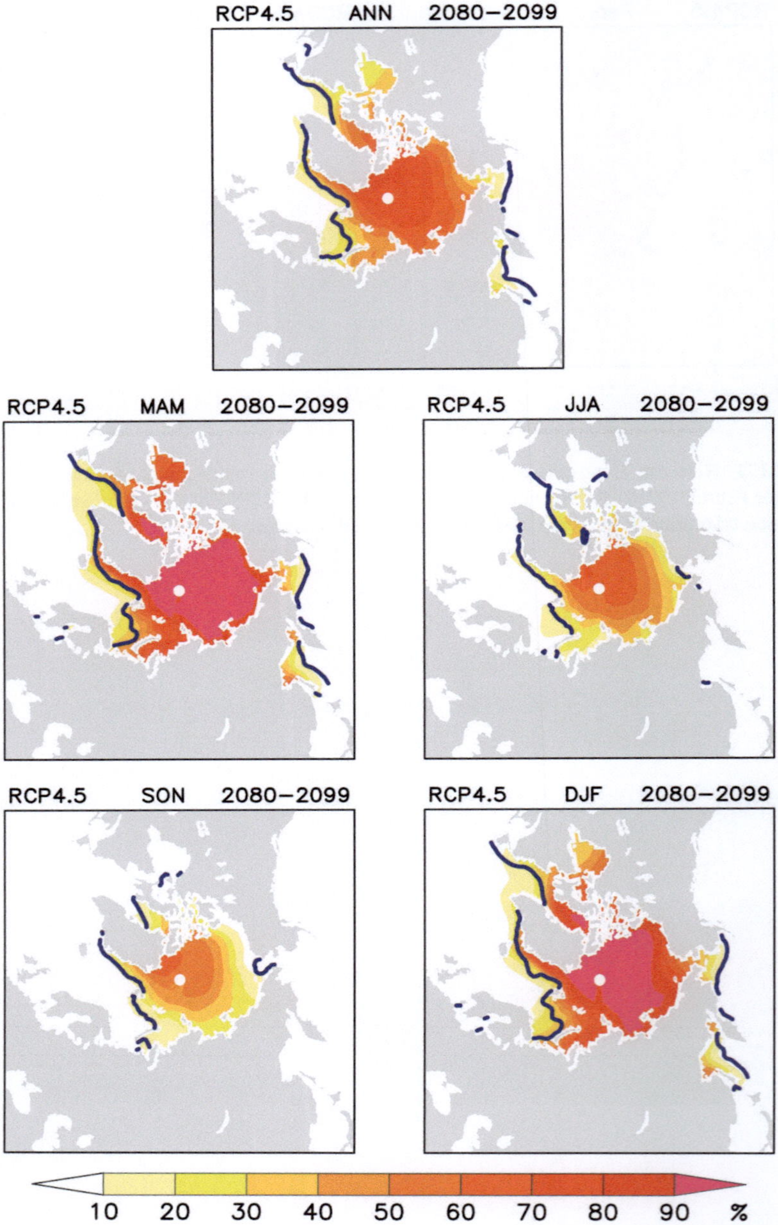

Fig. 2.172 Annual and seasonal mean of Northern Hemisphere sea ice concentration from multi-model average over 2080–2099 under the RCP4.5 scenario. The *blue line* indicates the observed outline of 20-year (1986–2005) averaged sea ice extent

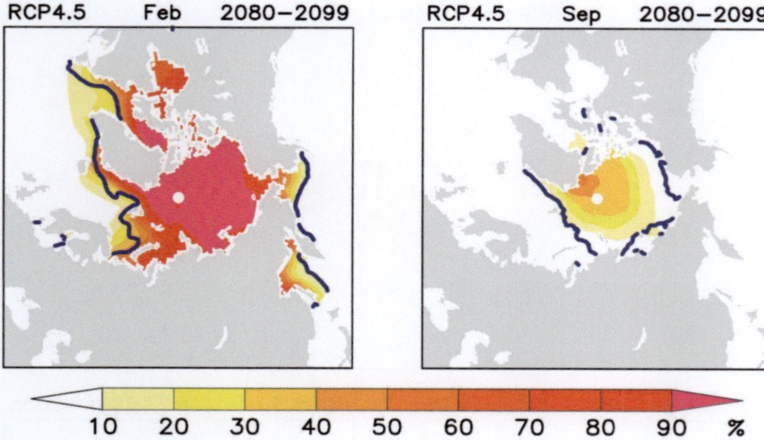

Fig. 2.173 Northern Hemisphere sea ice concentration from multi-model average over 2080–2099 for February (*left*) and September (*right*) under the RCP4.5 scenario. The *blue line* indicates the observed outline of 20-year (1986–2005) averaged sea ice extent

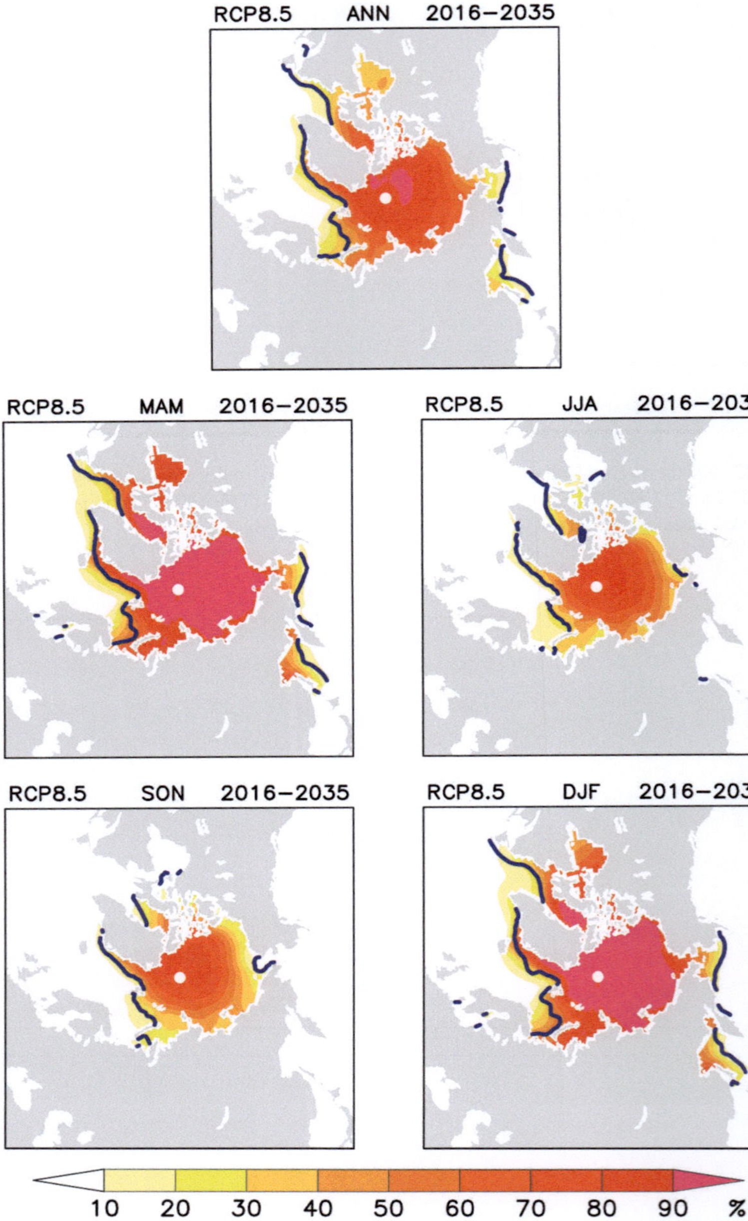

Fig. 2.174 Annual and seasonal mean of Northern Hemisphere sea ice concentration from multi-model average over 2016–2035 under the RCP8.5 scenario. The *blue line* indicates the observed outline of 20-year (1986–2005) averaged sea ice extent

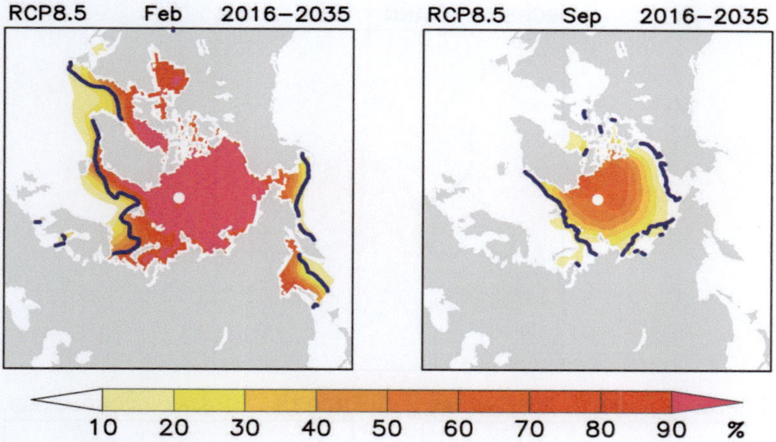

Fig. 2.175 Northern Hemisphere sea ice concentration from multi-model average over 2016–2035 for February (*left*) and September (*right*) under the RCP8.5 scenario. The *blue line* indicates the observed outline of 20-year (1986–2005) averaged sea ice extent

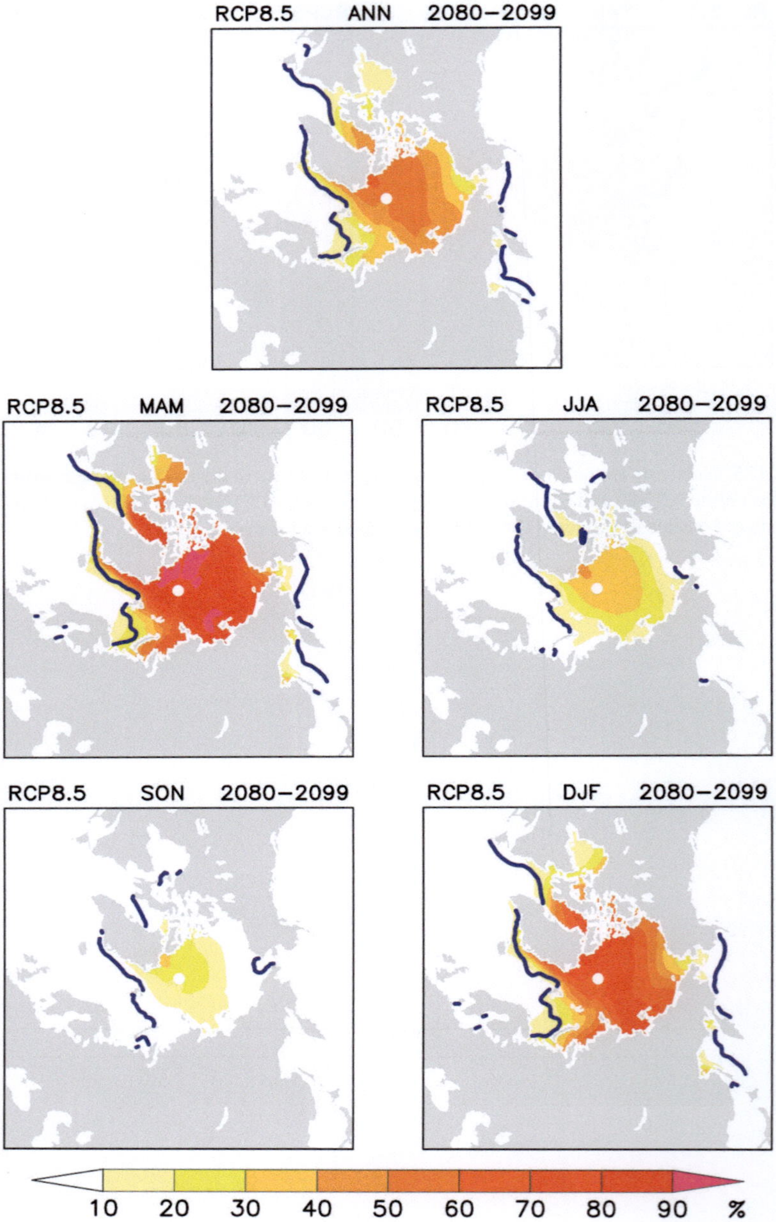

Fig. 2.176 Annual and seasonal mean of Northern Hemisphere sea ice concentration from multi-model average over 2080–2099 under the RCP8.5 scenario. The *blue line* indicates the observed outline of 20-year (1986–2005) averaged sea ice extent

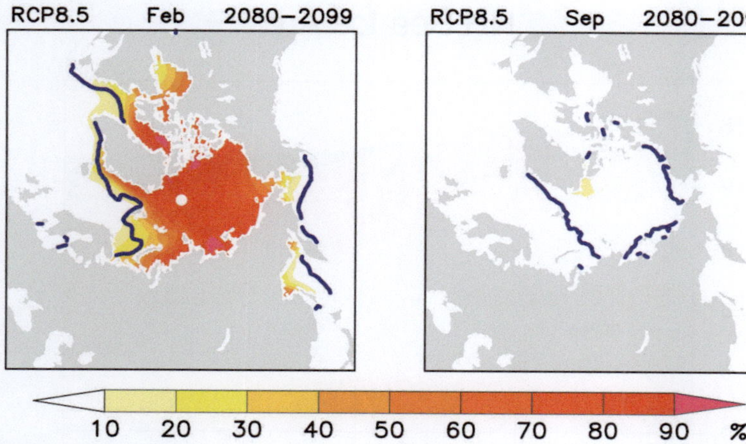

Fig. 2.177 Northern Hemisphere sea ice concentration from multi-model average over 2080–2099 for February (*left*) and September (*right*) under the RCP8.5 scenario. The *blue line* indicates the observed outline of 20-year (1986–2005) averaged sea ice extent

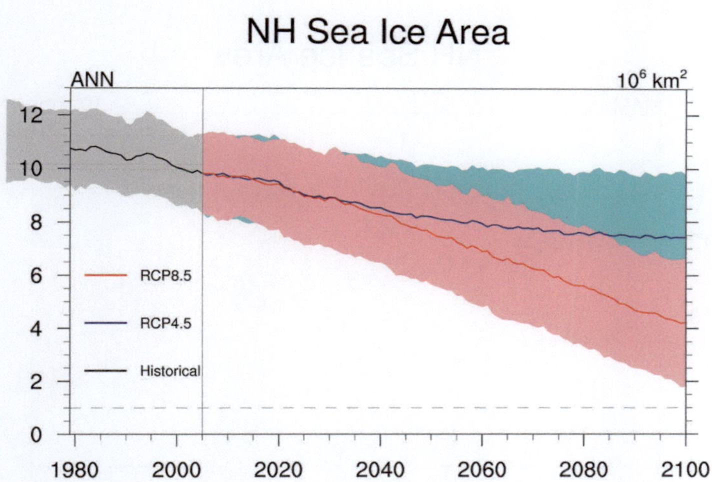

Fig. 2.178 Annual mean of Northern Hemisphere sea ice area from historical simulation and future projection under the RCP scenarios

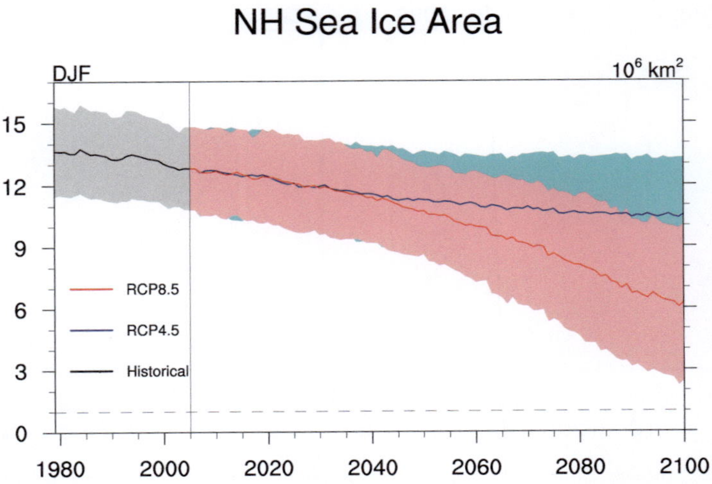

Fig. 2.179 Winter mean (December–February) of Northern Hemisphere sea ice area from historical simulation and future projection under the RCP scenarios

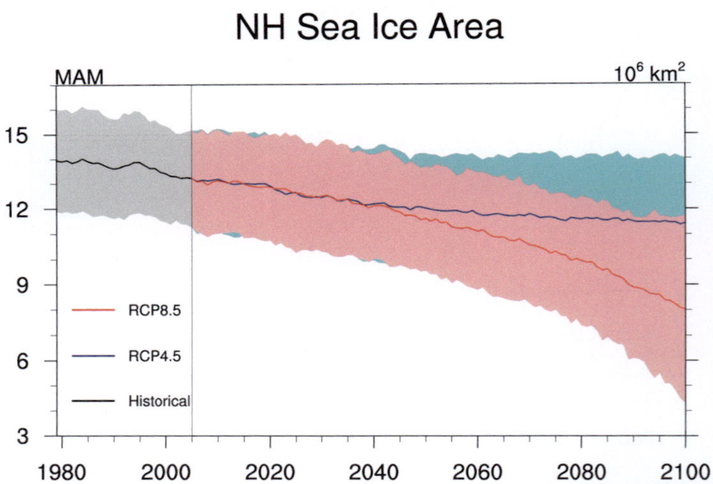

Fig. 2.180 Spring mean (March–May) of Northern Hemisphere sea ice area from historical simulation and future projection under the RCP scenarios

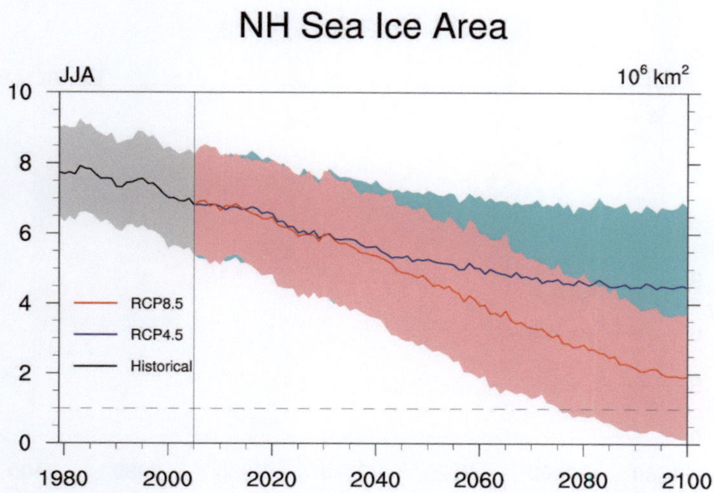

Fig. 2.181 Summer mean (June–August) of Northern Hemisphere sea ice area from historical simulation and future projection under the RCP scenarios

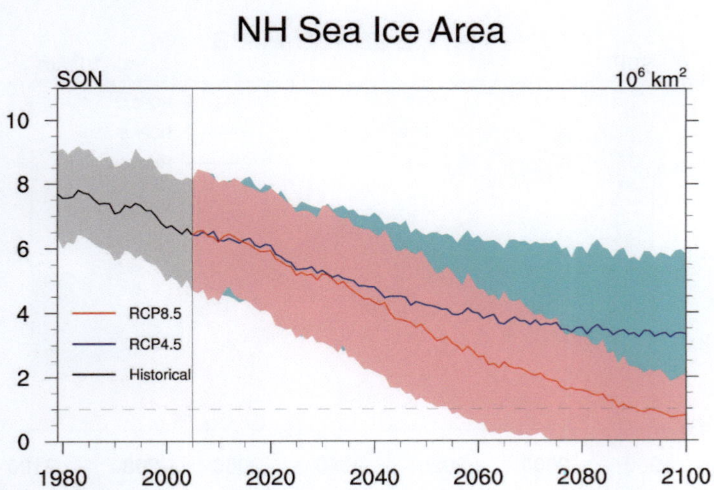

Fig. 2.182 Autumn mean (September–November) of Northern Hemisphere sea ice area from historical simulation and future projection under the RCP scenarios

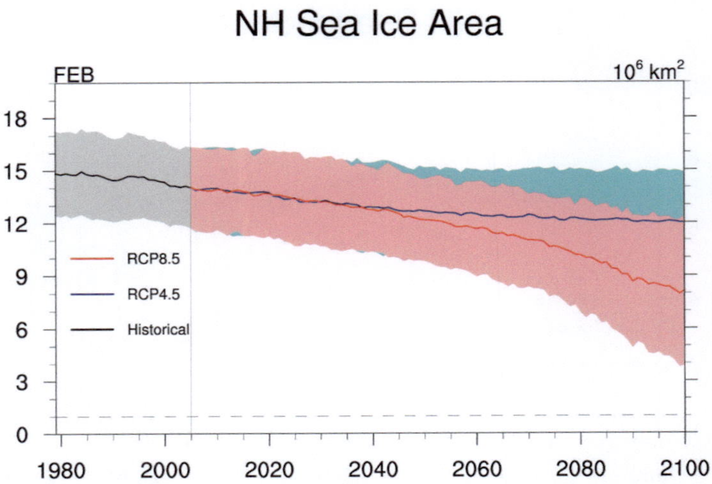

Fig. 2.183 Northern Hemisphere sea ice area from historical simulation and future projection under the RCP scenarios for February

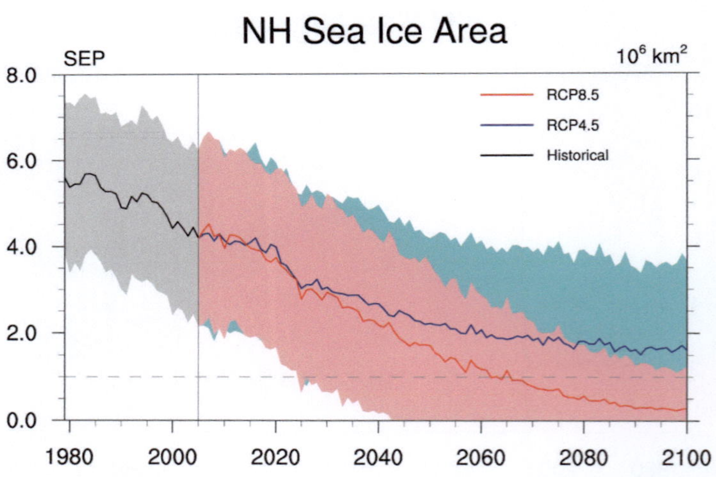

Fig. 2.184 Northern Hemisphere sea ice area from historical simulation and future projection under the RCP scenarios for September

References

Frisch P et al (2002) Observed coherent changes in climatic extremes during the second 16 half of the twentieth century. Clim Res 19:193–212

Guo QY (1983) The summer monsoon intensity index in East Asia and its variation. Acta Geograph Sin 38(3):208–217

Liu Y, Li W, Ai W, Li Q (2012) Reconstruction and application of the monthly Western Pacific subtropical high indices. J Appl Meteorol Sci 23(4):414–423

Thompson DWJ, Wallace JM (1998) The Arctic oscillation signature in the wintertime geopotential height and temperature fields. Geophys Res Lett 25:1297–1300

Wu B, Wang J (2002) Winter Arctic Oscillation, Siberian High and East Asian winter monsoon. Geophys Res Lett 29(19):1897. doi:10.1029/2002GL015373

Chapter 3
Comparisons Among Multi-model Ensemble Based on Different Ensemble Methods and Ensemble Member Sizes

Abstract In this chapter, the results of three different multi-model ensemble mean methods are compared. In Sect. 3.1, the global mean surface air temperature and precipitation from multi-model weighted mean and equal-weighted mean are compared. The weight is given according to the correlation coefficients and relative bias between model and observation. The results from the multi-model weighted mean and the equal-weighted mean show no significant difference. In Sect. 3.2, the surface air temperature and precipitation in China from multi-model equal-weighted mean and Bayesian model averaging (BMA) method are compared. BMA is a multi-model ensemble method which considered each model's posterior distributions and then weighted by posterior model probability. In our calculation, the training of BMA is based on observational data (derived from CRUTS 3.21) for the period of 1961–1990. The multi-model ensemble mean in Sects. 3.1 and 3.2 is calculated based on 22 CMIP5 models described in Chap. 1.

In Sect. 3.3, the global mean surface air temperature and precipitation simulated by equal-weighted ensemble mean of 8 (analyzed in another Atlas (Dong WJ et al, The atlas of climate change: based on SEAP-CMIP5. Springer. doi:10.1007/978-3-642-31773-6, 2013)) and 22 CMIP5 models are compared with the results from IPCC AR5 (Collins M et al., Long-term climate change: projections, commitments and irreversibility. In: Stocker TF et al., (eds) Climate change 2013: the physical science basis. Contribution of Working Group I to the Fifth assessment report of the Intergovernmental Panel on Climate Change. Cambridge University Press, Cambridge/New York, 2013; Flato G et al., Evaluation of climate models. In: Stocker TF et al., (eds) Climate change 2013: the physical science basis. Contribution of Working Group I to the Fifth assessment report of the Intergovernmental Panel on Climate Change. Cambridge University Press, Cambridge/New York, 2013). The ensemble of the 22 CMIP5 models show better performance than that of 8 models and is similar with that of IPCC AR5.

Keywords Multi-model ensemble mean • Equal-weighted mean • Weighted mean • BMA • Ensemble member sizes

W. Dong et al., *Atlas of Climate Change: Responsibility and Obligation of Human Society*, Springer Atmospheric Sciences,
DOI 10.1007/978-3-662-48444-9_3

3.1 Comparison between Multi-model Weighted Mean and Equal-Weighted Mean (Figs. 3.1, 3.2, 3.3, 3.4, 3.5, 3.6, 3.7, 3.8, 3.9, 3.10, 3.11, 3.12, 3.13, 3.14, 3.15, 3.16, 3.17, 3.18, 3.19, and 3.20)

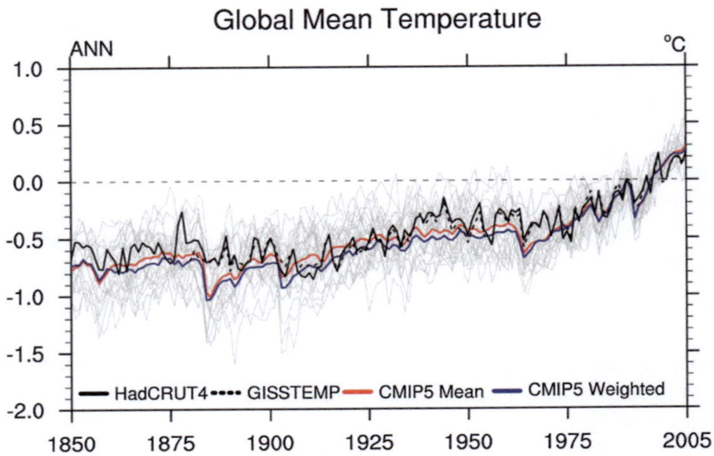

Fig. 3.1 Global annual mean surface air temperature anomalies, relative to 1986–2005 climatology from two observations and 22 CMIP5 models (*gray lines*) and their ensemble mean (*red line*) and weighted mean based on temporal correlations (unit: °C)

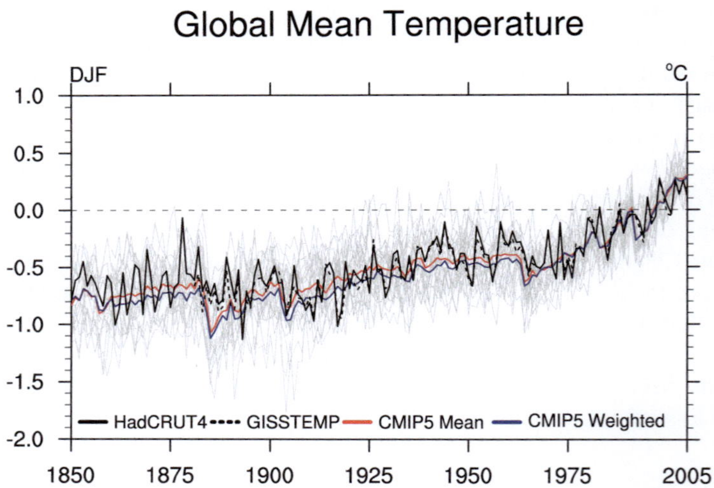

Fig. 3.2 Global mean surface air temperature anomalies for December–February, relative to 1986–2005 climatology from two observations and 22 CMIP5 models (*gray lines*) and their ensemble mean (*red line*) and weighted mean based on temporal correlations (unit: °C)

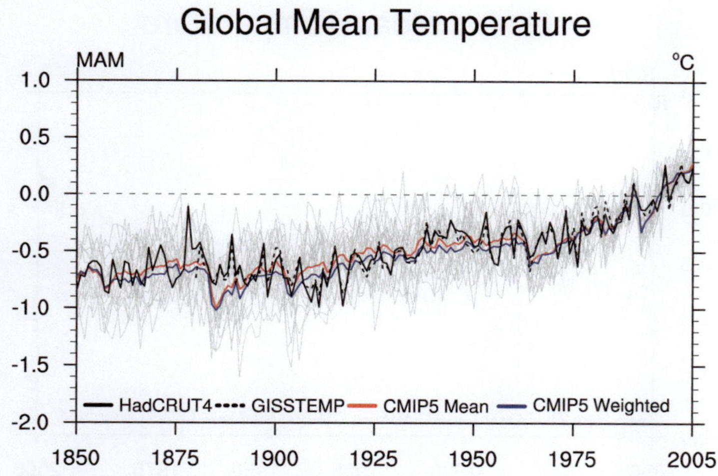

Fig. 3.3 Global mean surface air temperature anomalies for March–May, relative to 1986–2005 climatology from two observations and 22 CMIP5 models (*gray lines*) and their ensemble mean (*red line*) and weighted mean based on temporal correlations (unit: °C)

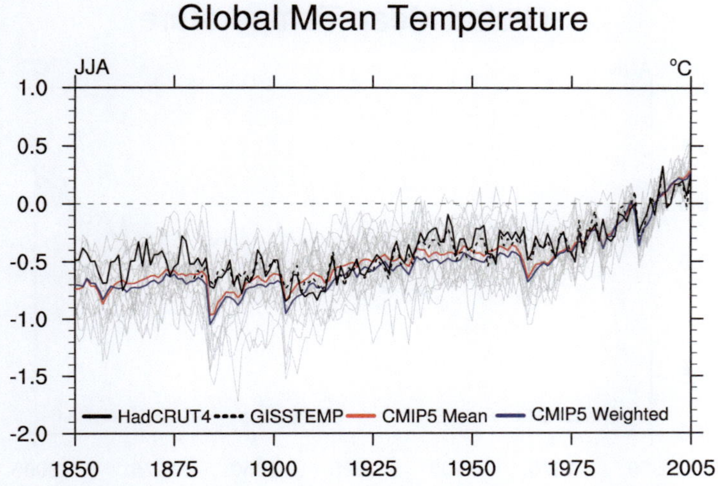

Fig. 3.4 Global mean surface air temperature anomalies for June–August, relative to 1986–2005 climatology from two observations and 22 CMIP5 models (*gray lines*) and their ensemble mean (*red line*) and weighted mean based on temporal correlations (unit: °C)

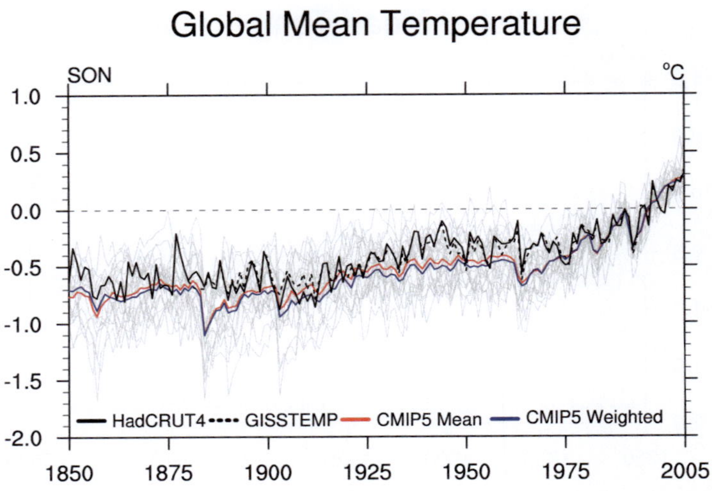

Fig. 3.5 Global mean surface air temperature anomalies for September–November, relative to 1986–2005 climatology from two observations and 22 CMIP5 models (*gray lines*) and their ensemble mean (*red line*) and weighted mean based on temporal correlations (unit: °C)

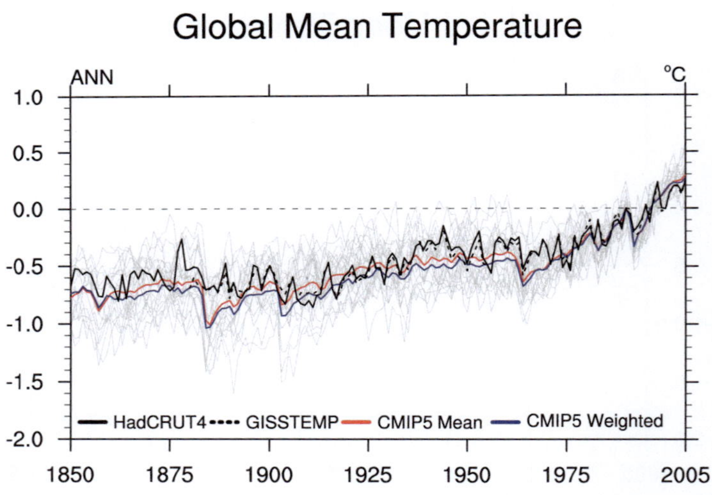

Fig. 3.6 Global mean annual surface air temperature anomalies relative to 1986–2005 climatology from two observations and 22 CMIP5 models (*gray lines*) and their ensemble mean (*red line*) and weighted mean based on relative bias (unit: °C)

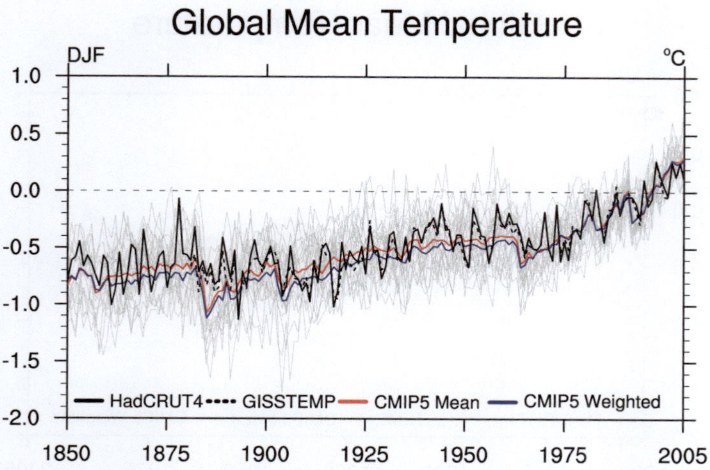

Fig. 3.7 Global mean surface air temperature anomalies for December–February, relative to 1986–2005 climatology from two observations and 22 CMIP5 models (*gray lines*) and their ensemble mean (*red line*) and weighted mean based on relative bias (unit: °C)

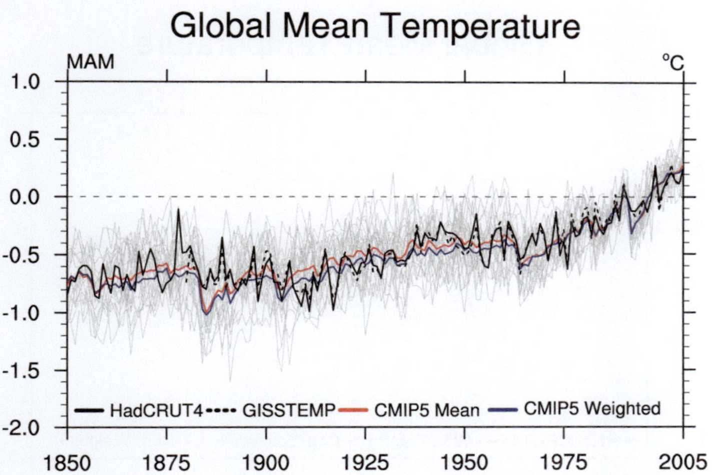

Fig. 3.8 Global mean surface air temperature anomalies for March–May, relative to 1986–2005 climatology from two observations and 22 CMIP5 models (*gray lines*) and their ensemble mean (*red line*) and weighted mean based on relative bias (unit: °C)

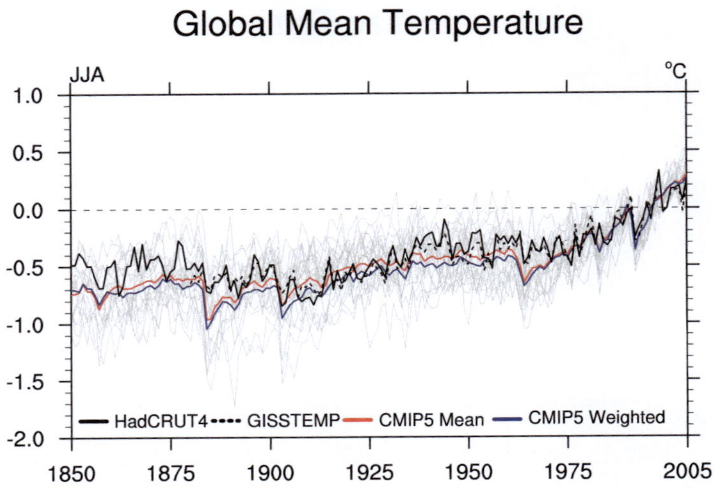

Fig. 3.9 Global mean surface air temperature anomalies for June–August, relative to 1986–2005 climatology from two observations and 22 CMIP5 models (*gray lines*) and their ensemble mean (*red line*) and weighted mean based on relative bias (unit: °C)

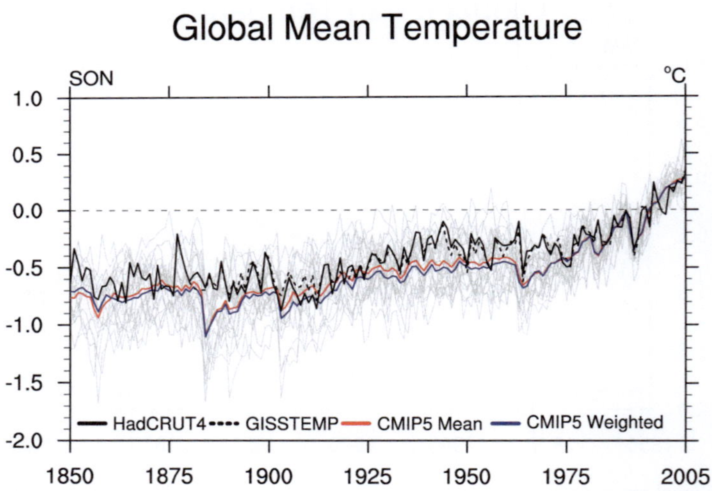

Fig. 3.10 Global mean surface air temperature anomalies for September–November, relative to 1986–2005 climatology from two observations and 22 CMIP5 models (*gray lines*) and their ensemble mean (*red line*) and weighted mean based on relative bias (unit: °C)

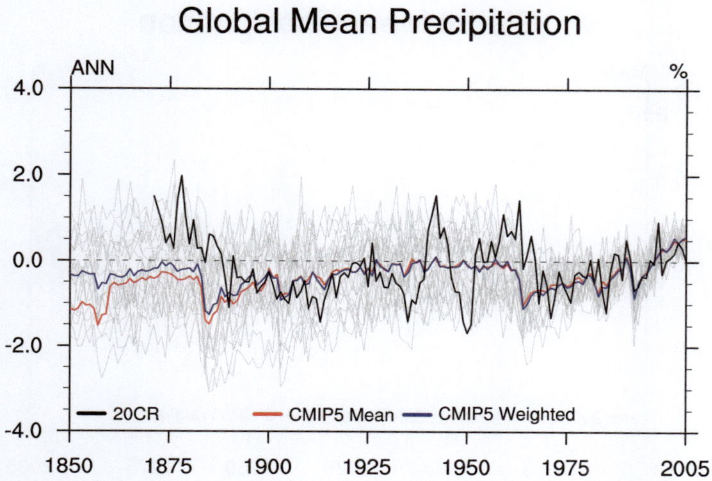

Fig. 3.11 Global annual mean precipitation anomalies, relative to 1986–2005 climatology from two observations and 22 CMIP5 models (*gray lines*) and their ensemble mean (*red line*) and weighted mean based on temporal correlations (unit: %)

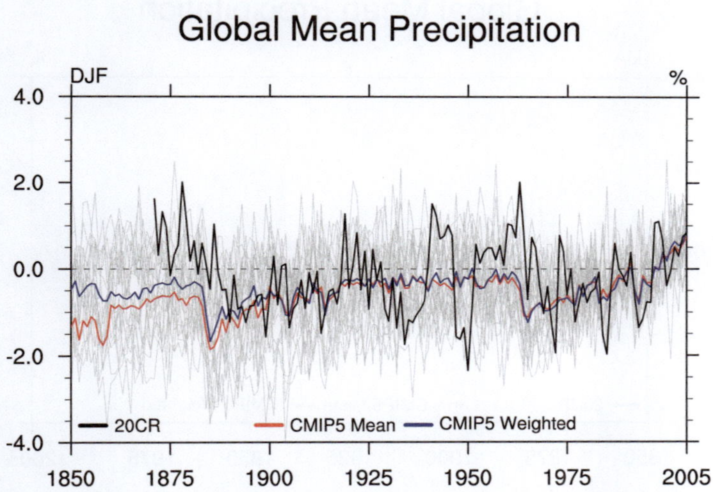

Fig. 3.12 Global mean precipitation anomalies for December–February, relative to 1986–2005 climatology from two observations and 22 CMIP5 models (*gray lines*) and their ensemble mean (*red line*) and weighted mean based on temporal correlations (unit: %)

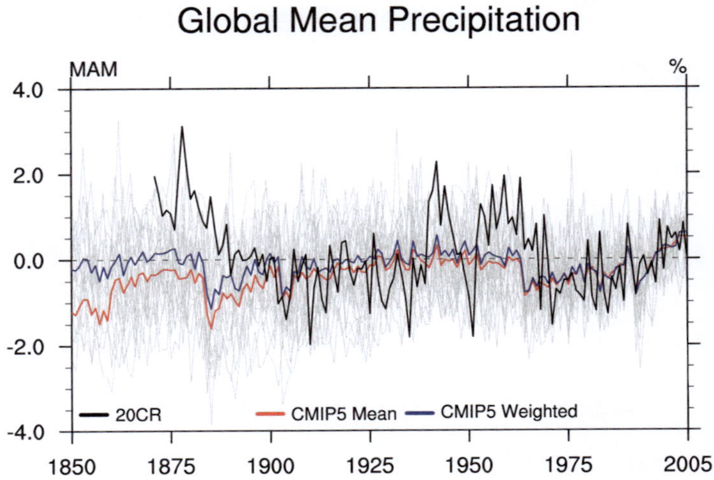

Fig. 3.13 Global mean precipitation anomalies for March–May, relative to 1986–2005 climatology from two observations and 22 CMIP5 models (*gray lines*) and their ensemble mean (*red line*) and weighted mean based on temporal correlations (unit: %)

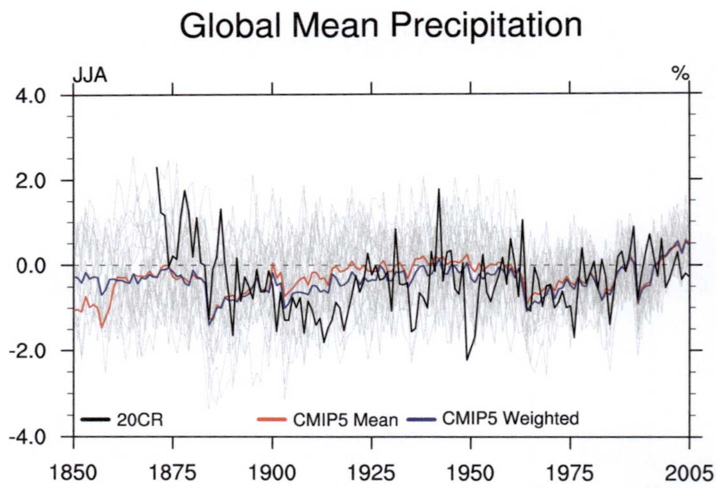

Fig. 3.14 Global mean precipitation anomalies for June–August, relative to 1986–2005 climatology from two observations and 22 CMIP5 models (*gray lines*) and their ensemble mean (*red line*) and weighted mean based on temporal correlations (unit: %)

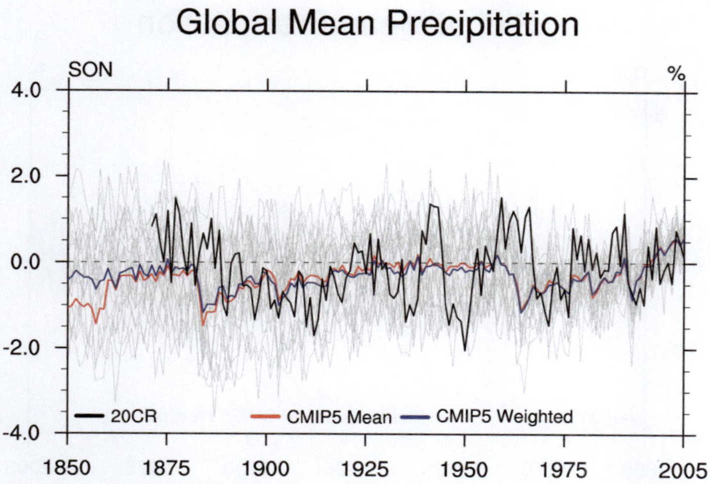

Fig. 3.15 Global mean precipitation anomalies for September–November, relative to 1986–2005 climatology from two observations and 22 CMIP5 models (*gray lines*) and their ensemble mean (*red line*) and weighted mean based on temporal correlations (unit: %)

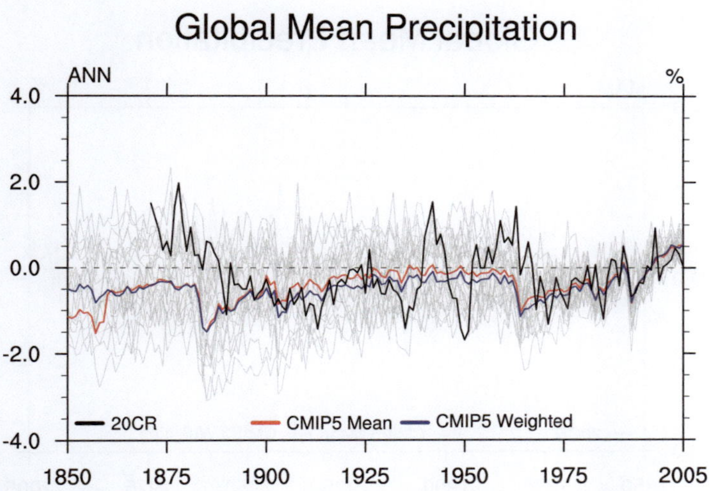

Fig. 3.16 Global annual mean precipitation anomalies, relative to 1986–2005 climatology from two observations and 22 CMIP5 models (*gray lines*) and their ensemble mean (*red line*) and weighted mean based on relative bias (unit: %)

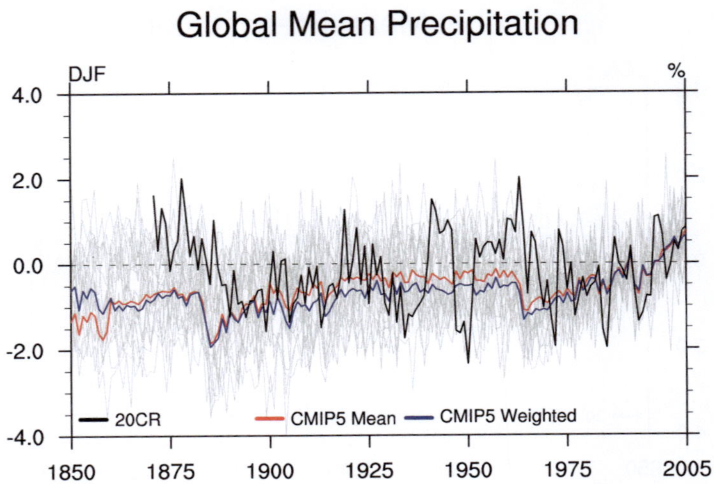

Fig. 3.17 Global mean precipitation anomalies for December–February, relative to 1986–2005 climatology from two observations and 22 CMIP5 models (*gray lines*) and their ensemble mean (*red line*) and weighted mean based on relative bias (unit: %)

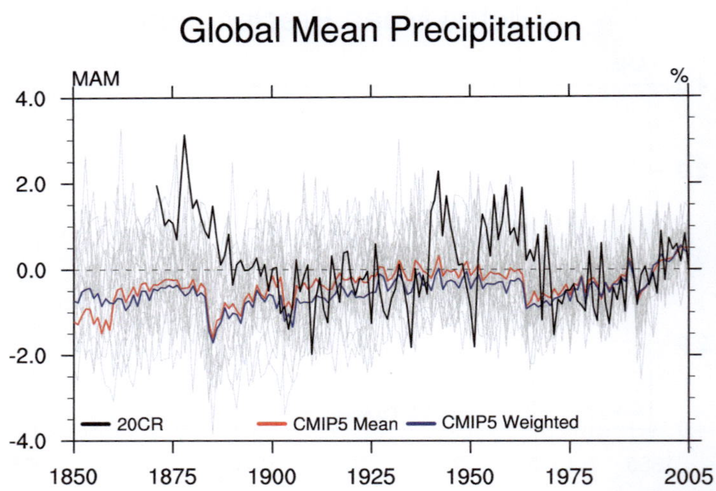

Fig. 3.18 Global mean precipitation anomalies for March–May, relative to 1986–2005 climatology from two observations and 22 CMIP5 models (*gray lines*) and their ensemble mean (*red line*) and weighted mean based on relative bias (unit: %)

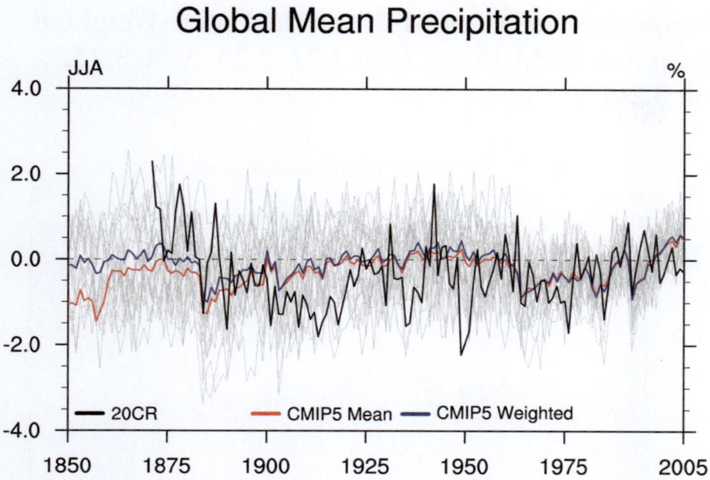

Fig. 3.19 Global mean precipitation anomalies for June–August, relative to 1986–2005 climatology from two observations and 22 CMIP5 models (*gray lines*) and their ensemble mean (*red line*) and weighted mean based on relative bias (unit: %)

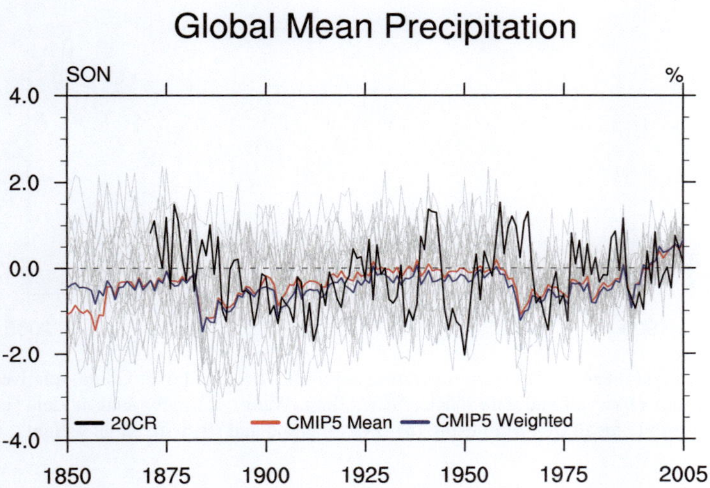

Fig. 3.20 Global mean precipitation anomalies for September–November, relative to 1986–2005 climatology from two observations and 22 CMIP5 models (*gray lines*) and their ensemble mean (*red line*) and weighted mean based on relative bias (unit: %)

3.2 Comparison between Multi-model Equal-Weighted Mean and BMA (Figs. 3.21, 3.22, 3.23, 3.24, 3.25, and 3.26)

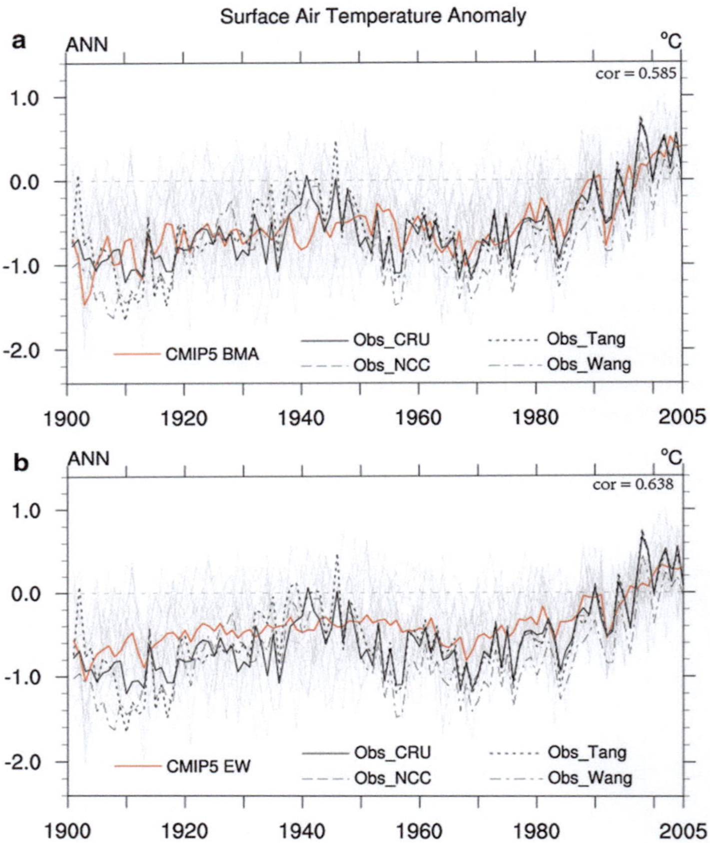

Fig. 3.21 Annual mean surface air temperature anomalies averaged over China, relative to 1986–2005, from four observational data (CRU, NCC, Tang, Wang), 22 individual models (*gray lines*) and their ensemble mean (*red line*) using (**a**) BMA method and (**b**) with equal weight

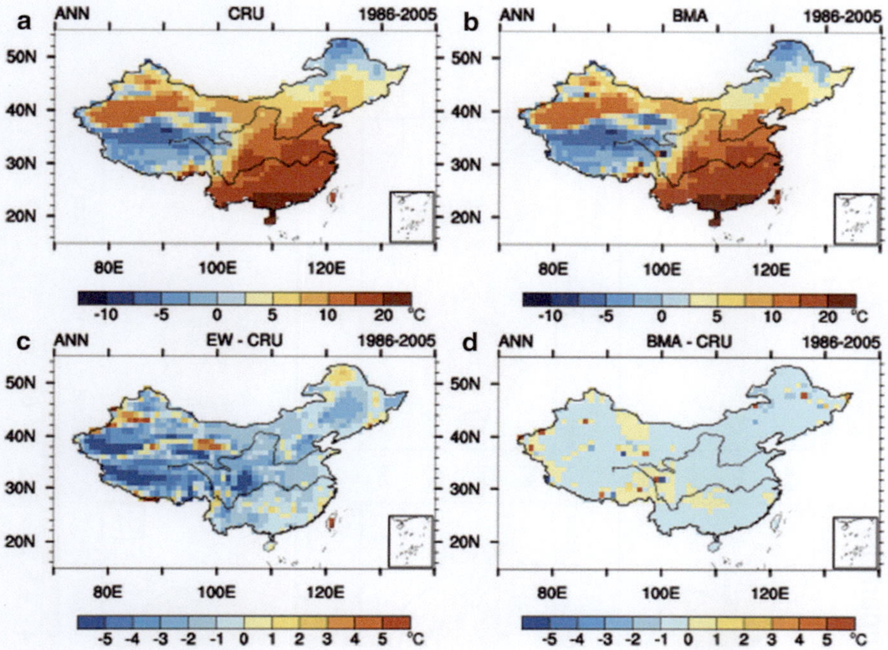

Fig. 3.22 Annual mean surface air temperature in China over 1986–2005: (**a**) observation (CRU), (**b**) multi-model ensemble mean using BMA method. The bias distribution of (**c**) multi-model ensemble mean with equal weight (EW), (**d**) multi-model ensemble mean using BMA method (unit: °C)

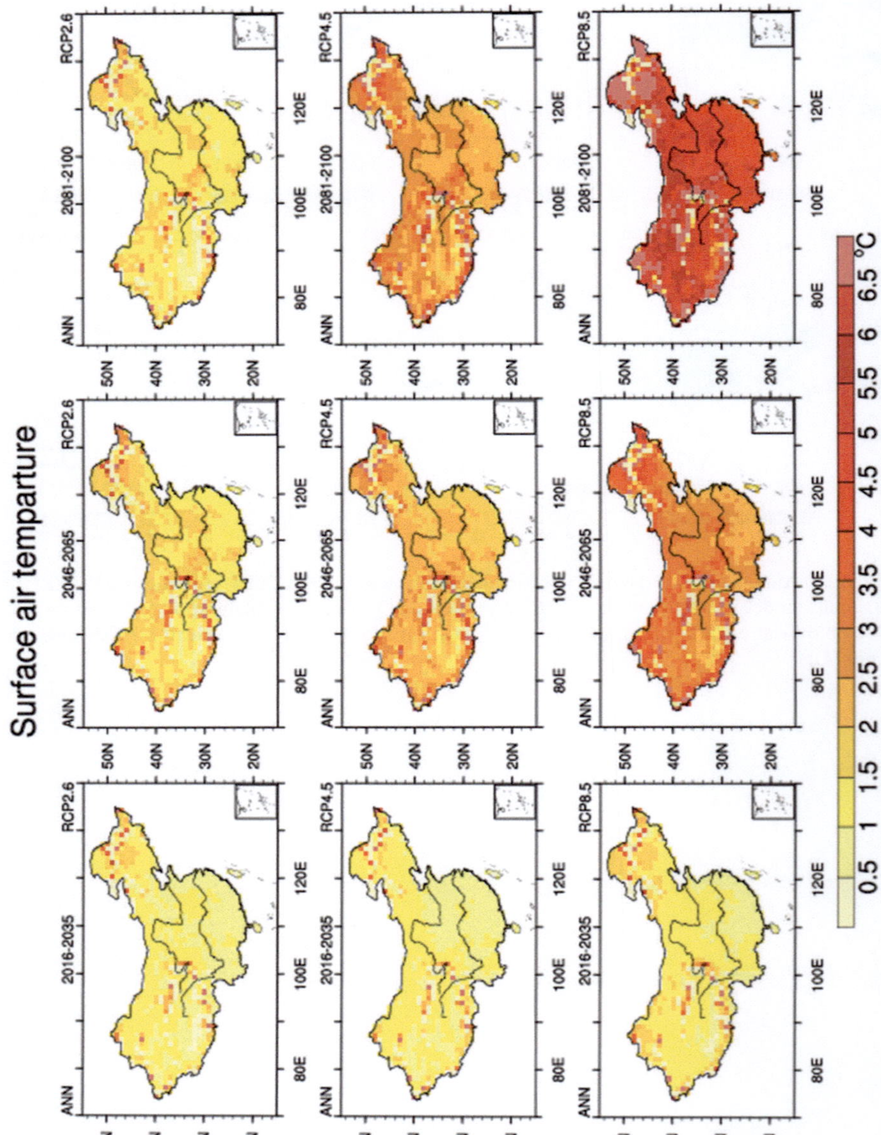

Fig. 3.23 Changes in annual mean surface air temperature in China over 2016–2035 (*left*), 2046–2065 (*middle*), and 2081–2100 (*right*), relative to 1986–2005, for RCP2.6 (*upper*), RCP4.5 (*middle*), and RCP8.5 (*low*) scenarios from multi-model ensemble mean using BMA method (unit: °C)

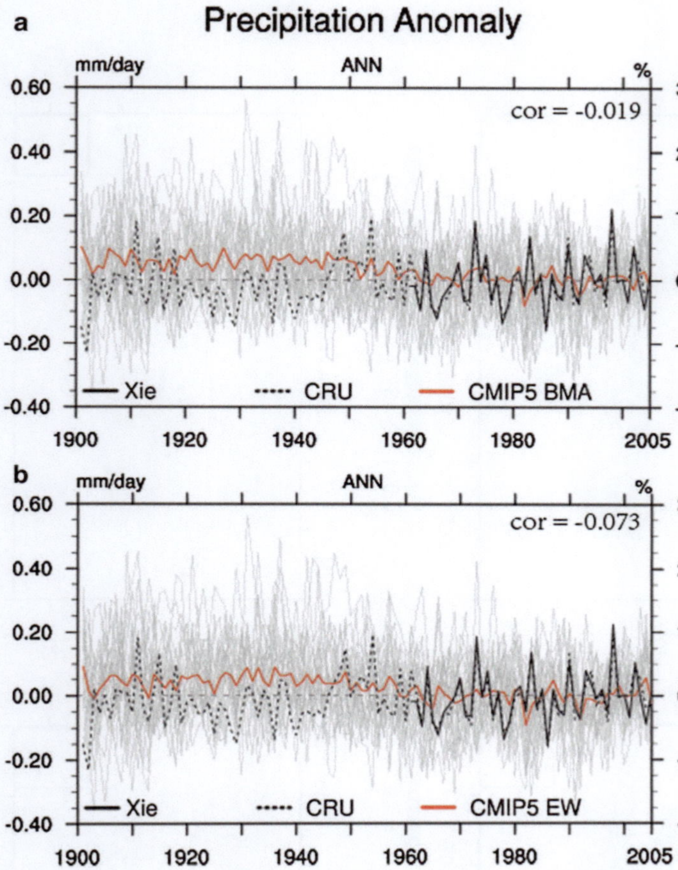

Fig. 3.24 Annual mean precipitation anomalies in China, relative to 1986–2005, from two observed data (CRU, Xie), 22 models (*gray lines*), and their ensemble mean (*red line*) (**a**) using BMA method and (**b**) with equal weight (units: mm/day; %)

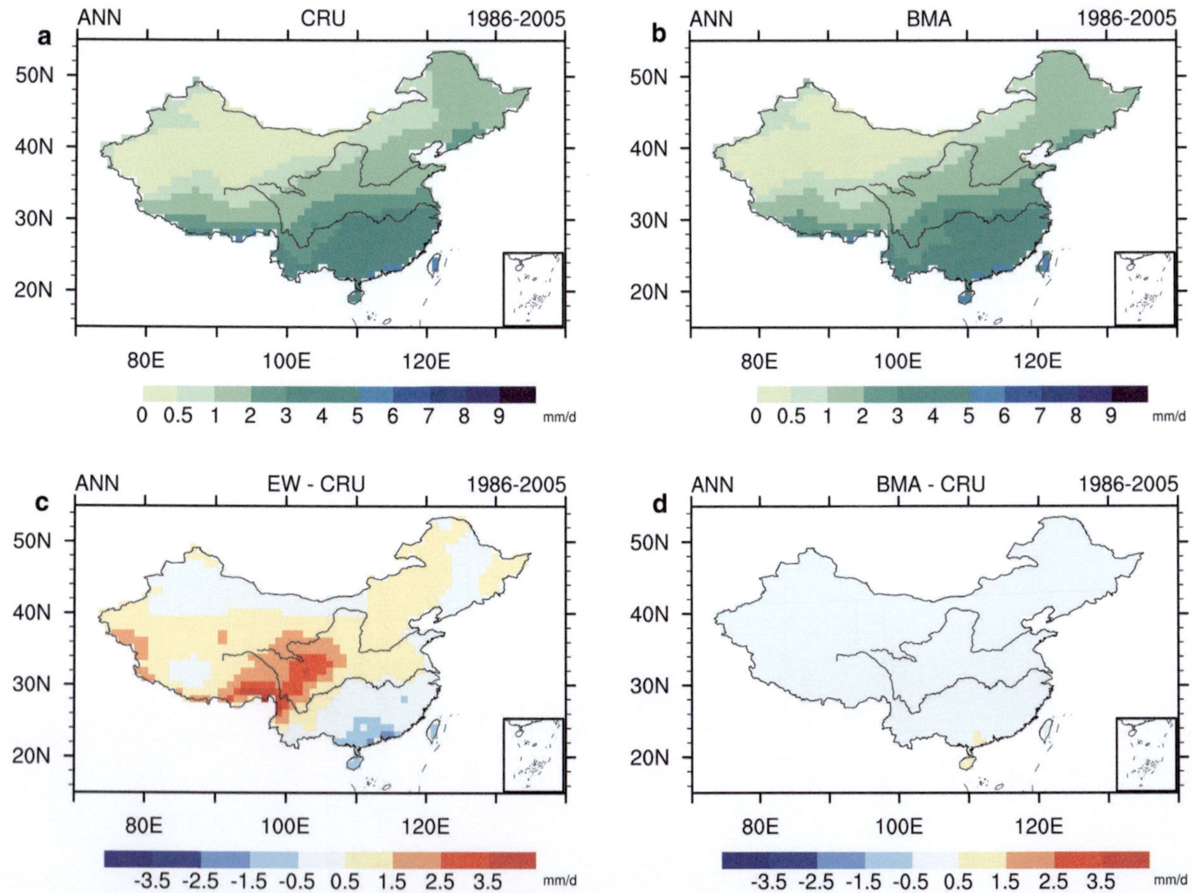

Fig. 3.25 Annual mean precipitation in China over 1986–2005: (**a**) observation (CRU), (**b**) multi-model ensemble mean using BMA method. The bias distribution of (**c**) multi-model ensemble mean with equal weight (EW), (**d**) multi-model ensemble mean using BMA method (unit: mm/d)

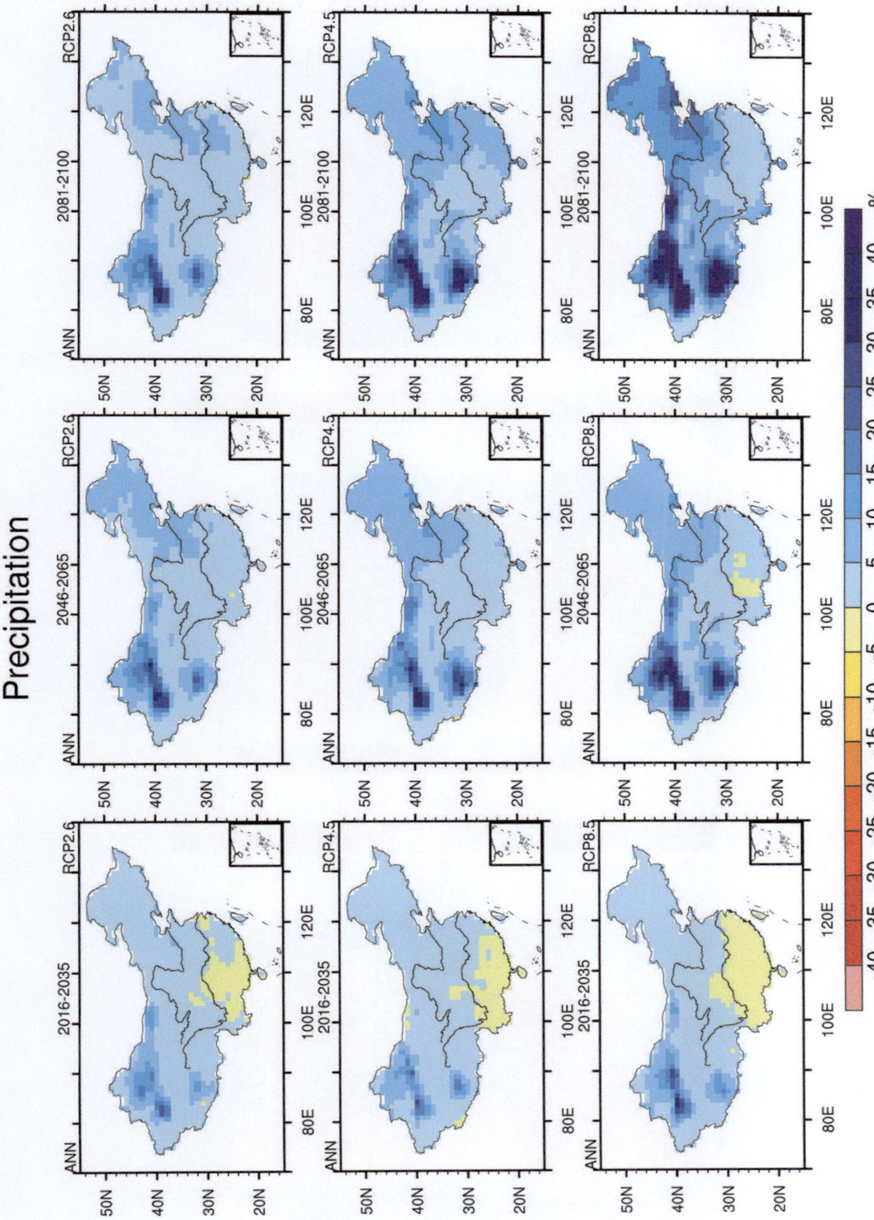

Fig. 3.26 Relative changes in annual mean precipitation in China over 2016–2035 (*left*), 2046–2065 (*middle*), and 2081–2100 (*right*), relative to 1986–2005, for RCP2.6 (*upper*), RCP4.5 (*middle*), and RCP8.5 (*low*) scenarios from multi-model ensemble mean using BMA method (unit: %)

3.3 Comparison among Multi-model Equal-Weighted Means with Different Ensemble Member Sizes (Figs. 3.27, 3.28, 3.29, 3.30, 3.31, 3.32, 3.33, 3.34, 3.35, 3.36, and 3.37)

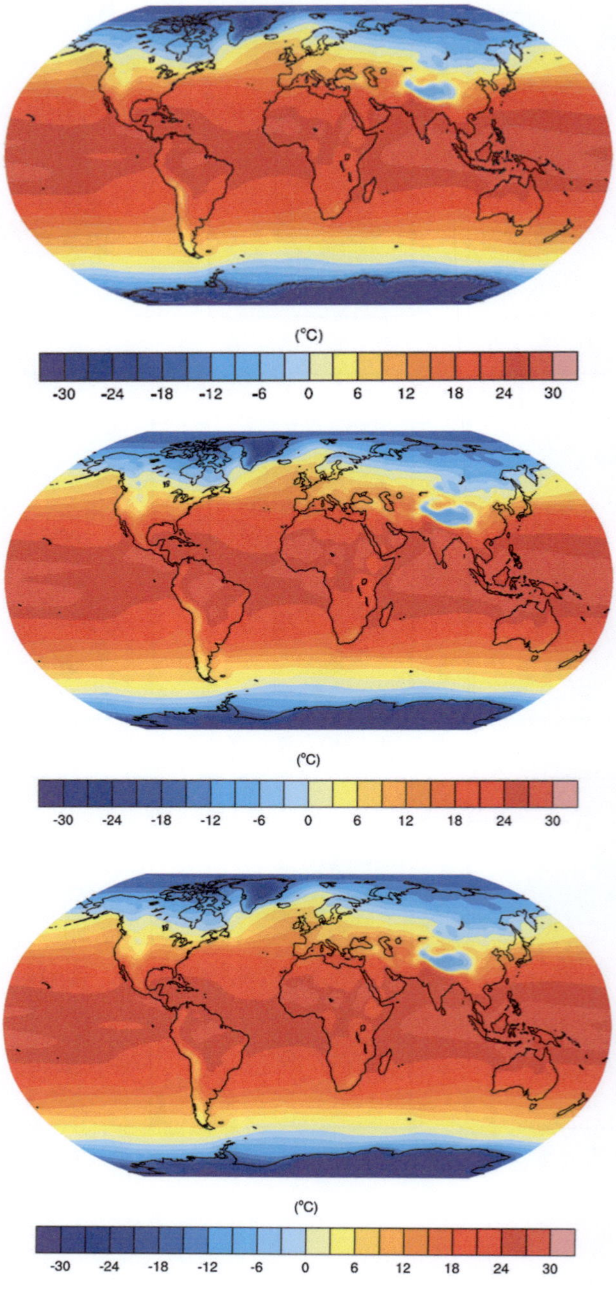

Fig. 3.27 Global mean surface air temperature in 1980–2005 simulated by all the CMIP5 models (*top*, from IPCC AR5 (Flato et al. 2013)), by 8 (*middle*, from Dong et al. (2013)) and 22 (*bottom*) CMIP5 models

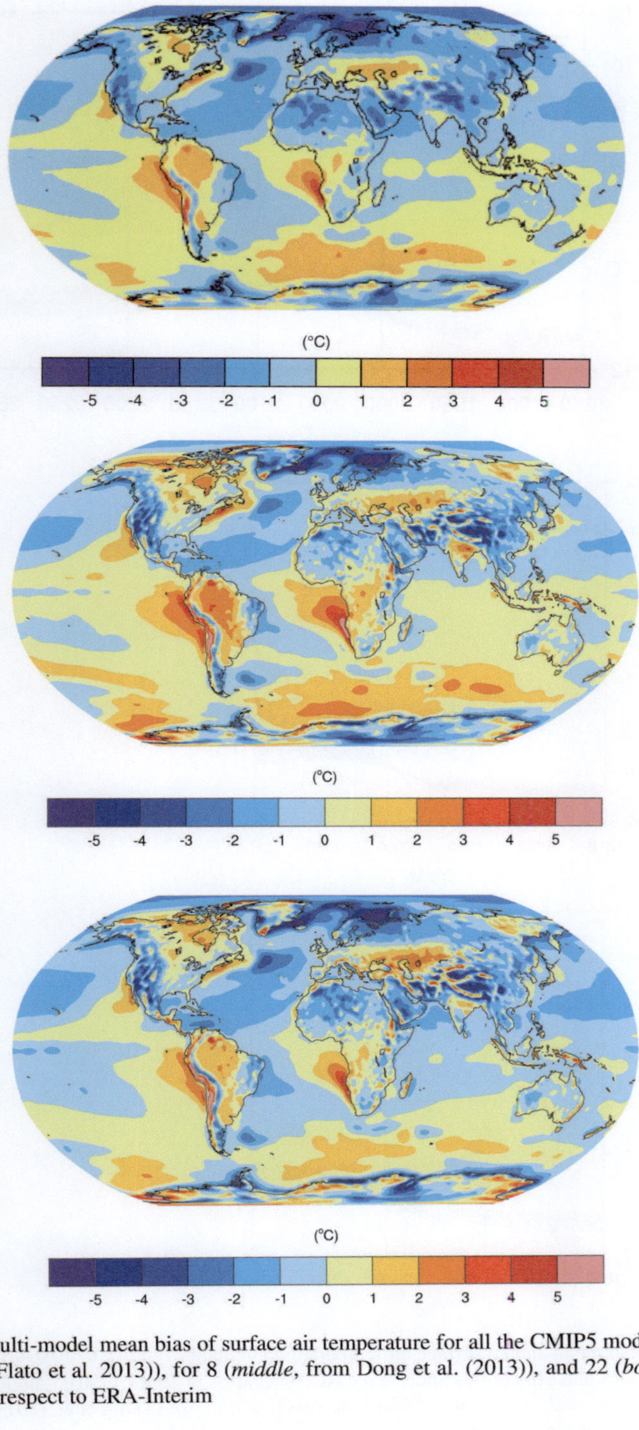

Fig. 3.28 Multi-model mean bias of surface air temperature for all the CMIP5 models (*top*, from IPCC AR5 (Flato et al. 2013)), for 8 (*middle*, from Dong et al. (2013)), and 22 (*bottom*) CMIP5 models with respect to ERA-Interim

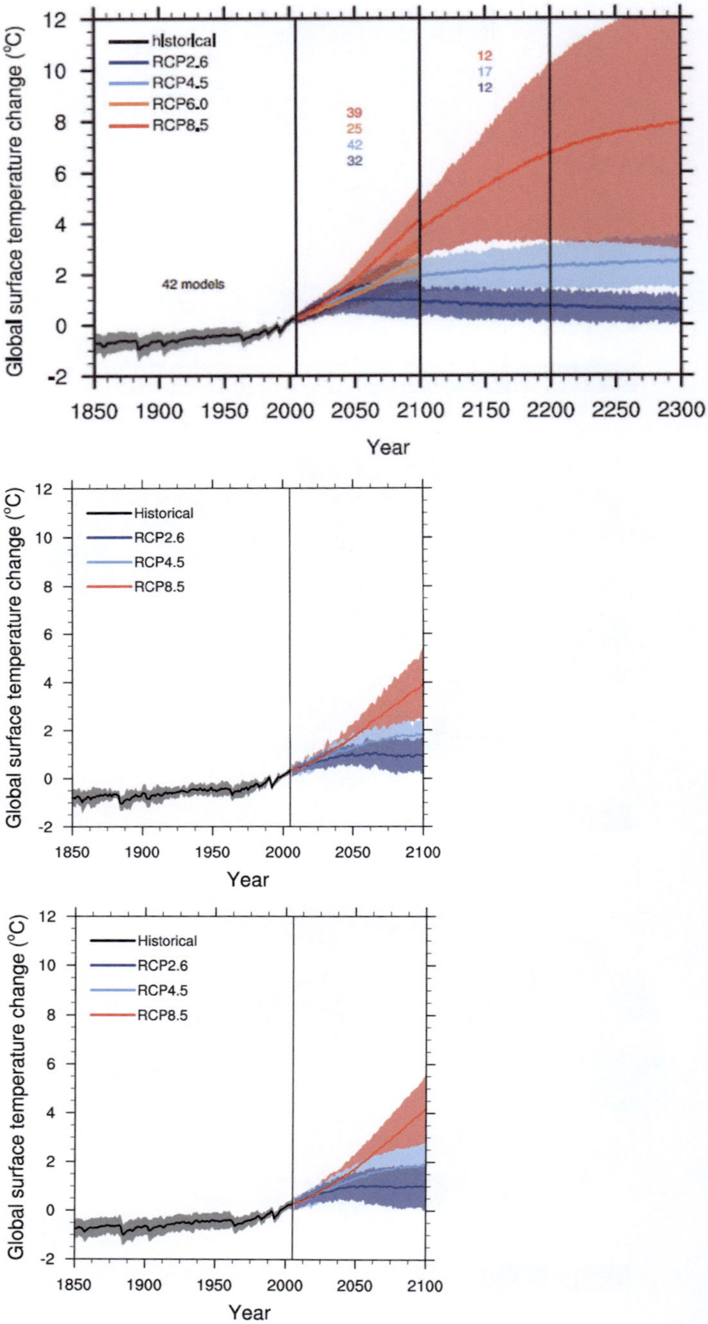

Fig. 3.29 Global mean annual surface air temperature anomalies relative to 1986–2005 from historical simulations and projections under three RCP scenarios in all the CMIP5 models (*top*, from IPCC AR5 (Collins et al. 2013)), in 8 CMIP5 models (*middle*, from Dong et al. (2013)), and in 22 (*bottom*) CMIP5 models (unit: °C)

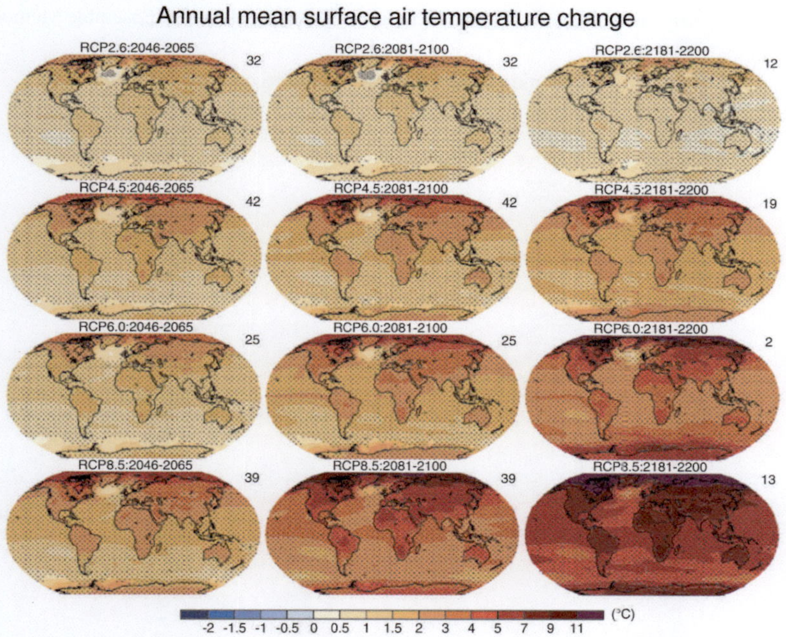

Fig. 3.30 Changes in annual mean surface air temperature in 2046–2065 (*left*), 2081–2100 (*middle*), and 2181–2200 (*right*) relative to the climatology in 1986–2005 projected by all the CMIP5 models (from IPCC AR5 (Collins et al. 2013))

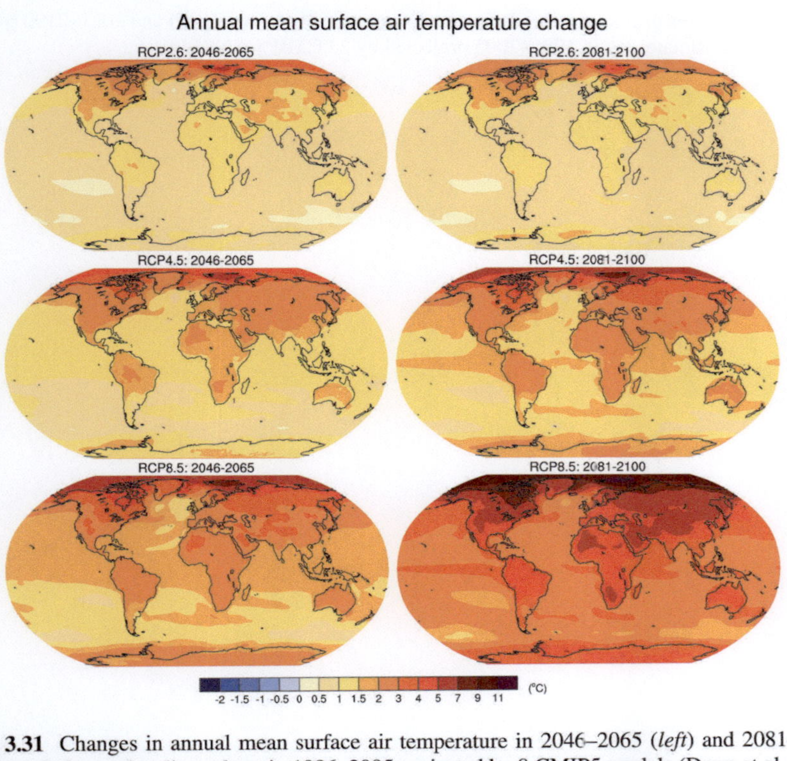

Fig. 3.31 Changes in annual mean surface air temperature in 2046–2065 (*left*) and 2081–2100 (*right*) relative to the climatology in 1986–2005 projected by 8 CMIP5 models (Dong et al. 2013)

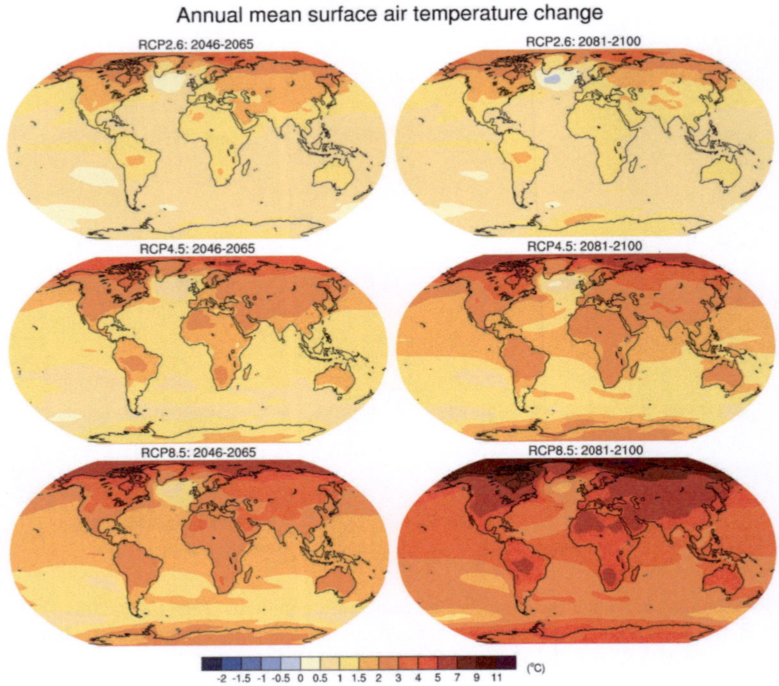

Fig. 3.32 Changes in annual mean surface temperature in 2046–2065 (*left*) and 2081–2100 (*right*) relative to the climatology in 1986–2005 projected by 22 CMIP5 models

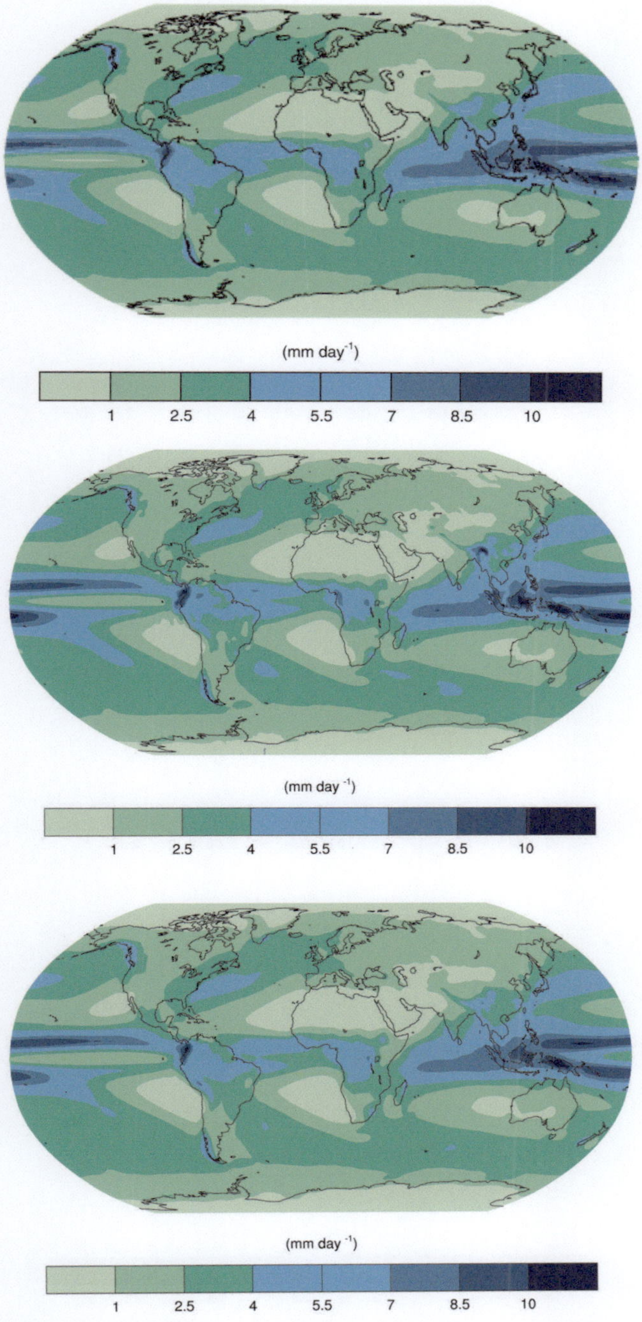

Fig. 3.33 Global mean precipitation in 1980–2005 simulated by all the CMIP5 models (*top*, from IPCC AR5 (Flato et al. 2013)), by 8 (*middle*, from Dong et al. (2013)), and 22 (*bottom*) CMIP5 models

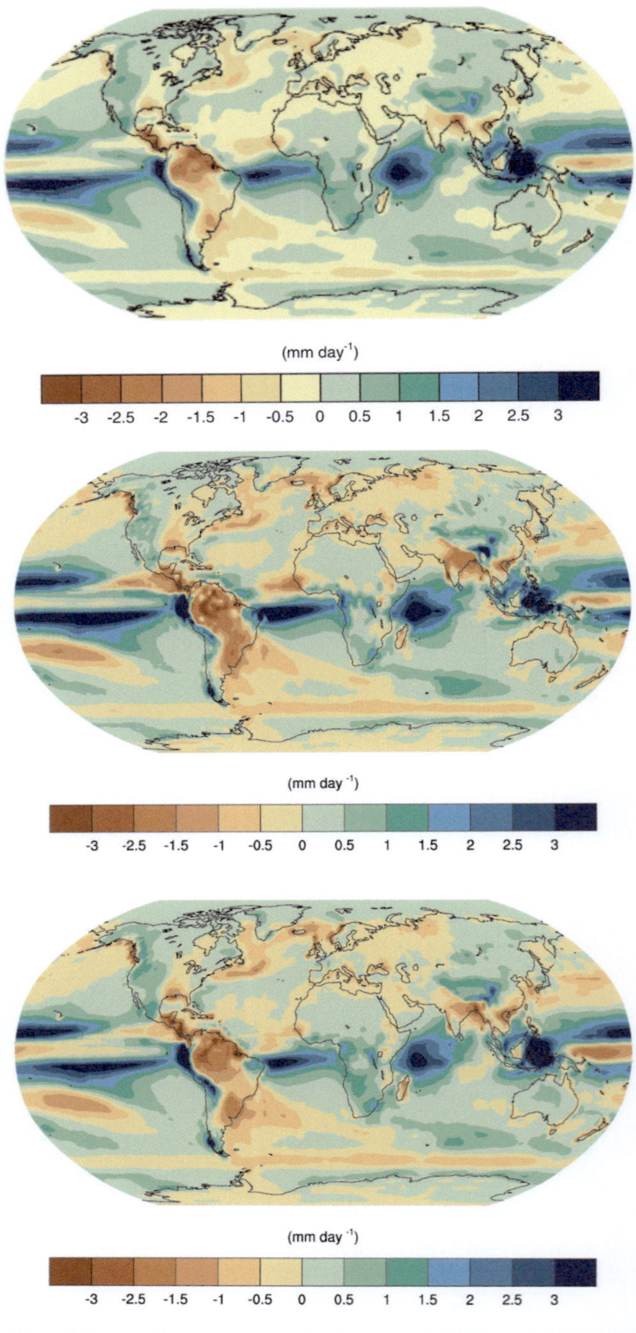

Fig. 3.34 Multi-model mean bias of precipitation for all the CMIP5 models (*top*, from IPCC AR5 (Flato et al. 2013)), 8 (*middle*, from Dong et al. (2013)), and 22 (*bottom*) CMIP5 models with respect to GPCP

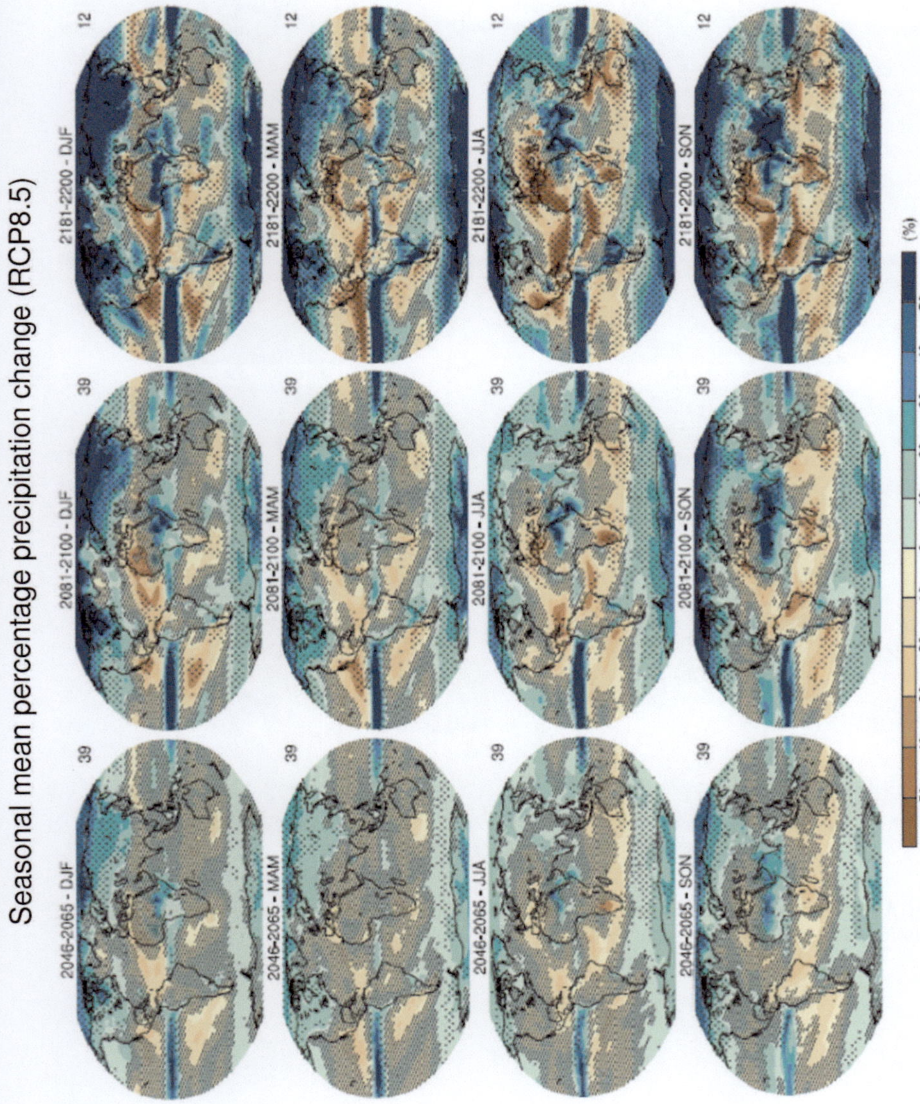

Fig. 3.35 Changes in precipitation in 2046–2065 (*left*), 2081–2100 (*middle*), and 2181–2200 (*right*) relative to the climatology in 1986–2005 projected by all the CMIP5 models (from IPCC AR5 (Collins et al. 2013))

Fig. 3.36 Changes in precipitation in 2046–2065 (*left*) and 2081–2100 (*right*) relative to the climatology in 1986–2005 projected by 8 CMIP5 models (Dong et al. 2013)

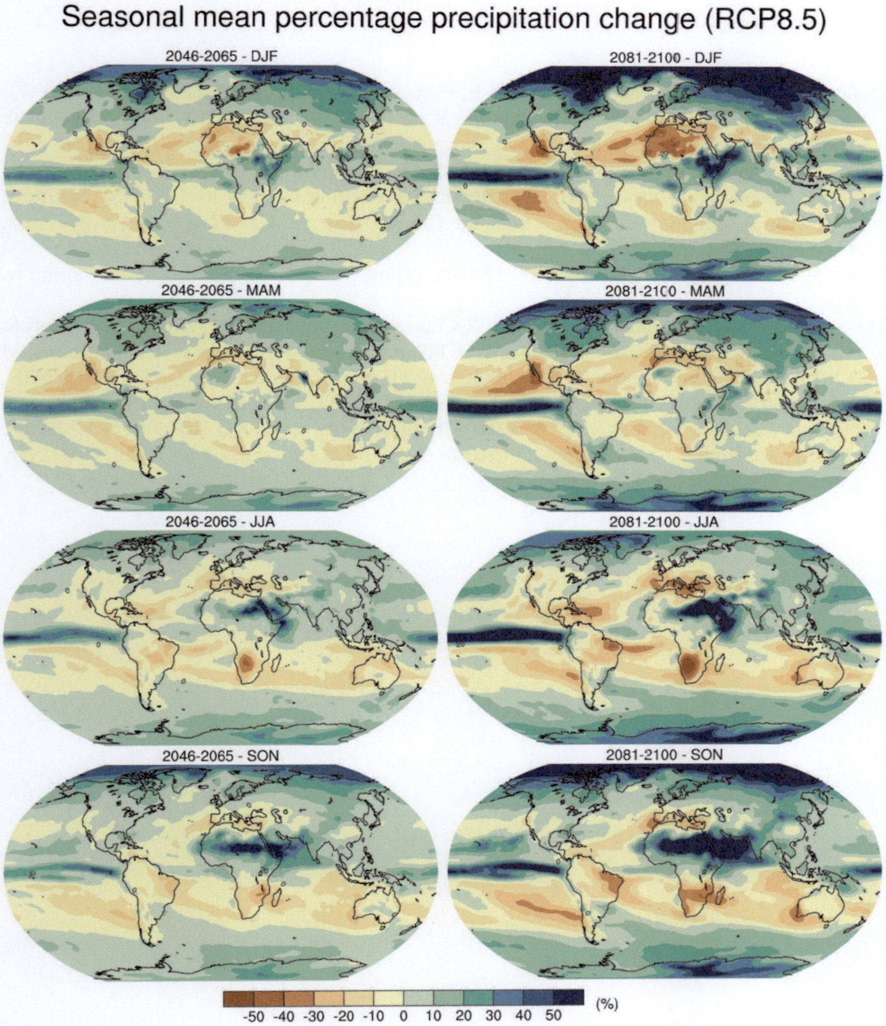

Fig. 3.37 Changes in precipitation in 2046–2065 (*left*) and 2081–2100 (*right*) relative to the climatology in 1986–2005 projected by 22 CMIP5 models

References

Collins M, Knutti R, Arblaster J, Dufresne J-L, Fichefet T, Friedlingstein P, Gao X, Gutowski WJ, Johns T, Krinner G, Shongwe M, Tebaldi C, Weaver AJ, Wehner M (2013) Long-term climate change: projections, commitments and irreversibility. In: Stocker TF, Qin D, Plattner G-K, Tignor M, Allen SK, Boschung J, Nauels A, Xia Y, Bex V, Midgley PM (eds) Climate change 2013: the physical science basis. Contribution of Working Group I to the Fifth assessment report of the Intergovernmental Panel on Climate Change. Cambridge University Press, Cambridge/New York

Dong WJ et al (2013) The atlas of climate change: based on SEAP-CMIP5. Springer. doi:10.1007/978-3-642-31773-6

Flato G, Marotzke J, Abiodun B, Braconnot P, Chou SC, Collins W, Cox P, Driouech F, Emori S, Eyring V, Forest C, Gleckler P, Guilyardi E, Jakob C, Kattsov V, Reason C, Rummukainen M (2013) Evaluation of climate models. In: Stocker TF, Qin D, Plattner G-K, Tignor M, Allen SK, Boschung J, Nauels A, Xia Y, Bex V, Midgley PM (eds) Climate change 2013: the physical science basis. Contribution of Working Group I to the Fifth assessment report of the Intergovernmental Panel on Climate Change. Cambridge University Press, Cambridge/New York

Chapter 4
Responsibility for Historical Climate Change Induced from Developed and Developing World Anthropogenic Carbon Emissions

Abstract This chapter shows the results of attribution experiments which assess the contributions of historical anthropogenic carbon emissions to global warming from developed and developing countries. These numerical experiments are performed based on NCAR/CESM1_0_2 model for the period of 1850–2011.

Keywords Developing/developed countries • Responsibility • Recent CO_2 emission

W. Dong et al., *Atlas of Climate Change: Responsibility and Obligation of Human Society*, Springer Atmospheric Sciences,
DOI 10.1007/978-3-662-48444-9_4

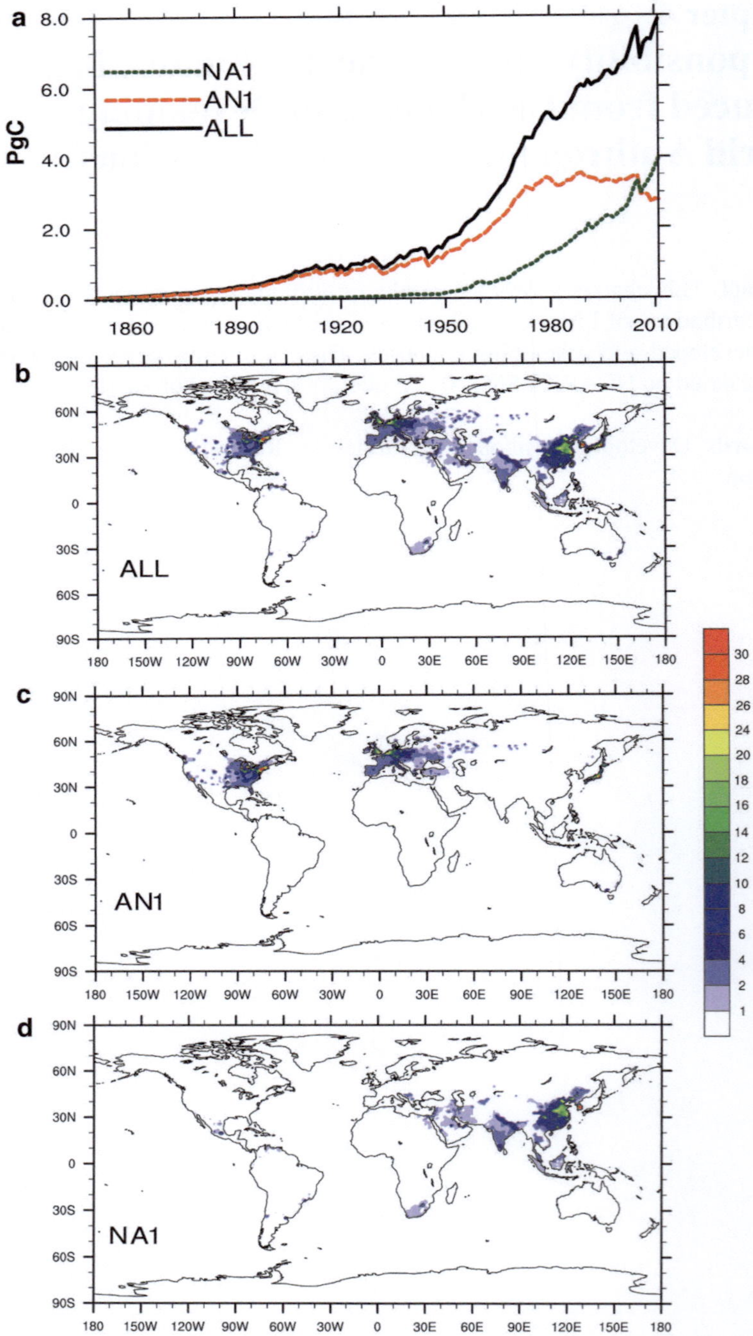

Fig. 4.1 (**a**) The annual CO_2 emissions rates (unit: pgC), (**b**)–(**d**) gridded CO_2 emission fluxes from 2006 to 2011 used by ALL, AN1, and NA1 experiments (unit: $KgCO_2/m^2 \cdot 16y$)

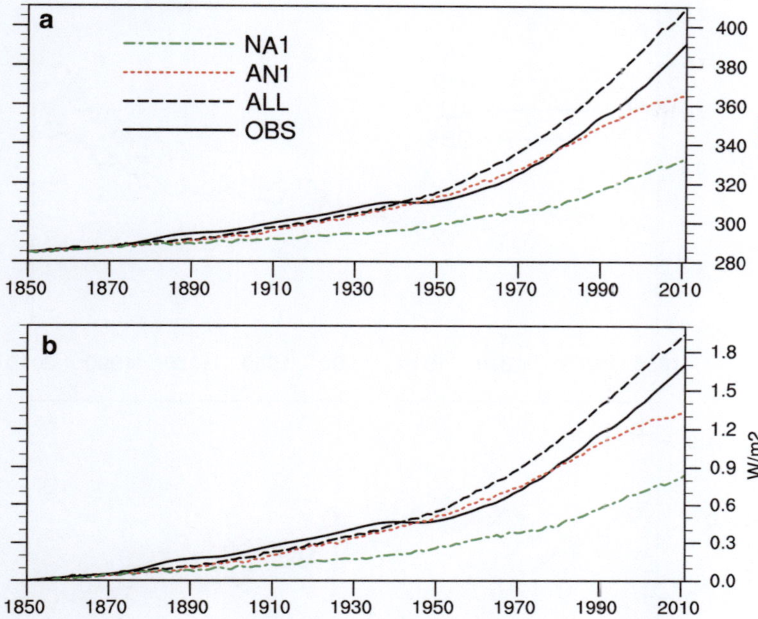

Fig. 4.2 (**a**) Global averaged CO_2 concentration (unit: ppmv), (**b**) CO_2 radiative forcings (unit: W/m^2) simulated by ALL, AN1, and NA1 experiments and observation

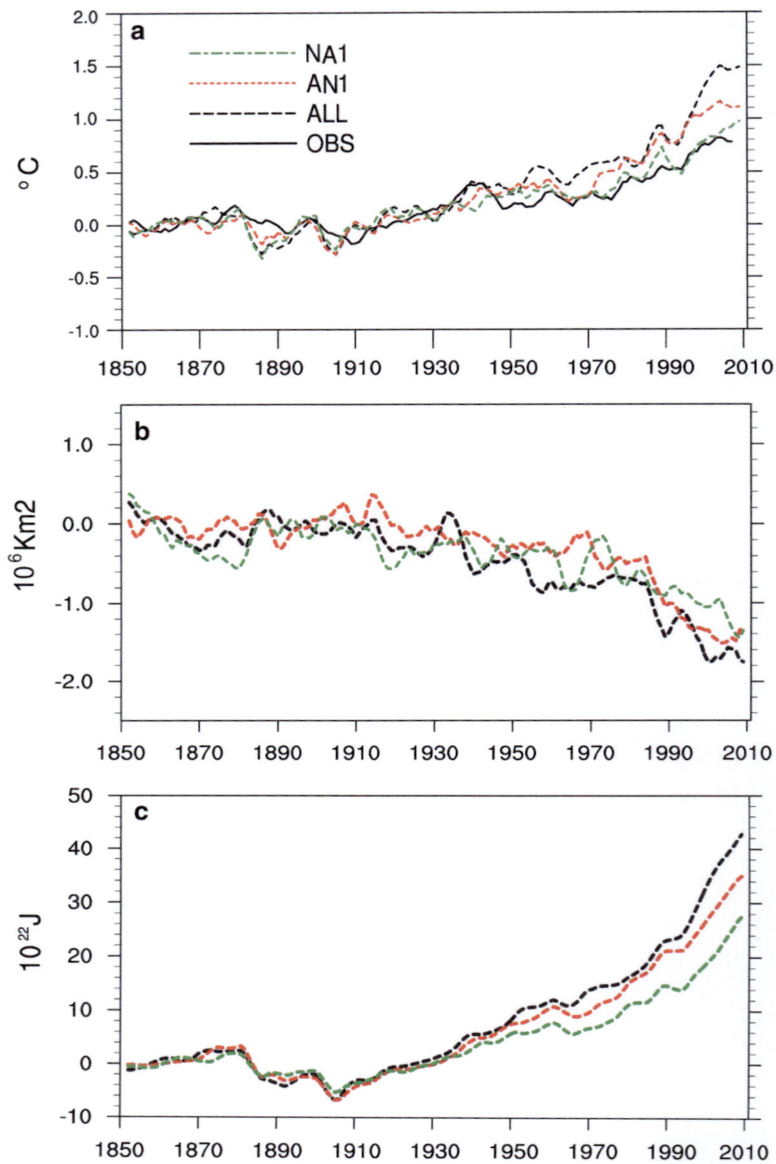

Fig. 4.3 (**a**) Global averaged annual average surface air temperature (unit:°C), (**b**) annual average North Hemisphere sea-ice area (unit: 10^6 km^2), and (**c**) global averaged annual average sea heat content from surface to 700 m depth (unit: 10^{22} J), simulated by ALL, AN1, and NA1 experiments. The *lines* denote variability with period exceeding 10 years

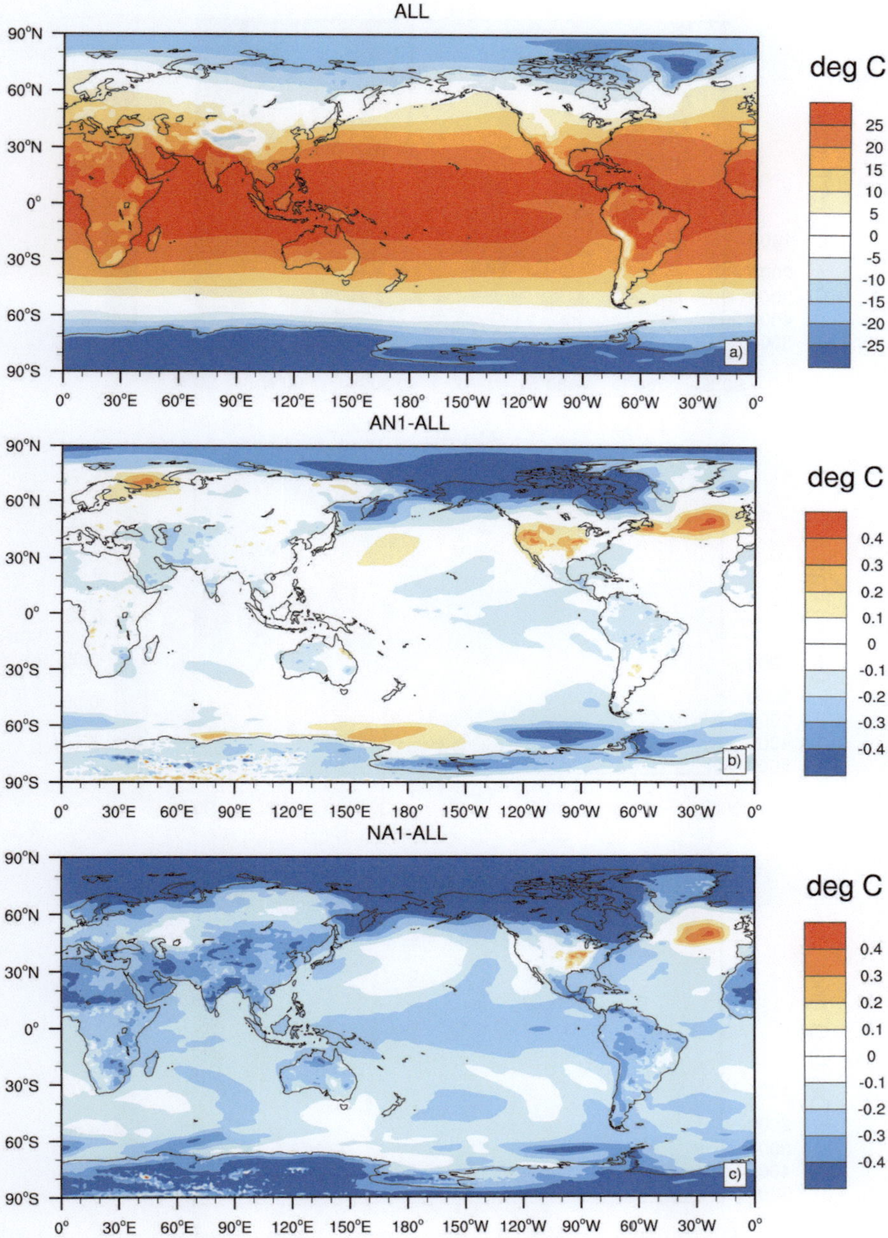

Fig. 4.4 Annual average surface air temperature averaged over 1941–2011, simulated by (**a**) ALL, (**b**) AN1 minus ALL, and (**c**) NA1 minus ALL (unit: °C)

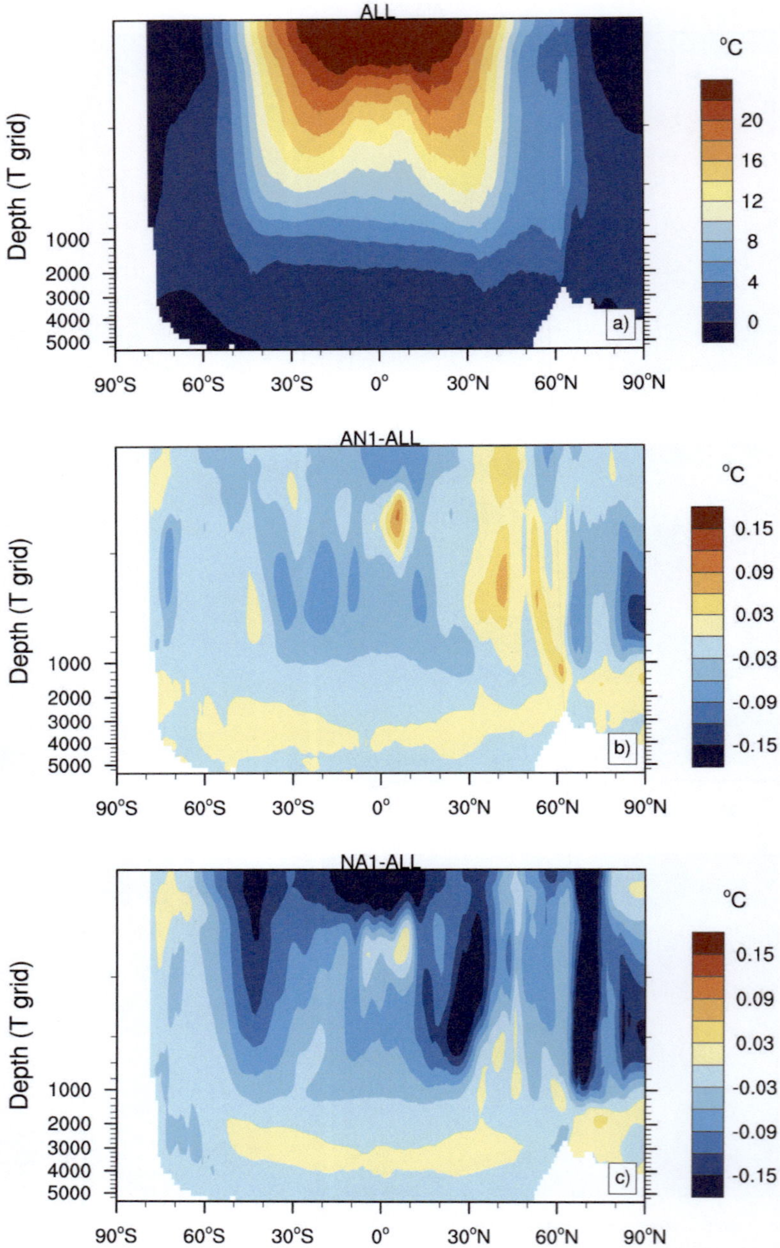

Fig. 4.5 Latitude-depth section of annual average sea potential temperature averaged over 1941–2011, simulated by (**a**) ALL, (**b**) AN1 minus ALL, and (**c**) NA1 minus ALL (unit: °C)

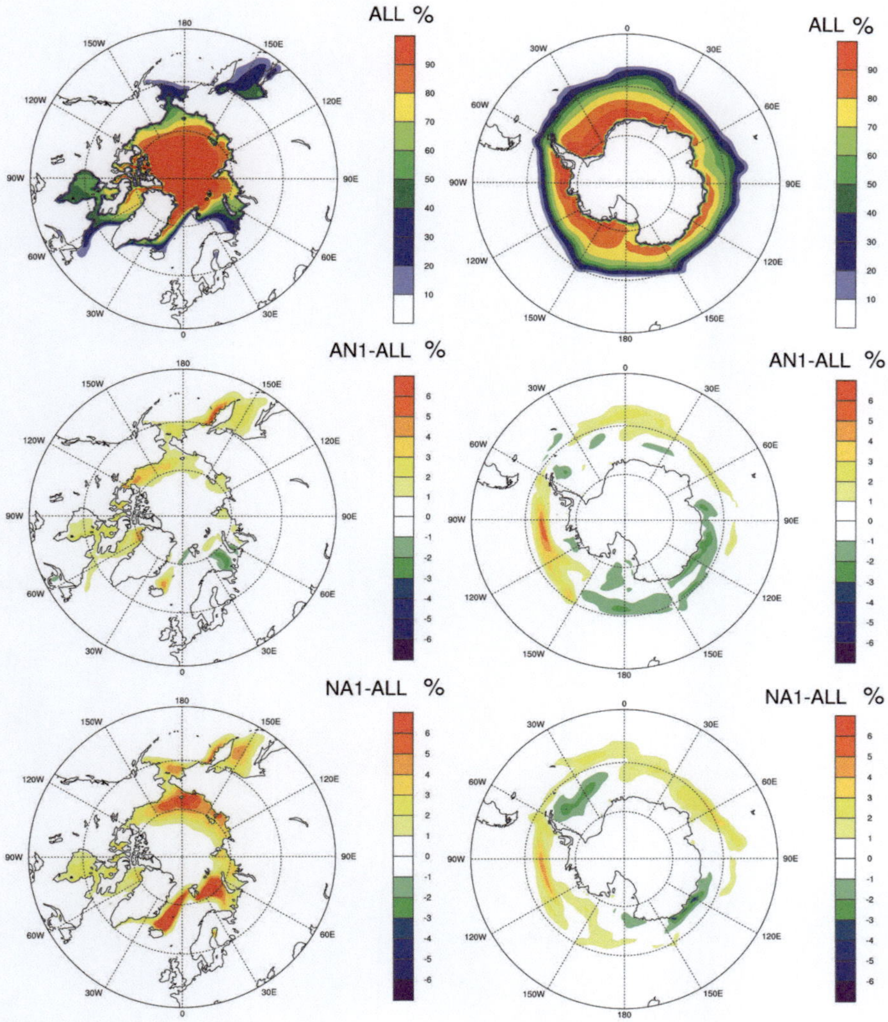

Fig. 4.6 Annual average sea ice concentration averaged over 1941–2011 at North Hemisphere (*left*) and South Hemisphere (*right*), simulated by ALL (*upper*), AN1 minus ALL (*middle*), and NA1 minus ALL (low; unit: %)

Chapter 5
Responsibilities for Climate Change Induced from Developed and Developing World Anthropogenic Sulfur Emissions

Abstract This chapter shows the results of the impacts of anthropogenic sulfur emissions from Annex I and non-Annex I countries for the period of 1850–2005. These numerical experiments named Both, Ann1, Non1, and Zero are performed based on NCAR/CESM1_0_2 model. The analyzed climate elements involve aerosol concentration, aerosol optical depth, air temperature, precipitation, cloud, solar flux, sea level pressure, zonal circulation, sea ice extent, etc.

Keywords Sulfur emission • Climate change • Responsibility

W. Dong et al., *Atlas of Climate Change: Responsibility and Obligation of Human Society*, Springer Atmospheric Sciences,
DOI 10.1007/978-3-662-48444-9_5

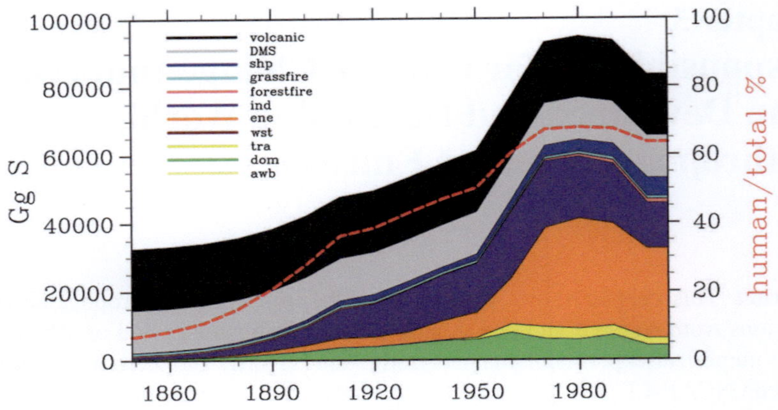

Fig. 5.1 Sources of natural and anthropogenic sulfur emissions (unit: Gg S), and the ratio of anthropogenic emissions to the whole is shown by the red dashed (unit: %). Emissions of sulfur dioxide are update of Smith et al. (2001, 2004, 2011)

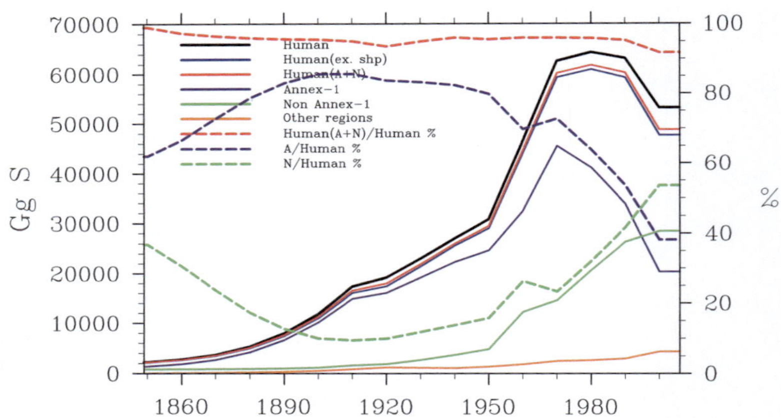

Fig. 5.2 Anthropogenic sulfur emissions from different regions (unit: Gg S), and the ratios of Annex I and non-Annex I country groups to the anthropogenic sulfur emissions are shown by purple and green dashed lines, respectively

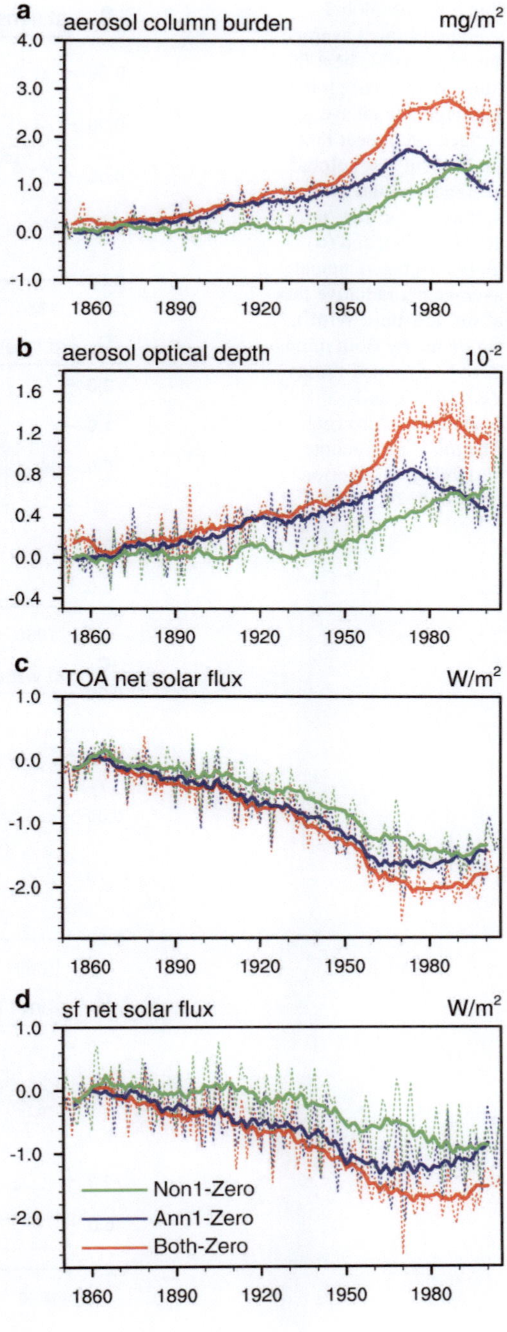

Fig. 5.3 (**a**) Global averaged annual average aerosol column burden (unit: mg/m^2), (**b**) global averaged annual average aerosol optical depth (unit: 10^{-2}), (**c**) global averaged annual average net solar flux at the top of atmosphere (unit: W/m^2), and (**d**) global averaged annual average net solar flux at surface (unit: W/m^2), simulated by Both minus Zero (*red*), Ann1 minus Zero (*blue*), and Non1minus Zero (*green*). The *thick lines* denote variability with period exceeding 10 years

Fig. 5.4 (**a**) Global averaged annual average surface sensible heat flux (unit: W/m^2), (**b**) global averaged annual average surface latent heat flux (unit: W/m^2), (**c**) global averaged annual average surface water flux (unit: 10^{-5} kg/m^2/s), and (**d**) global averaged annual average net radiative flux at surface (unit: W/m^2), simulated by Both minus Zero (*red*), Ann1 minus Zero (*blue*), and Non1minus Zero (*green*). The *thick lines* denote variability with period exceeding 10 years

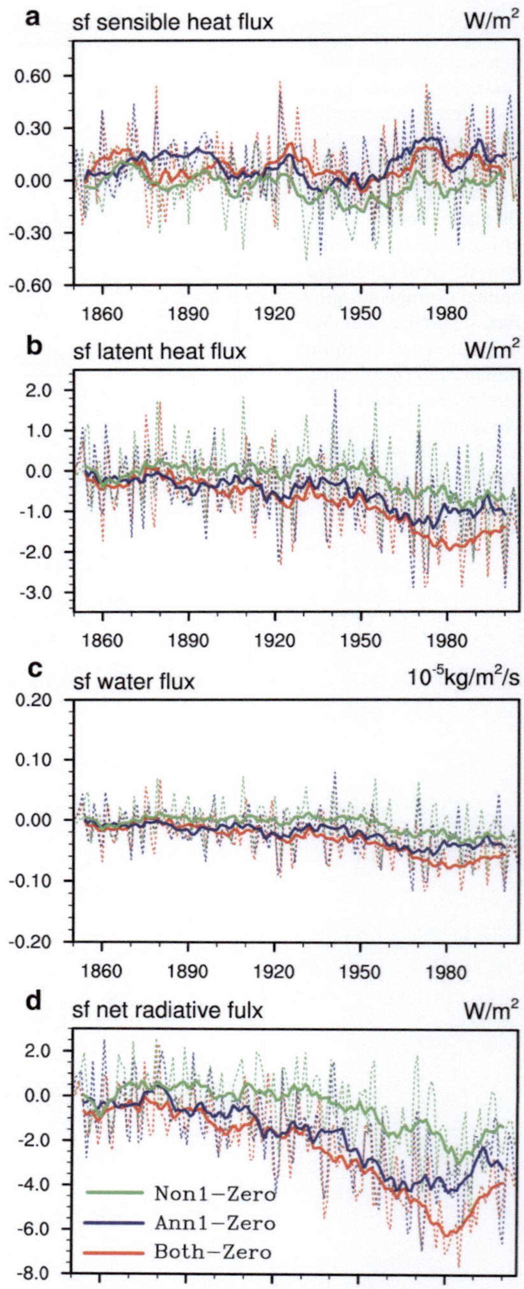

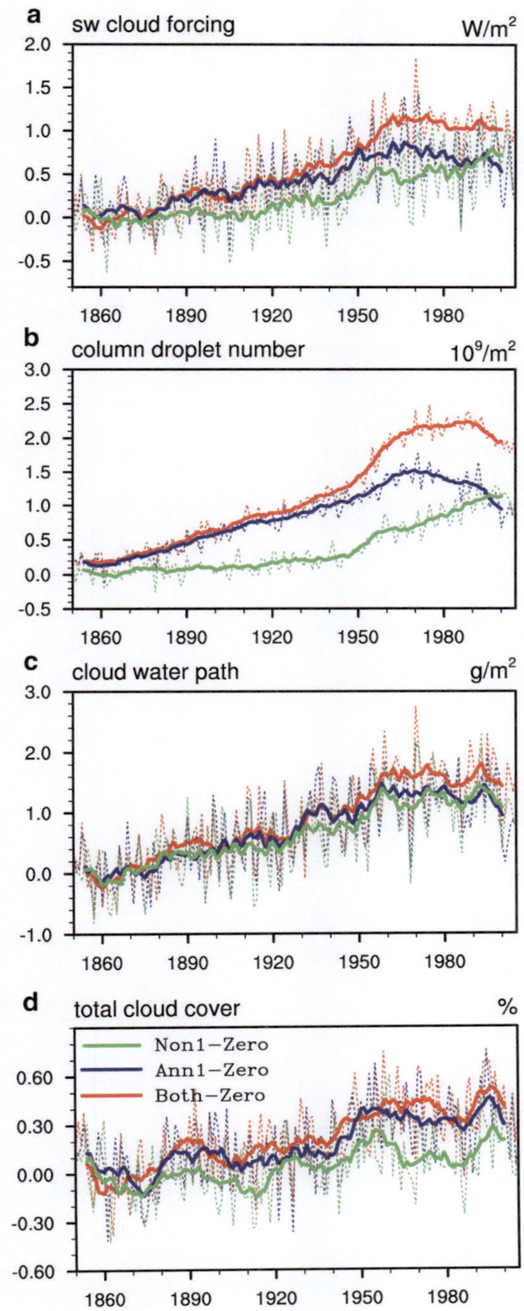

Fig. 5.5 (**a**) Global averaged annual average shortwave cloud forcing, positive value means an increase of solar radiation reflected by cloud (unit: W/m2), (**b**) global averaged annual average column droplet number (unit: 109/m2), (**c**) global averaged annual average cloud water path (unit: g/m2), and (**d**) global averaged annual average total cloud cover (unit: %), simulated by Both minus Zero (*red*), Ann1 minus Zero (*blue*), and Non1minus Zero (*green*). The *thick lines* denote variability with period exceeding 10 years

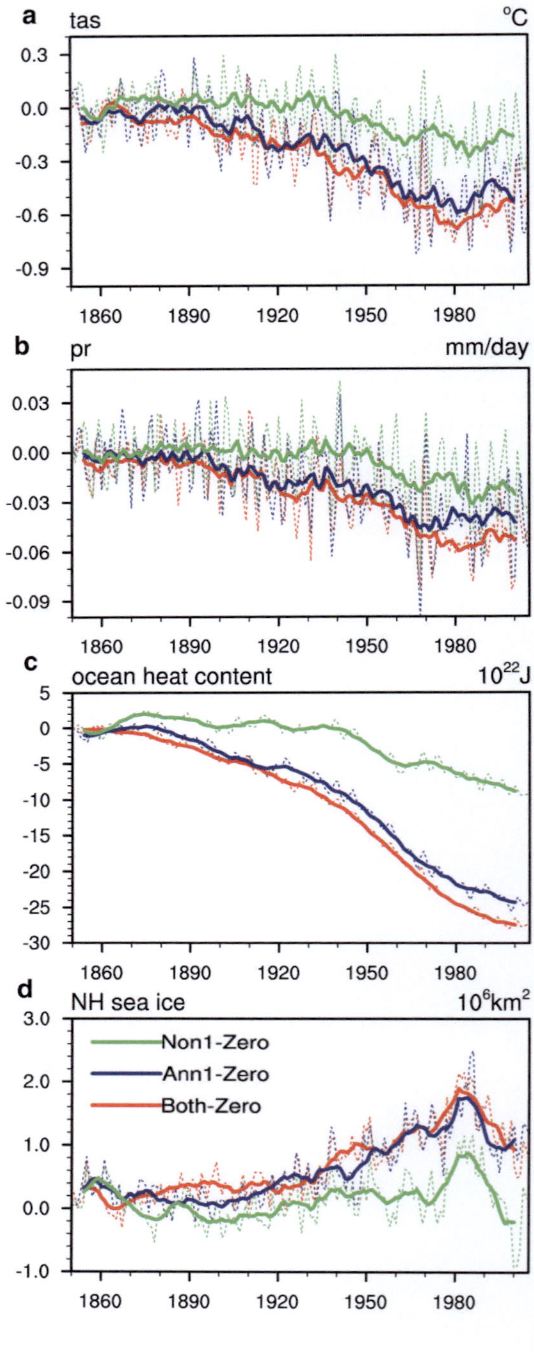

Fig. 5.6 (**a**) Global averaged annual average air temperature (unit: °C), (**b**) global averaged annual average precipitation (unit: mm/day), (**c**) global averaged annual average sea heat content from surface to 700 m depth (unit: 10^{22} J), and (**d**) annual average North Hemisphere sea ice area (unit: million km^2), simulated by Both minus Zero (*red*), Ann1 minus Zero (*blue*), and Non1minus Zero (*green*). The *thick lines* denote variability with period exceeding 10 years

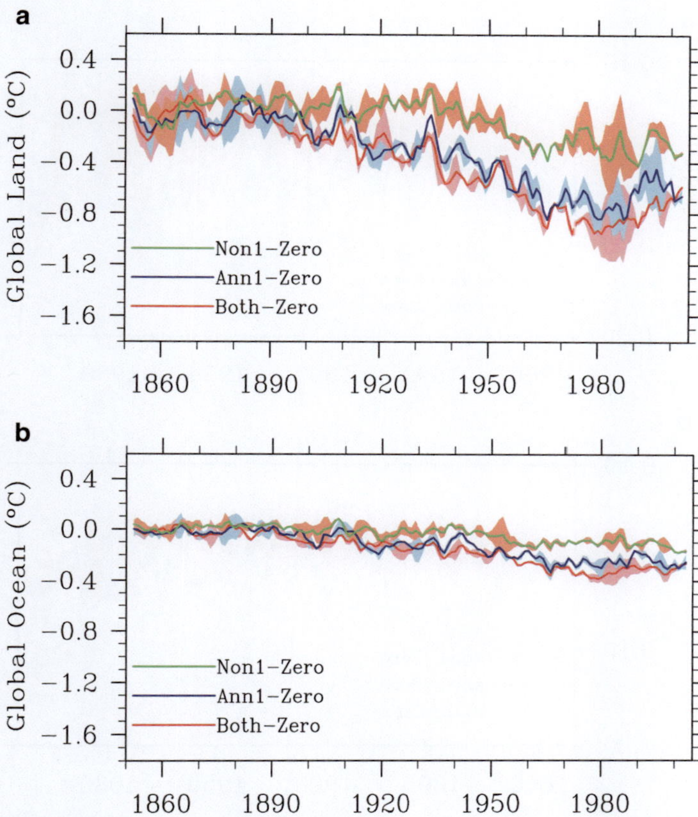

Fig. 5.7 Five-year running mean annual average surface air temperature averaged over global land (**a**) and ocean (**b**) (unit: °C), simulated by Both minus Zero (*red*), Ann1 minus Zero (*blue*), and Non1minus Zero (*green*). The *shades* denote variation range

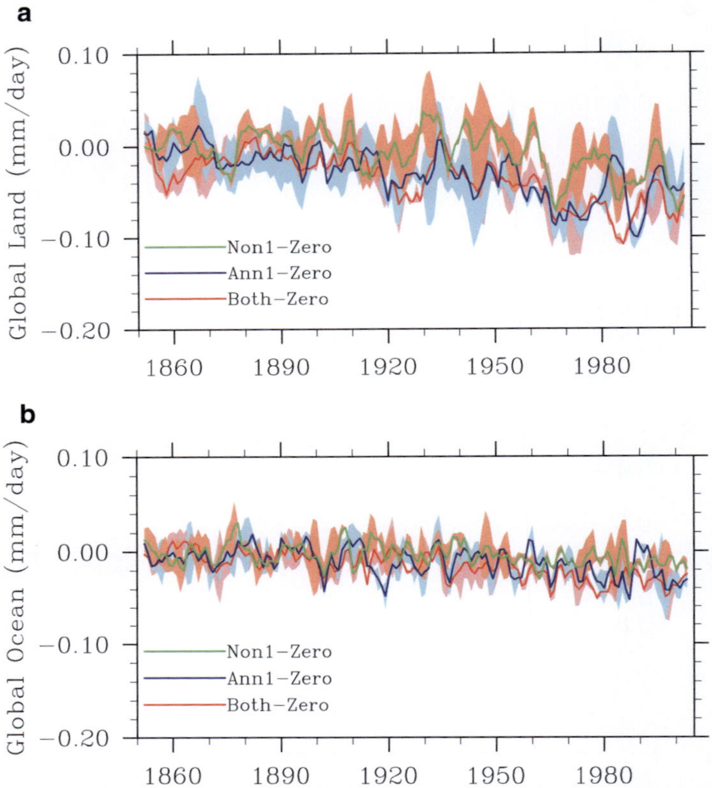

Fig. 5.8 Five-year running mean annual average precipitation averaged over global land (**a**) and ocean (**b**) (unit: mm/day), simulated by Both minus Zero (*red*), Ann1 minus Zero (*blue*), and Non1minus Zero (*green*). The *shades* denote variation range

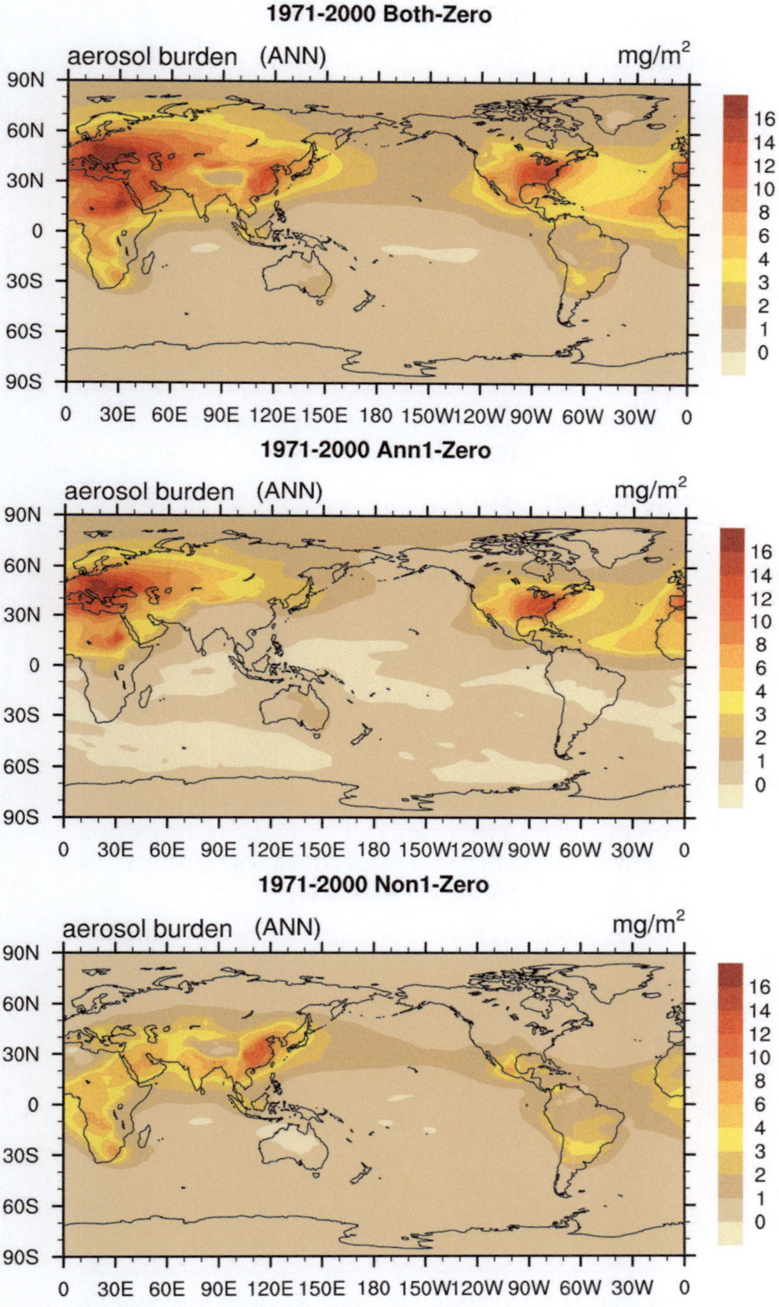

Fig. 5.9 Annual average aerosol column burden averaged over 1971–2000, simulated by Both minus Zero (*upper*), Ann1 minus Zero (*middle*), and Non1 minus Zero (*lower*) (unit: mg/m^2)

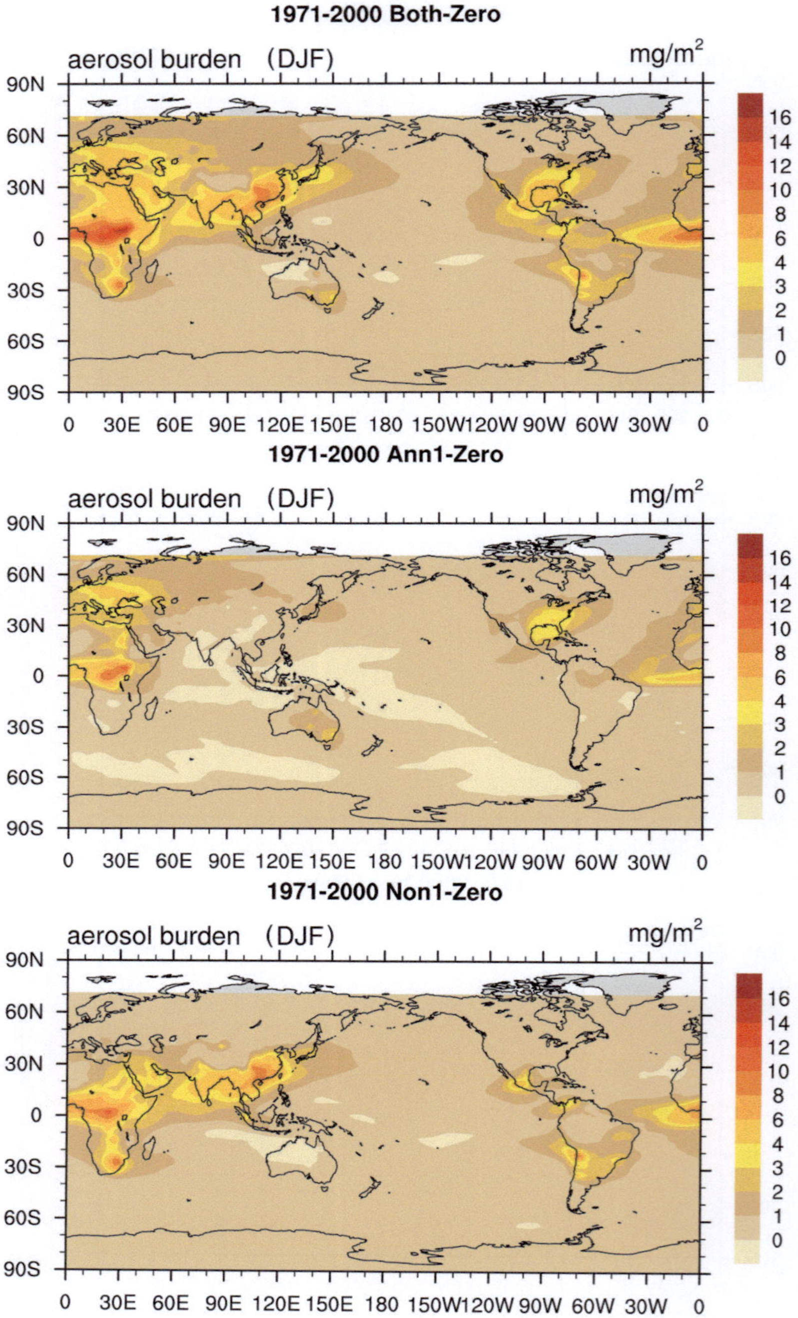

Fig. 5.10 Aerosol column burden for December–February averaged over 1971–2000, simulated by Both minus Zero (*upper*), Ann1 minus Zero (*middle*), and Non1 minus Zero (*lower*) (unit: mg/m^2)

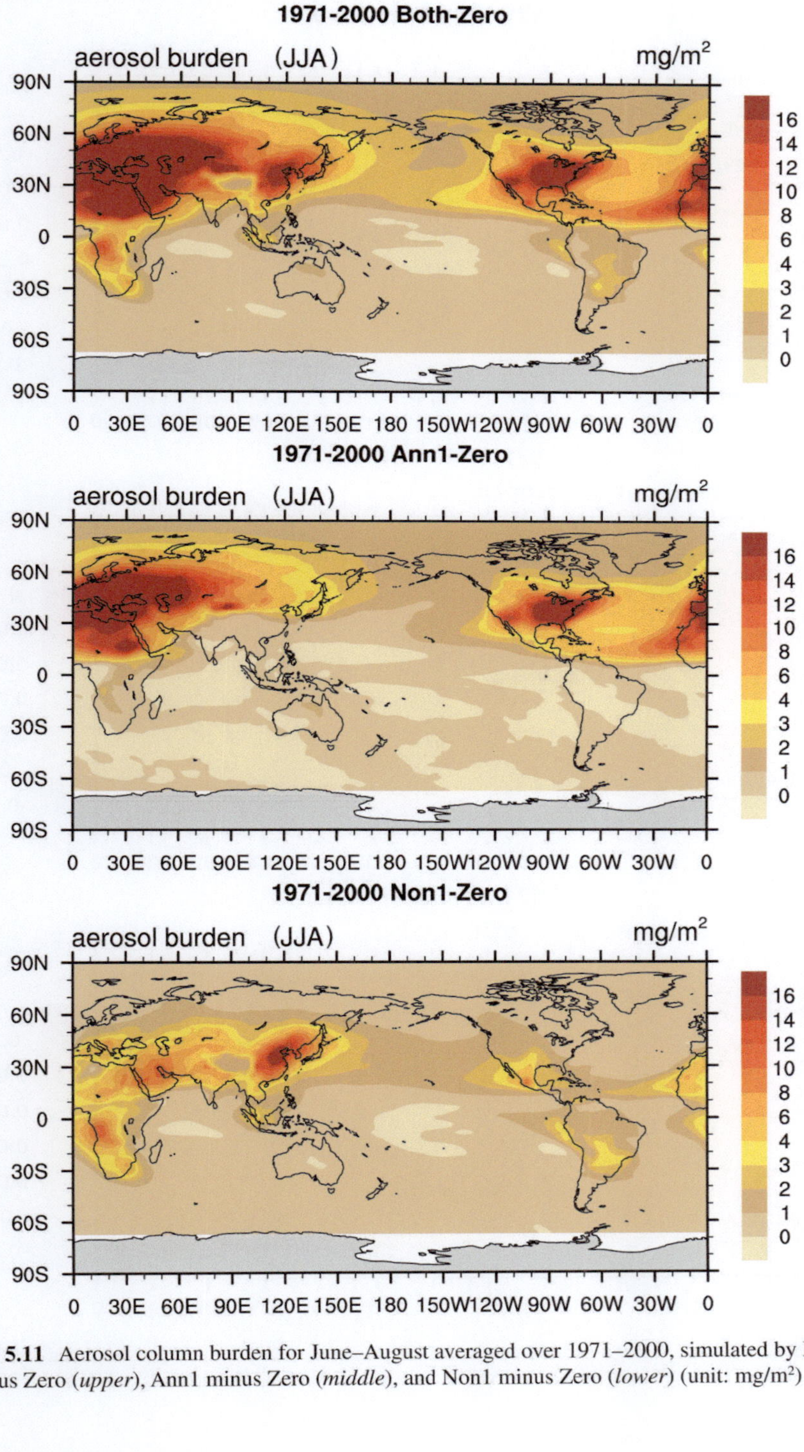

Fig. 5.11 Aerosol column burden for June–August averaged over 1971–2000, simulated by Both minus Zero (*upper*), Ann1 minus Zero (*middle*), and Non1 minus Zero (*lower*) (unit: mg/m²)

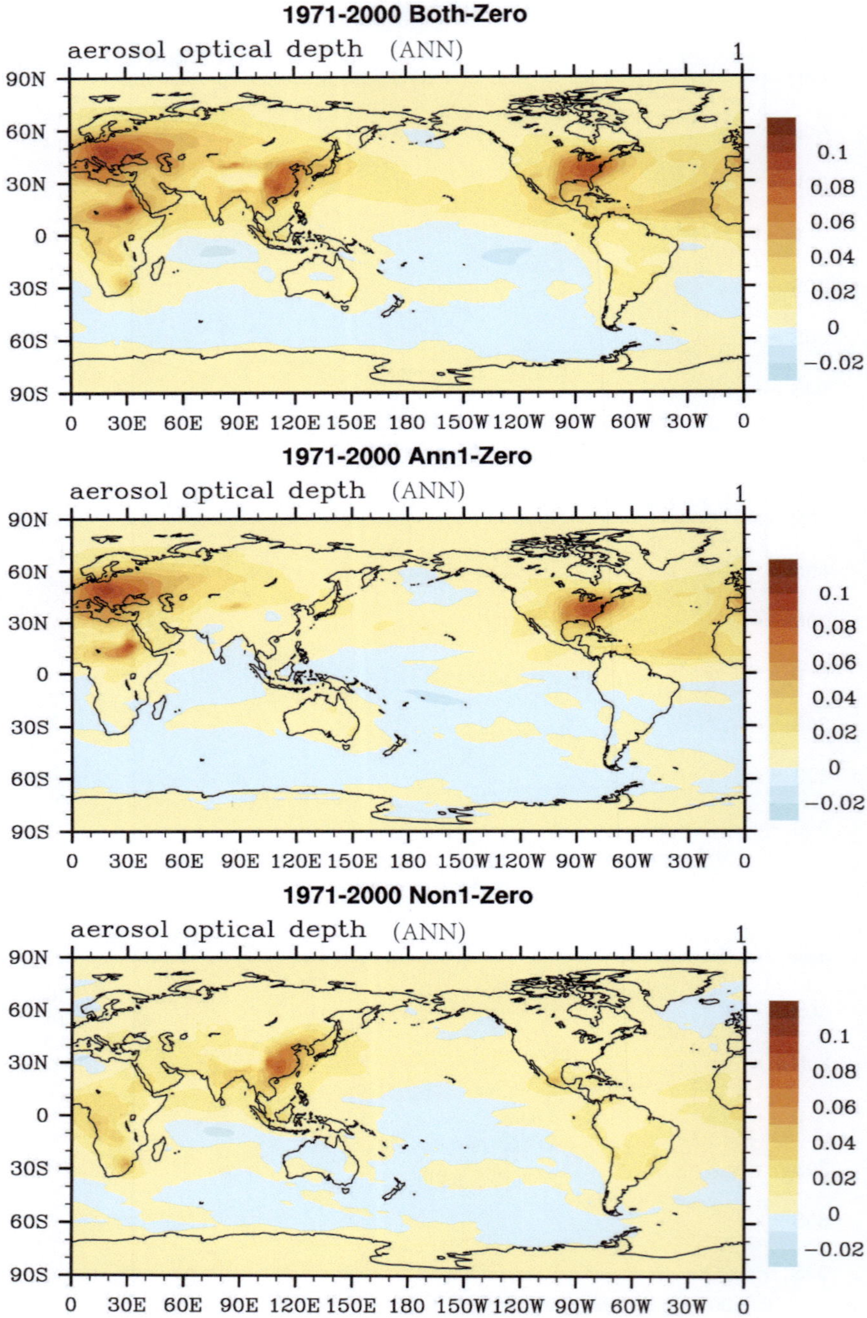

Fig. 5.12 Annual average aerosol optical depth averaged over 1971–2000, simulated by Both minus Zero (*upper*), Ann1 minus Zero (*middle*), and Non1 minus Zero (*lower*) (unit: 1)

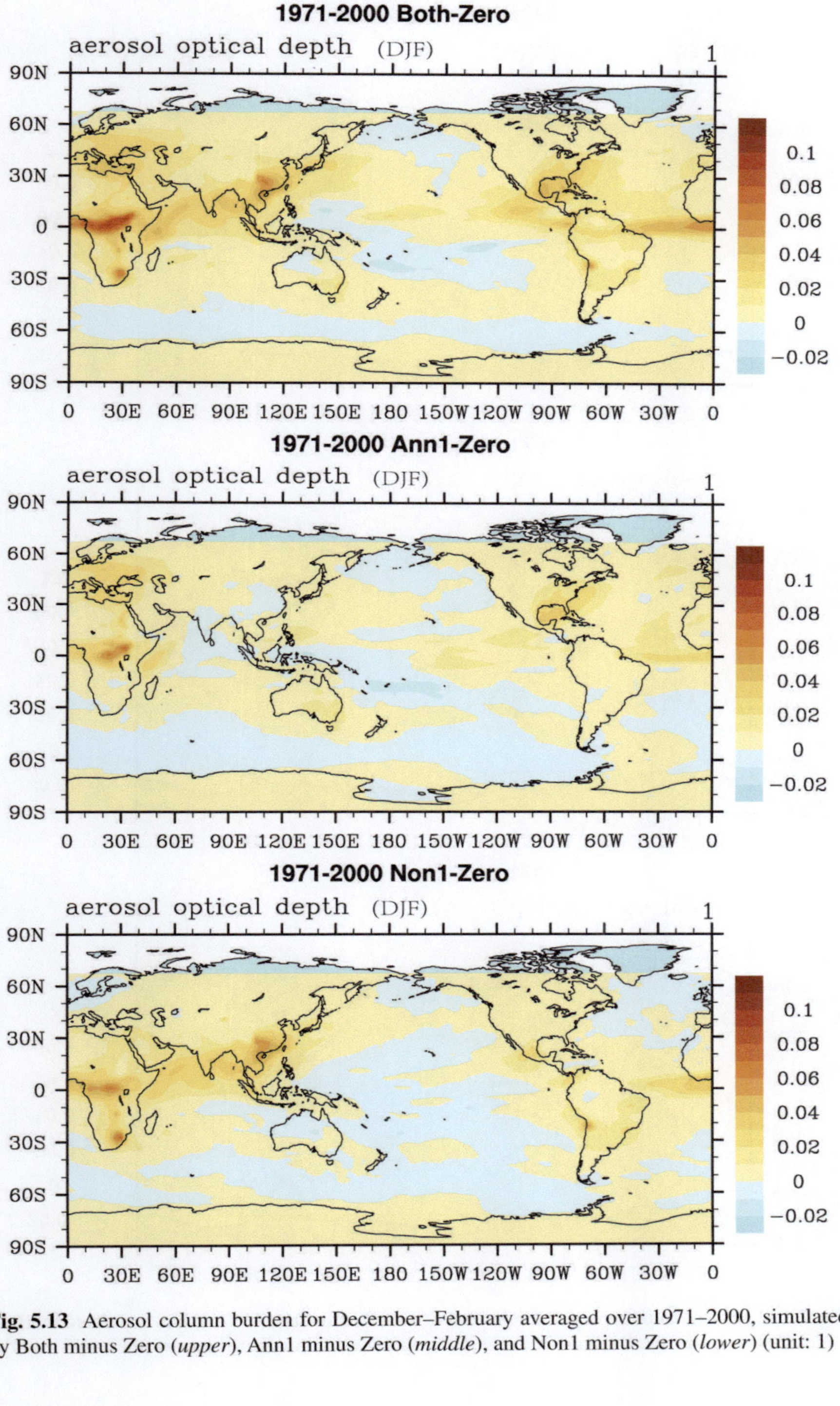

Fig. 5.13 Aerosol column burden for December–February averaged over 1971–2000, simulated by Both minus Zero (*upper*), Ann1 minus Zero (*middle*), and Non1 minus Zero (*lower*) (unit: 1)

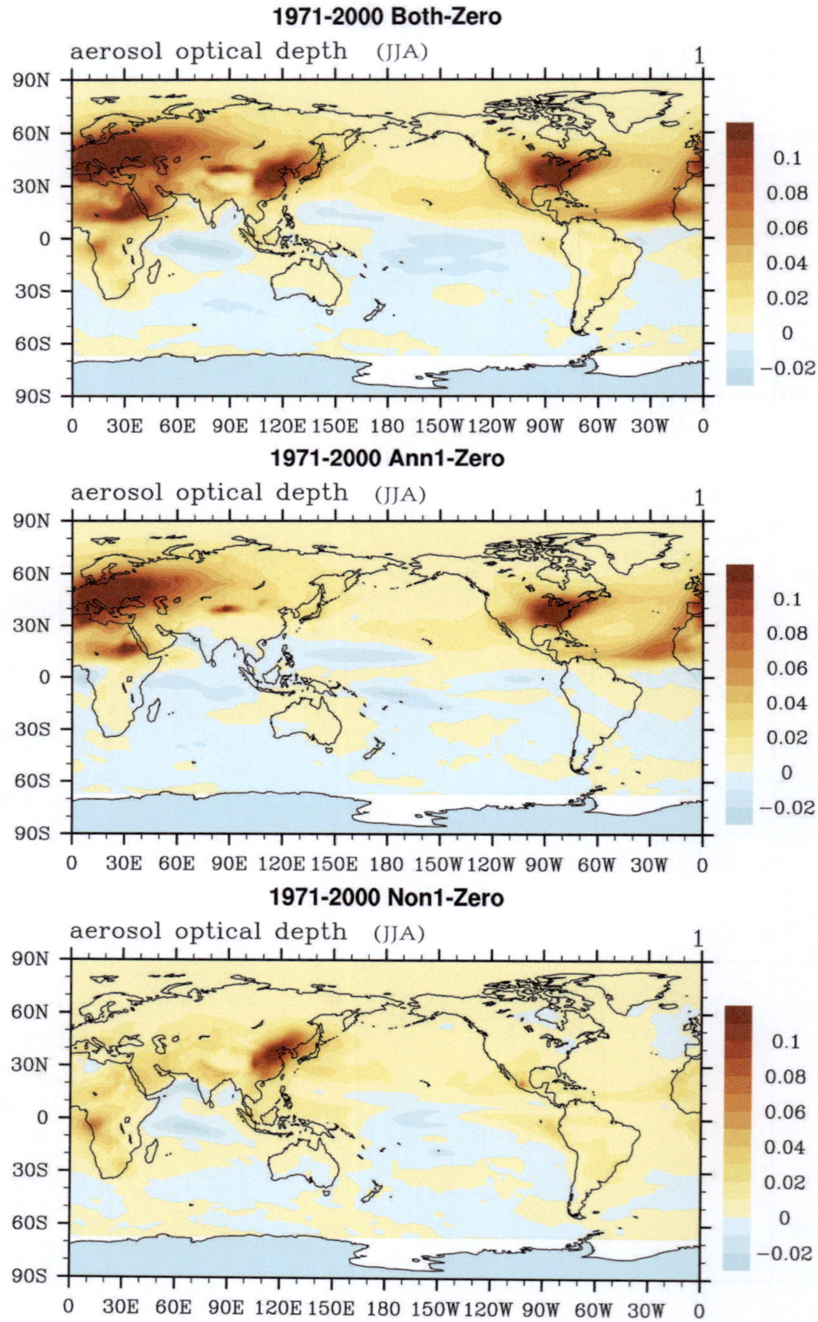

Fig. 5.14 Aerosol column burden for June–August averaged over 1971–2000, simulated by Both minus Zero (*upper*), Ann1 minus Zero (*middle*), and Non1 minus Zero (*lower*) (unit: 1)

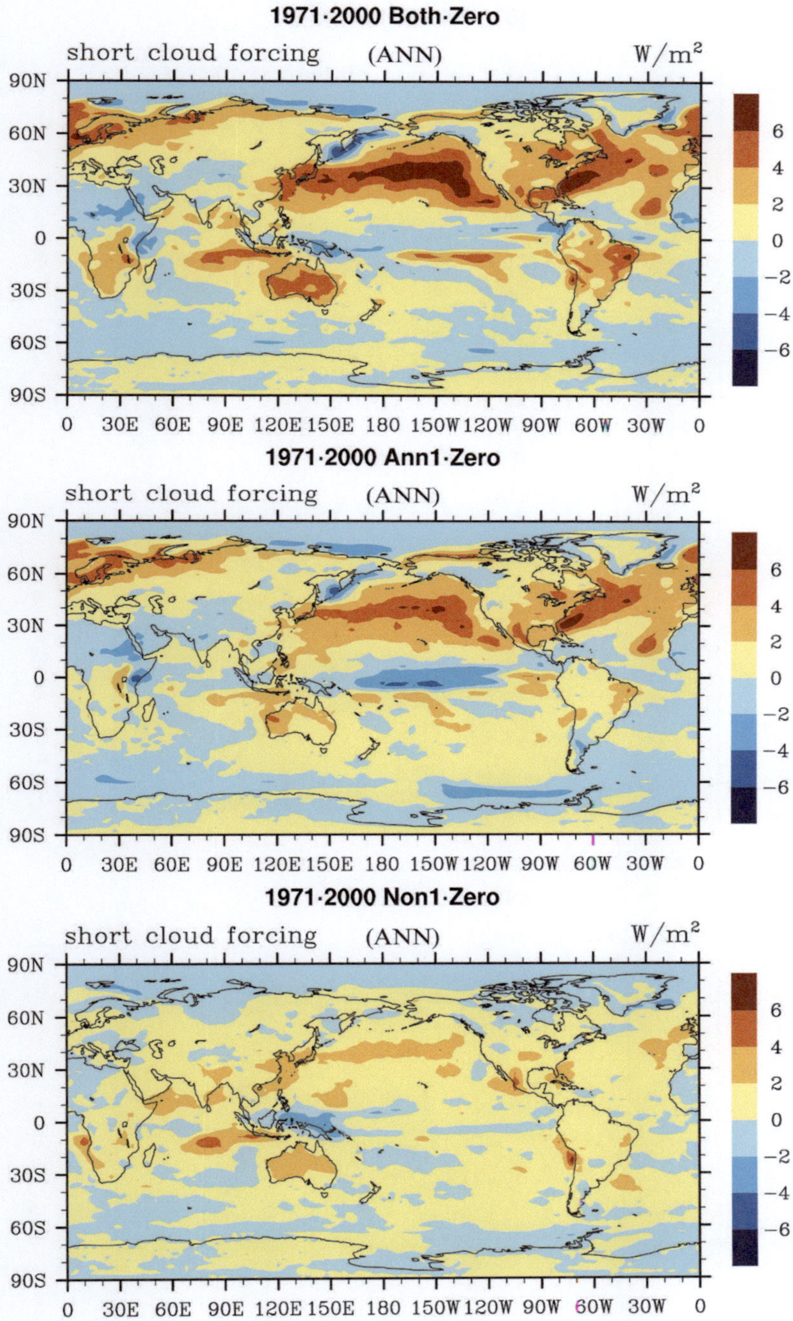

Fig. 5.15 Annual average shortwave cloud forcing averaged over 1971–2000, simulated by Both minus Zero (*upper*), Ann1 minus Zero (*middle*), and Non1 minus Zero (*lower*); positive value means an increase of solar radiation reflected by cloud (unit: W/m^2)

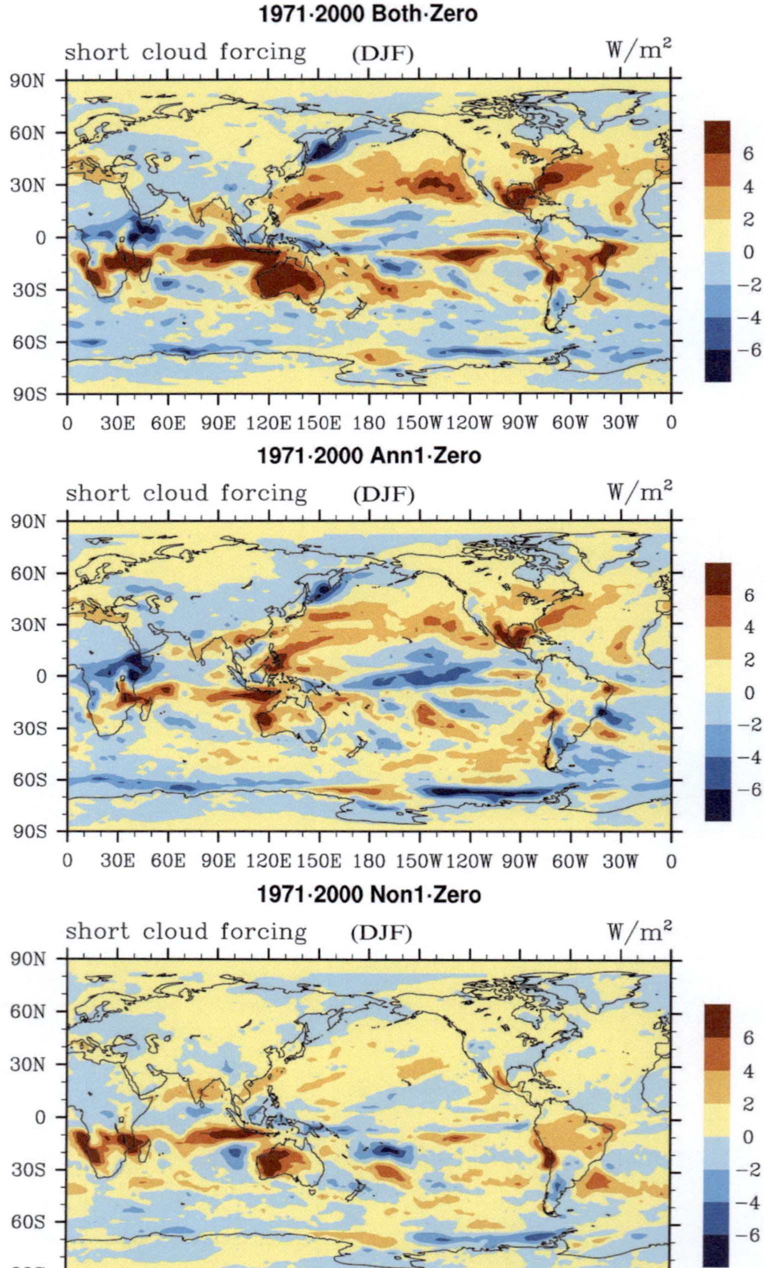

Fig. 5.16 Shortwave cloud forcing for December–February averaged over 1971–2000, simulated by Both minus Zero (*upper*), Ann1 minus Zero (*middle*), and Non1 minus Zero (*lower*) (unit: W/m^2)

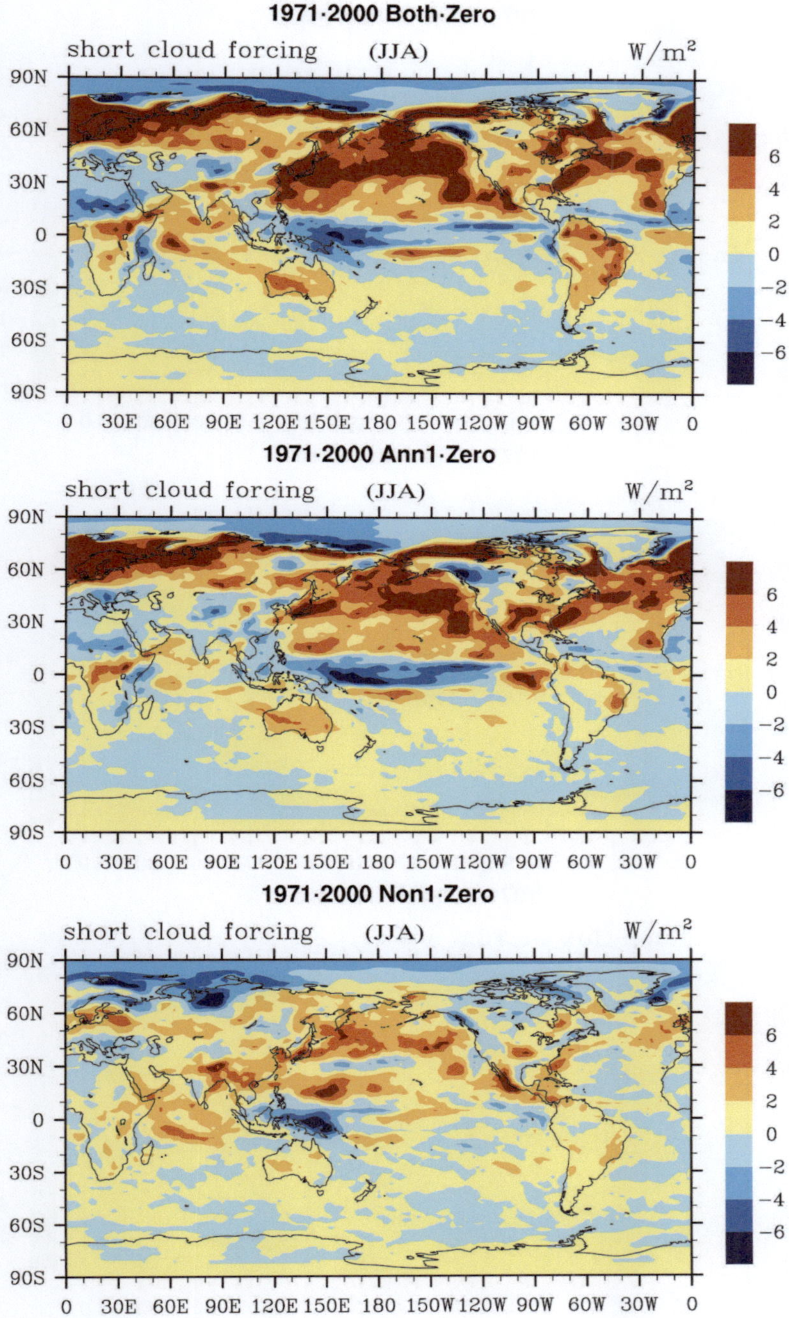

Fig. 5.17 Shortwave cloud forcing for June–August averaged over 1971–2000, simulated by Both minus Zero (*upper*), Ann1 minus Zero (*middle*), and Non1 minus Zero (*lower*) (unit: W/m^2)

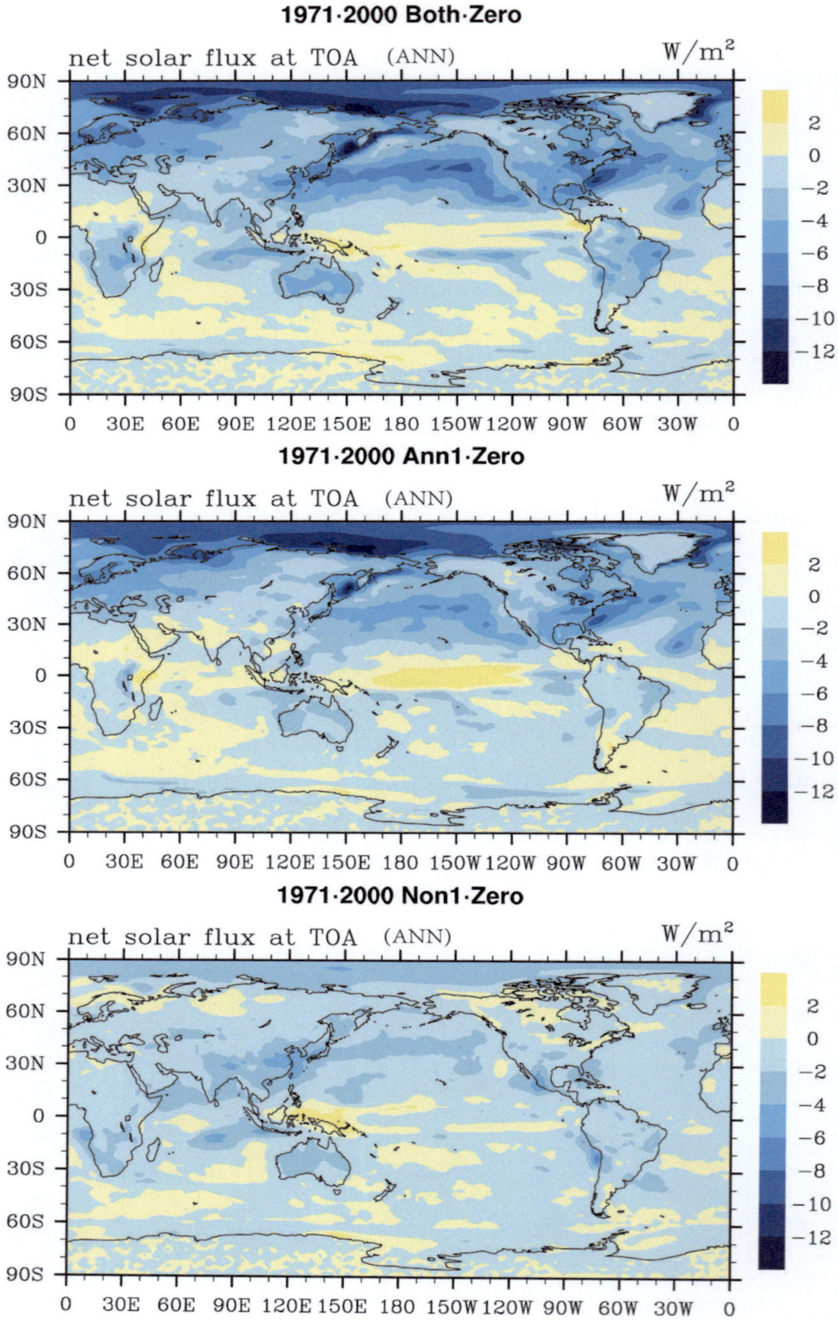

Fig. 5.18 Annual average net solar flux at TOA averaged over 1971–2000, simulated by Both minus Zero (*upper*), Ann1 minus Zero (*middle*), and Non1 minus Zero (*lower*), positive value means an increase of solar radiation received by earth (unit: W/m^2)

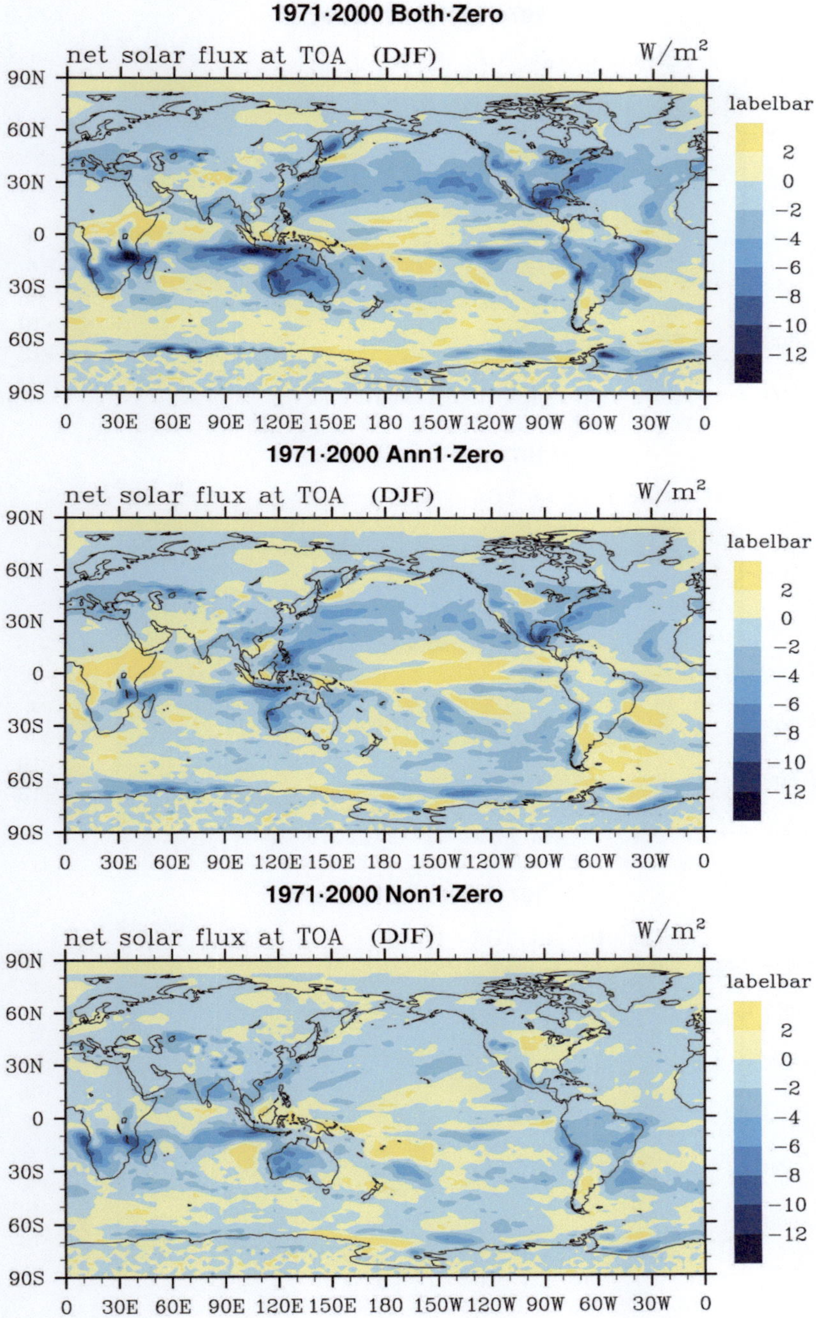

Fig. 5.19 Net solar flux at TOA for December–February averaged over 1971–2000, simulated by Both minus Zero (*upper*), Ann1 minus Zero (*middle*), and Non1 minus Zero (*lower*) (unit: W/m^2)

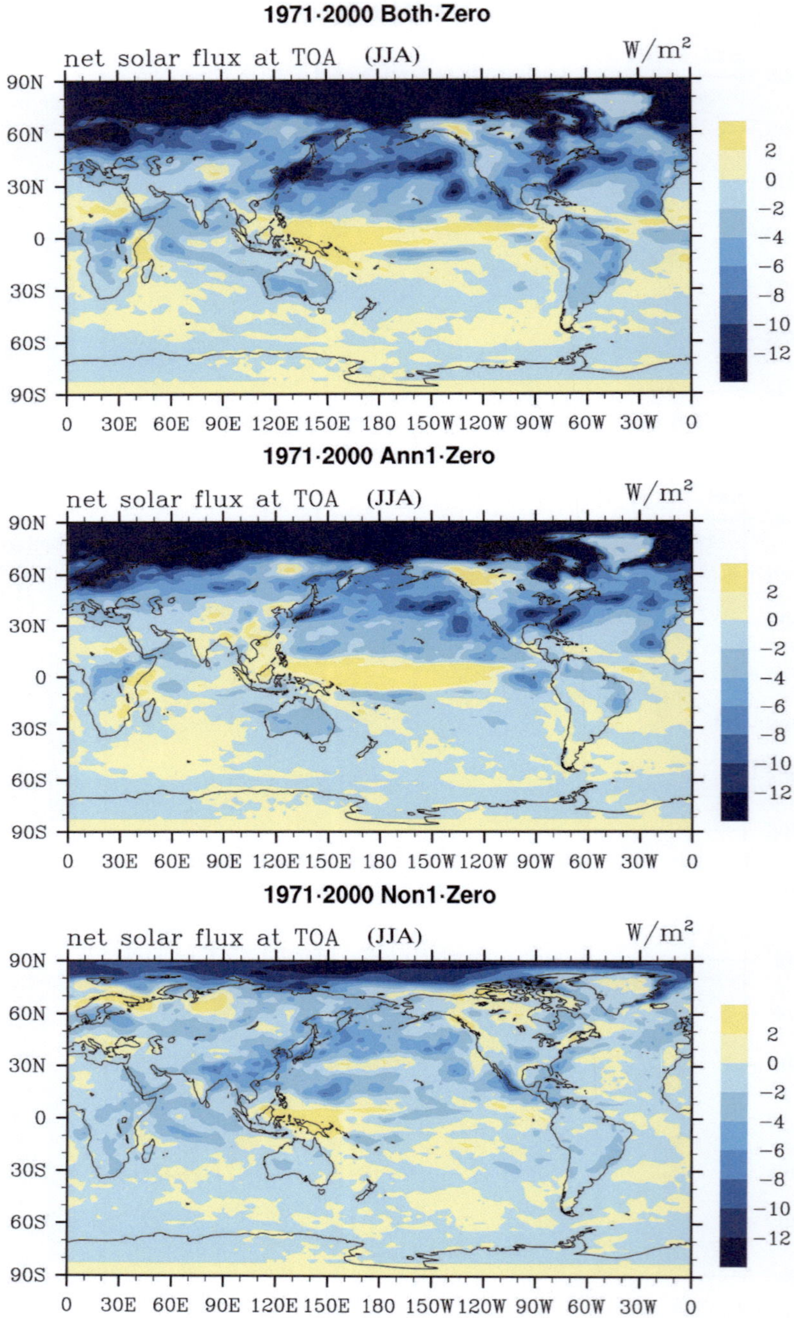

Fig. 5.20 Net solar flux at TOA for June–August averaged over 1971–2000, simulated by Both minus Zero (*upper*), Ann1 minus Zero (*middle*), and Non1 minus Zero (*lower*) (unit: W/m²)

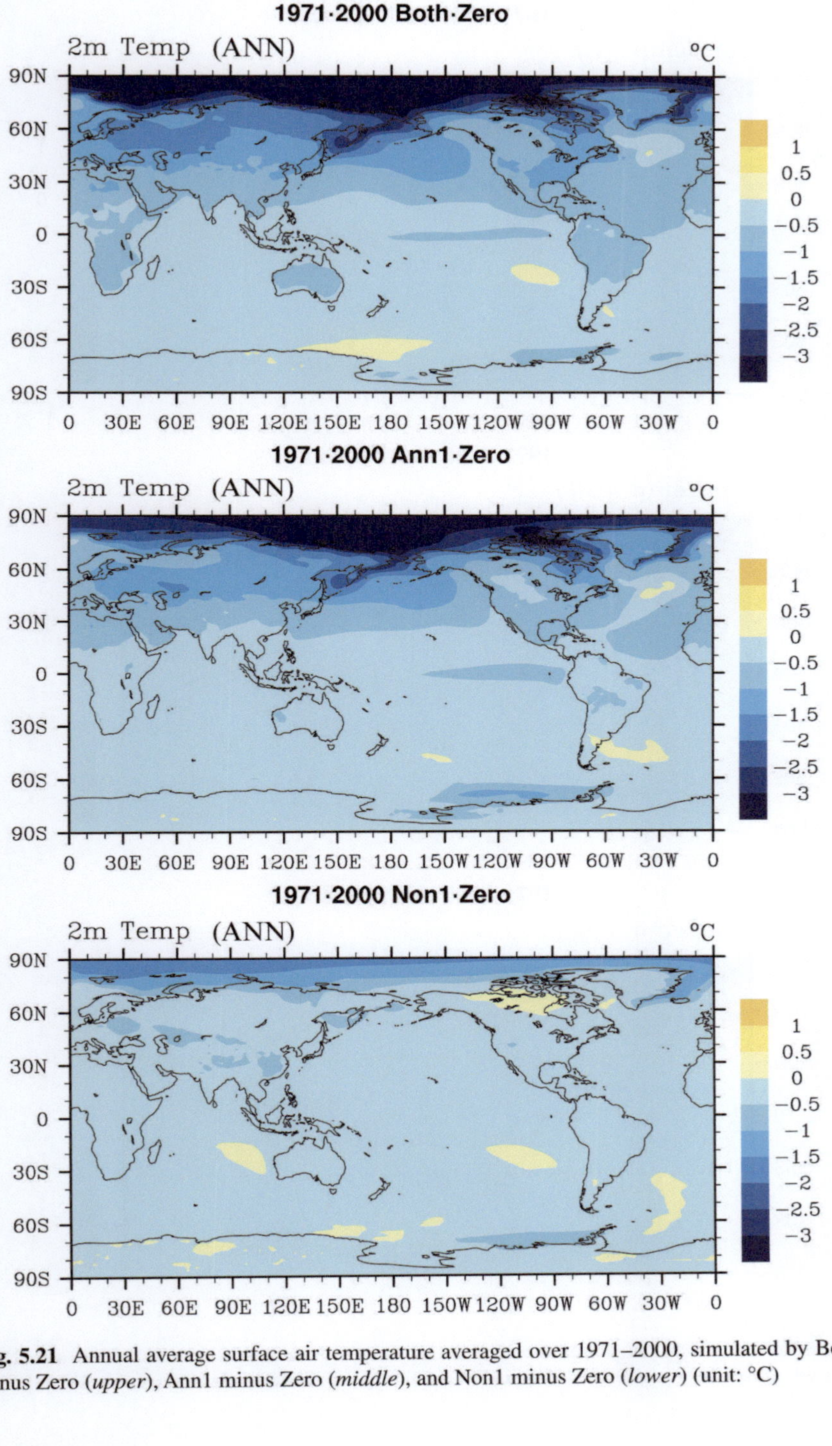

Fig. 5.21 Annual average surface air temperature averaged over 1971–2000, simulated by Both minus Zero (*upper*), Ann1 minus Zero (*middle*), and Non1 minus Zero (*lower*) (unit: °C)

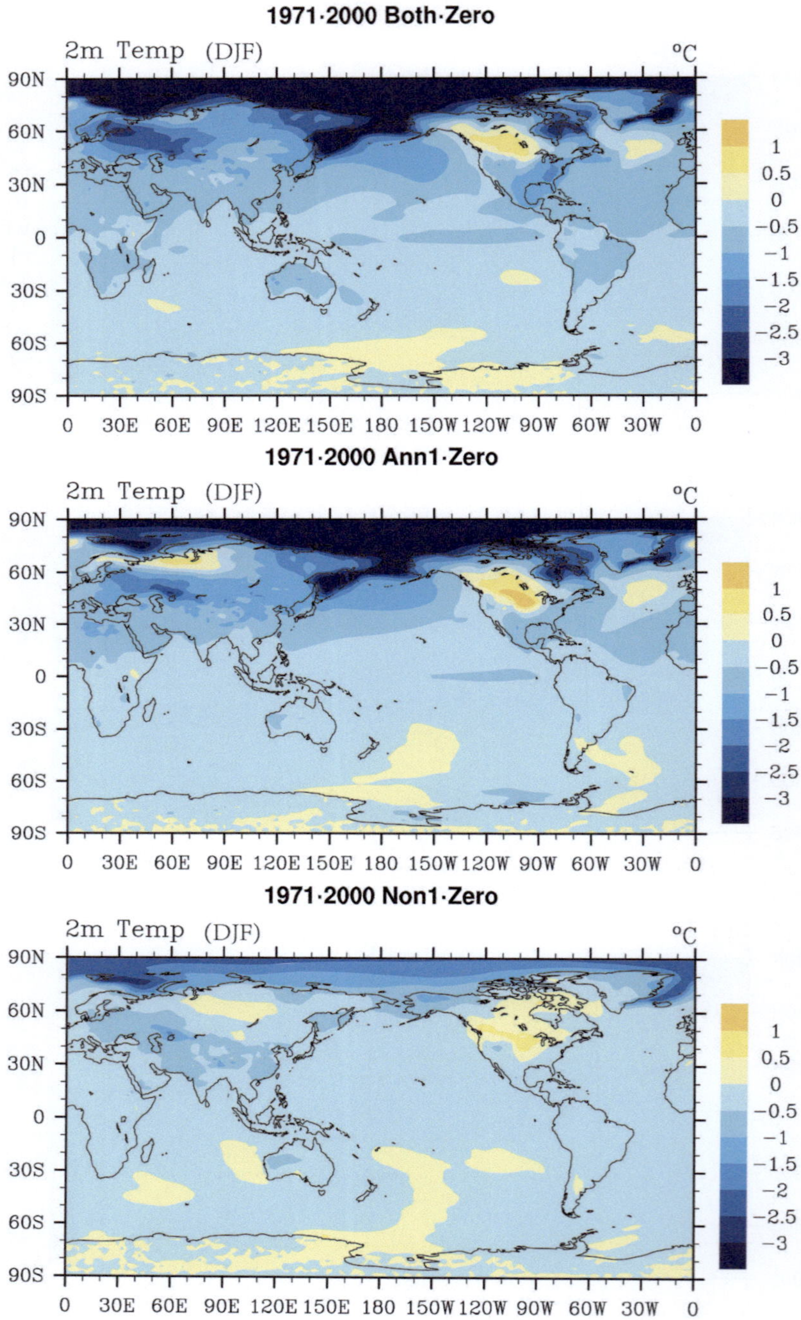

Fig. 5.22 Surface air temperature for December–February averaged over 1971–2000, simulated by Both minus Zero (*upper*), Ann1 minus Zero (*middle*), and Non1 minus Zero (*lower*) (unit: °C)

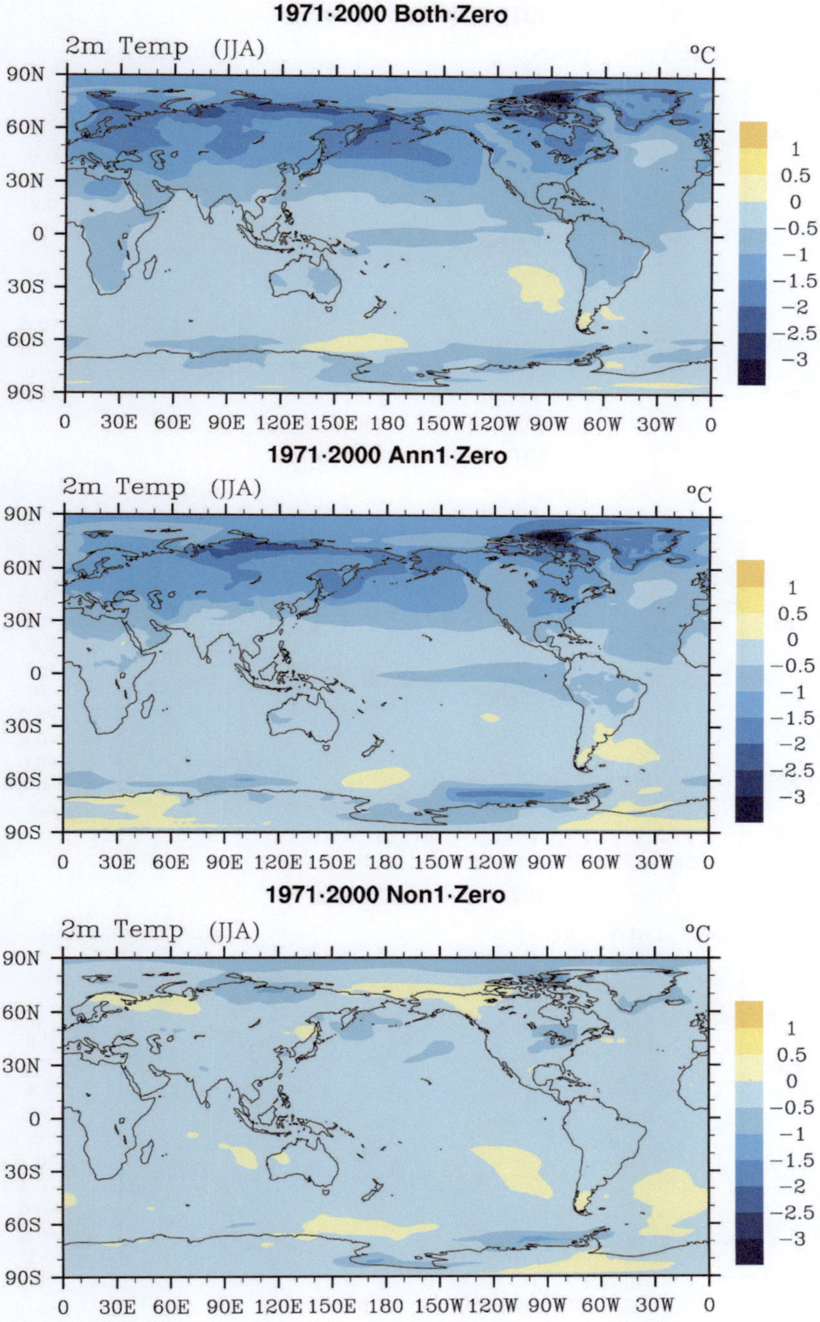

Fig. 5.23 Surface air temperature for June–August averaged over 1971–2000, simulated by Both minus Zero (*upper*), Ann1 minus Zero (*middle*), and Non1 minus Zero (*lower*) (unit: °C)

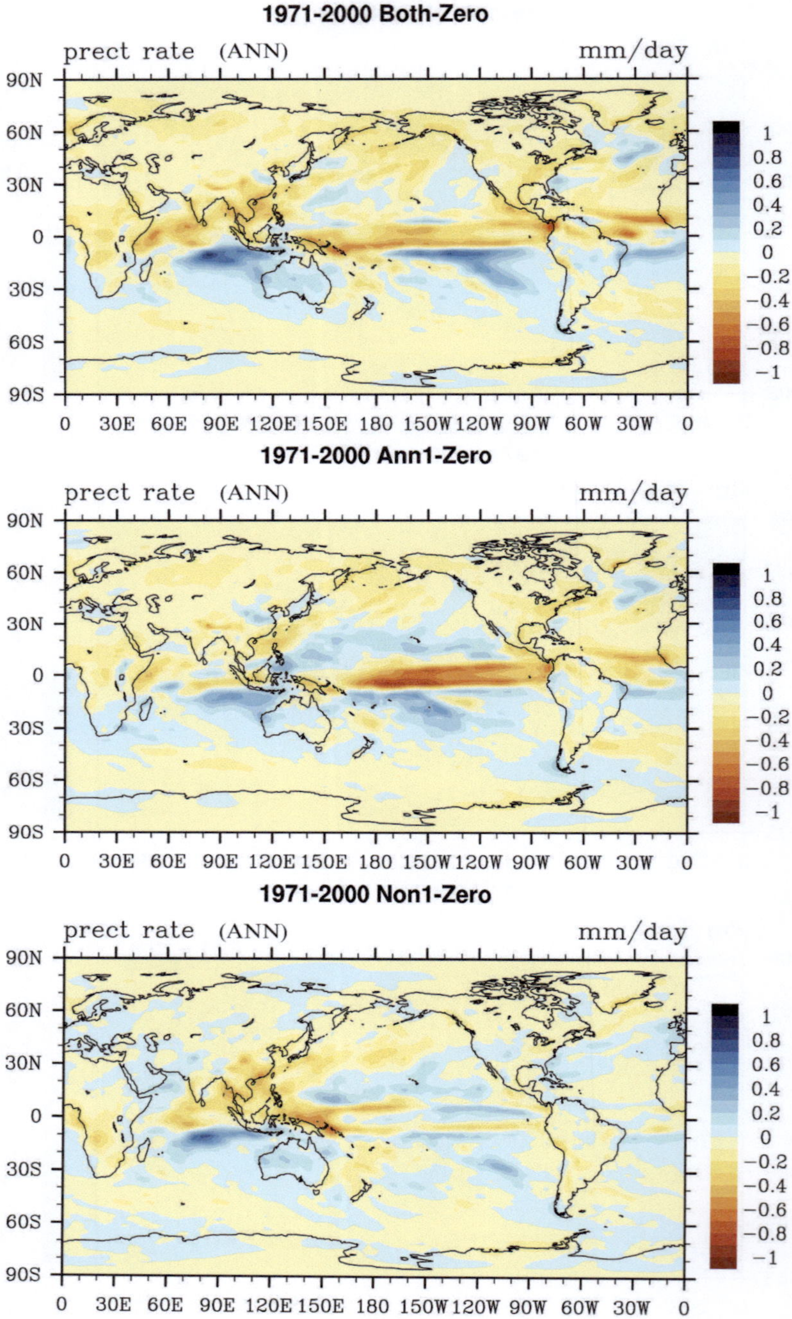

Fig. 5.24 Annual average precipitation averaged over 1971–2000, simulated by Both minus Zero (*upper*), Ann1 minus Zero (*middle*), and Non1 minus Zero (*lower*) (unit: mm/day)

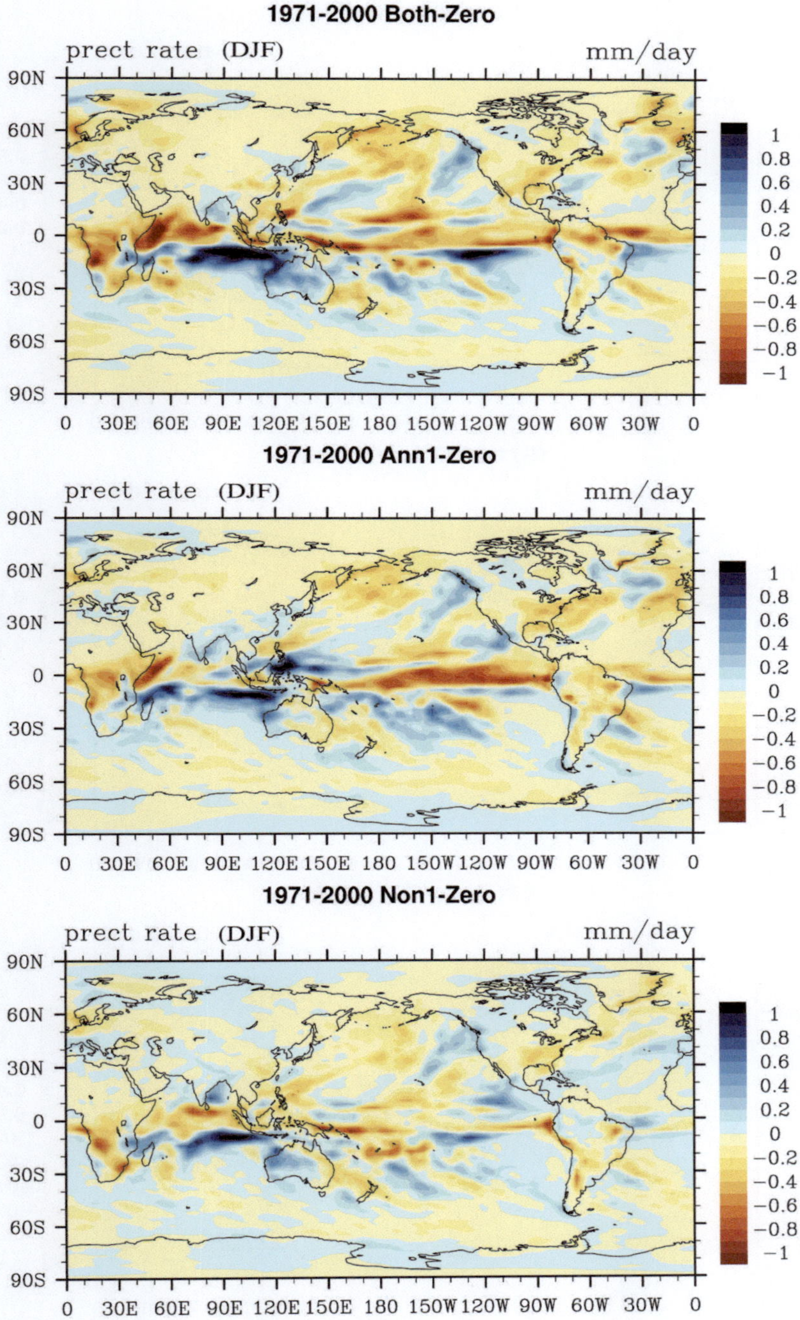

Fig. 5.25 Precipitation for December–February averaged over 1971–2000, simulated by Both minus Zero (*upper*), Ann1 minus Zero (*middle*), and Non1 minus Zero (*lower*) (unit: mm/day)

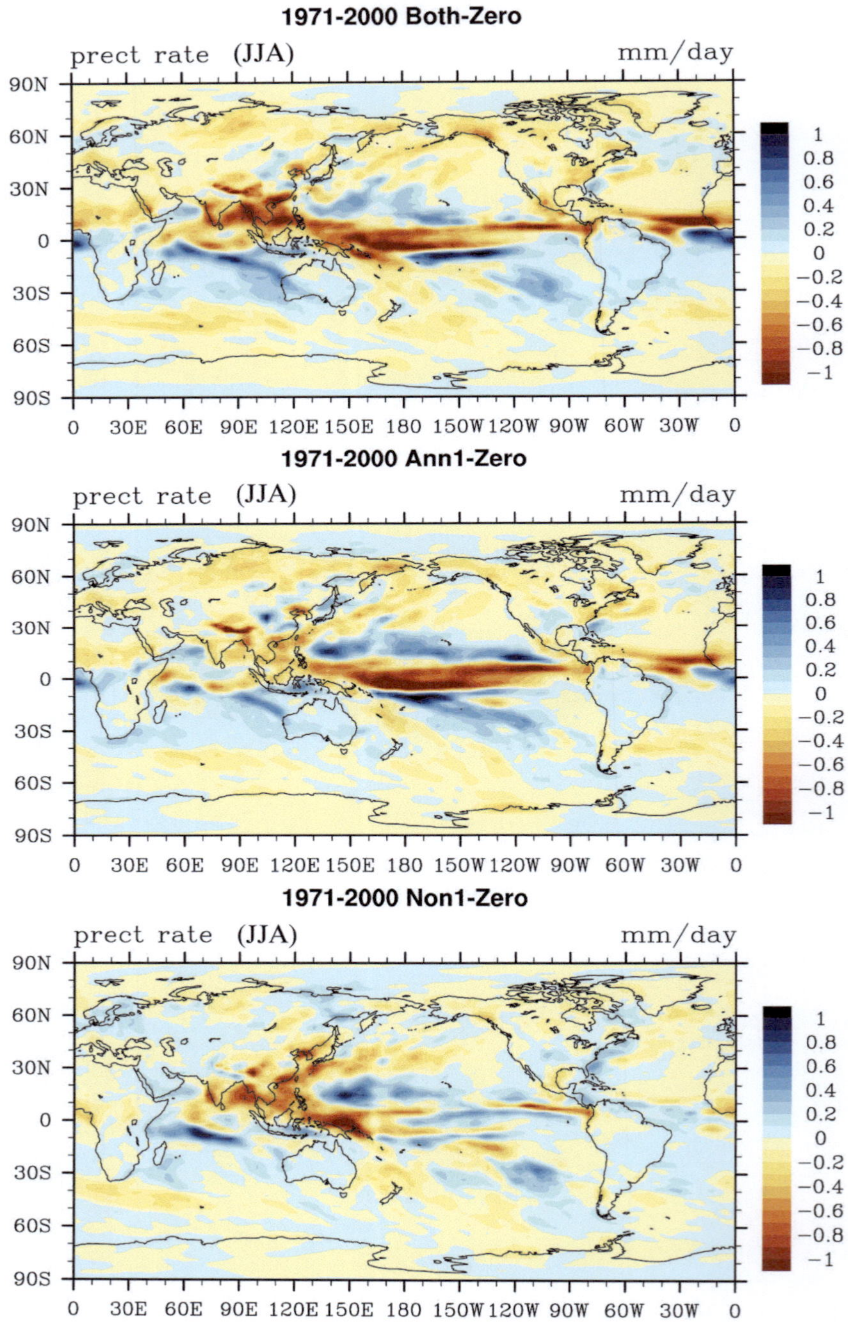

Fig. 5.26 Precipitation for June–August averaged over 1971–2000, simulated by Both minus Zero (*upper*), Ann1 minus Zero (*middle*), and Non1 minus Zero (*lower*) (unit: mm/day)

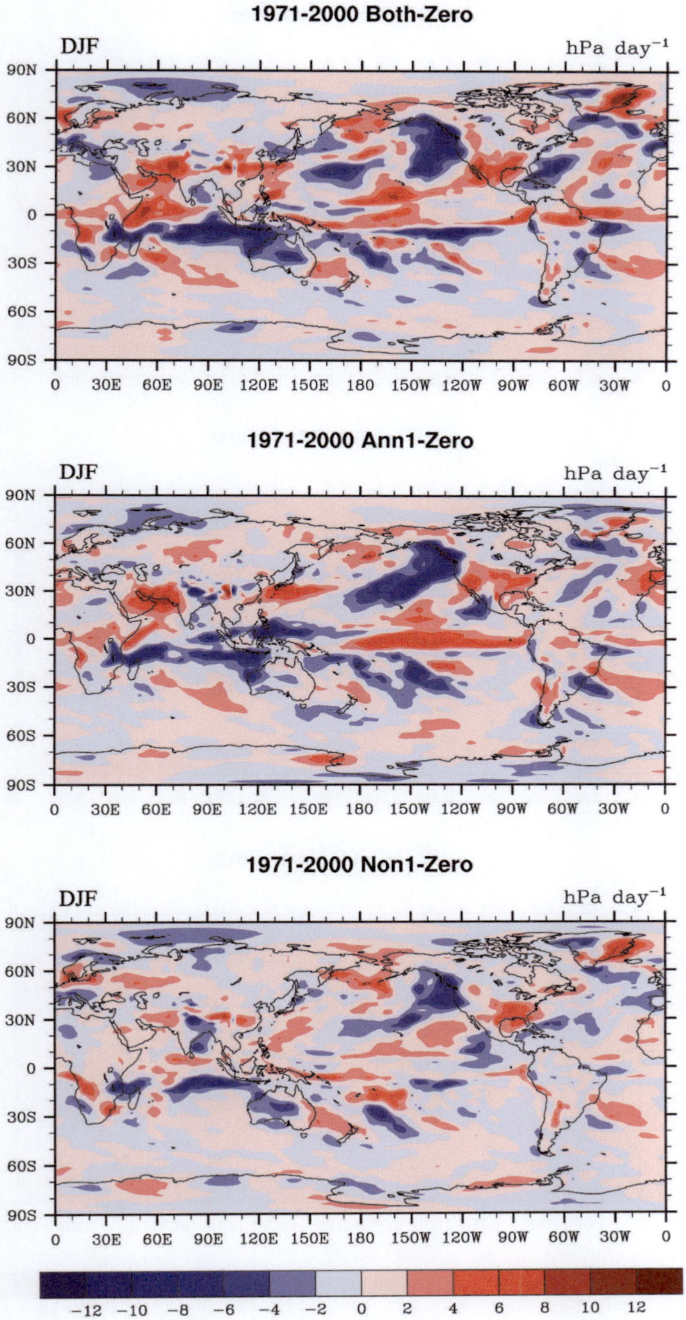

Fig. 5.27 500 hPa vertical velocity for December–February averaged over 1971–2000, simulated by Both minus Zero (*upper*), Ann1 minus Zero (*middle*), and Non1 minus Zero (*lower*) (unit: hPa/day^{-1})

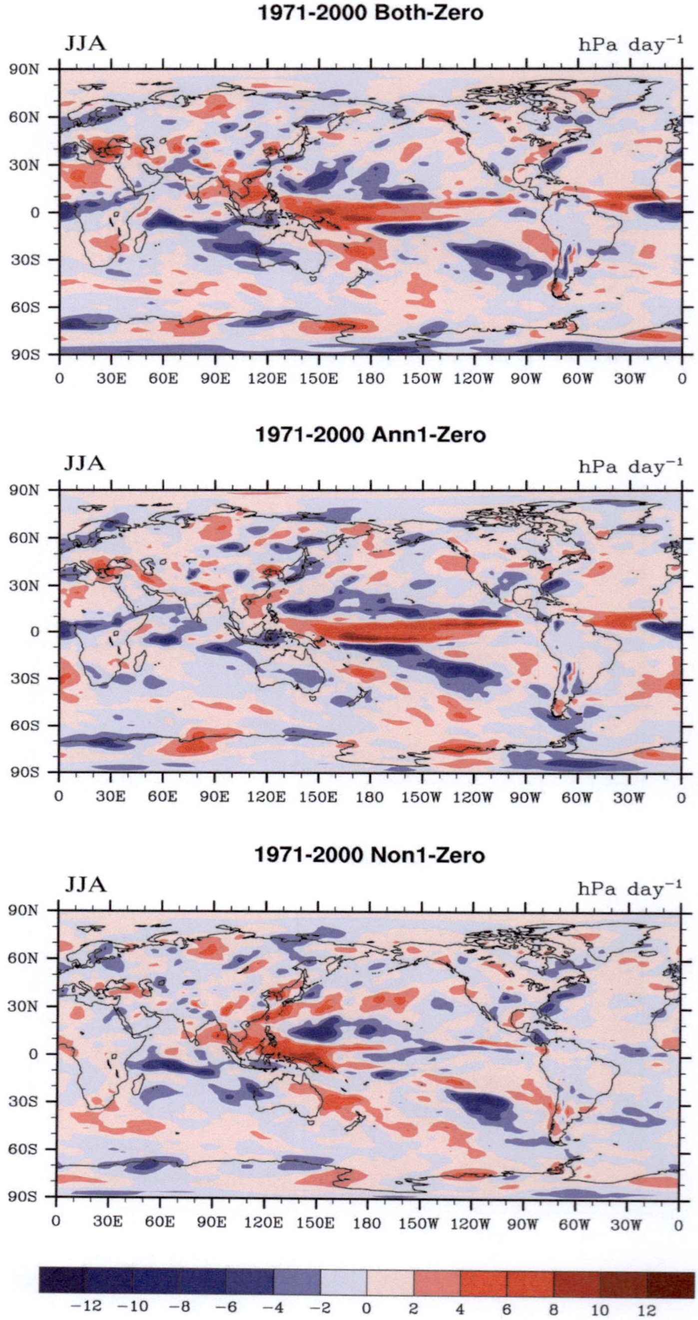

Fig. 5.28 500 hPa vertical velocity for June–August averaged over 1971–2000, simulated by Both minus Zero (*upper*), Ann1 minus Zero (*middle*), and Non1 minus Zero (*lower*) (unit: hPa/day^{-1})

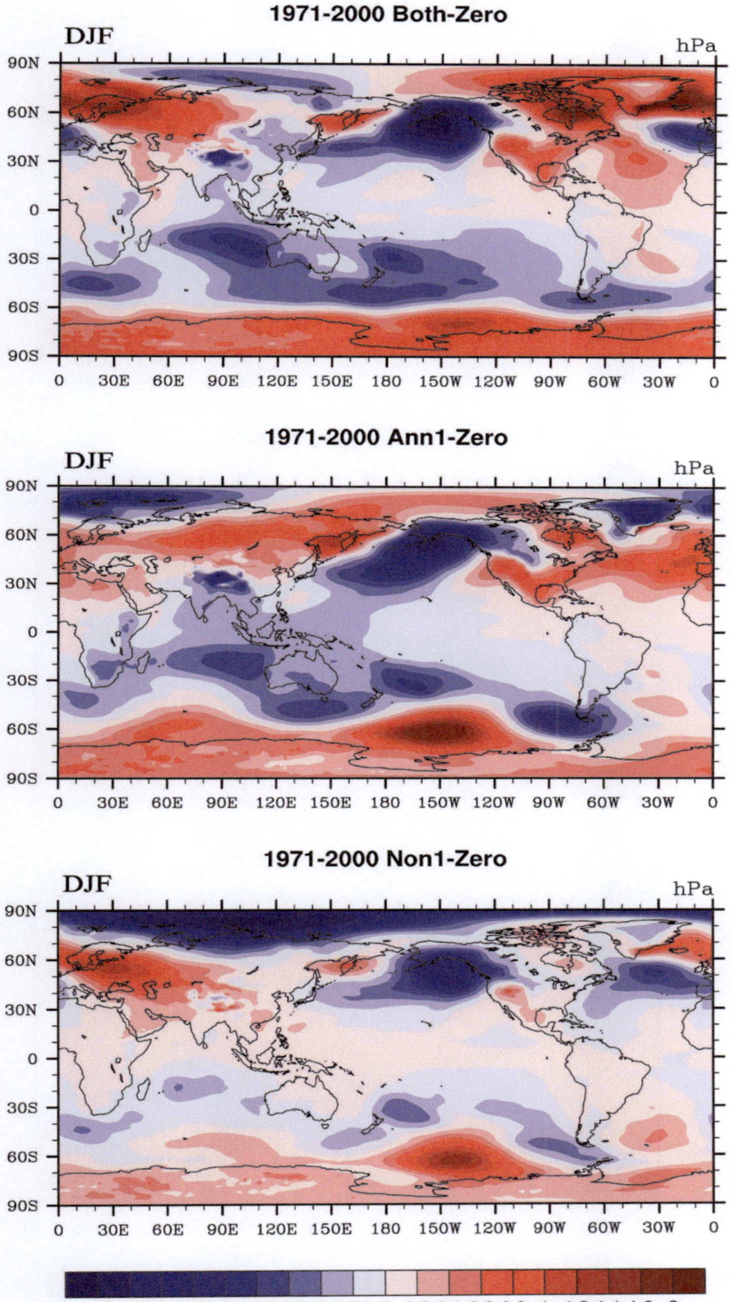

Fig. 5.29 Sea level pressure for December–February averaged over 1971–2000, simulated by Both minus Zero (*upper*), Ann1 minus Zero (*middle*), and Non1 minus Zero (*lower*) (unit: hPa)

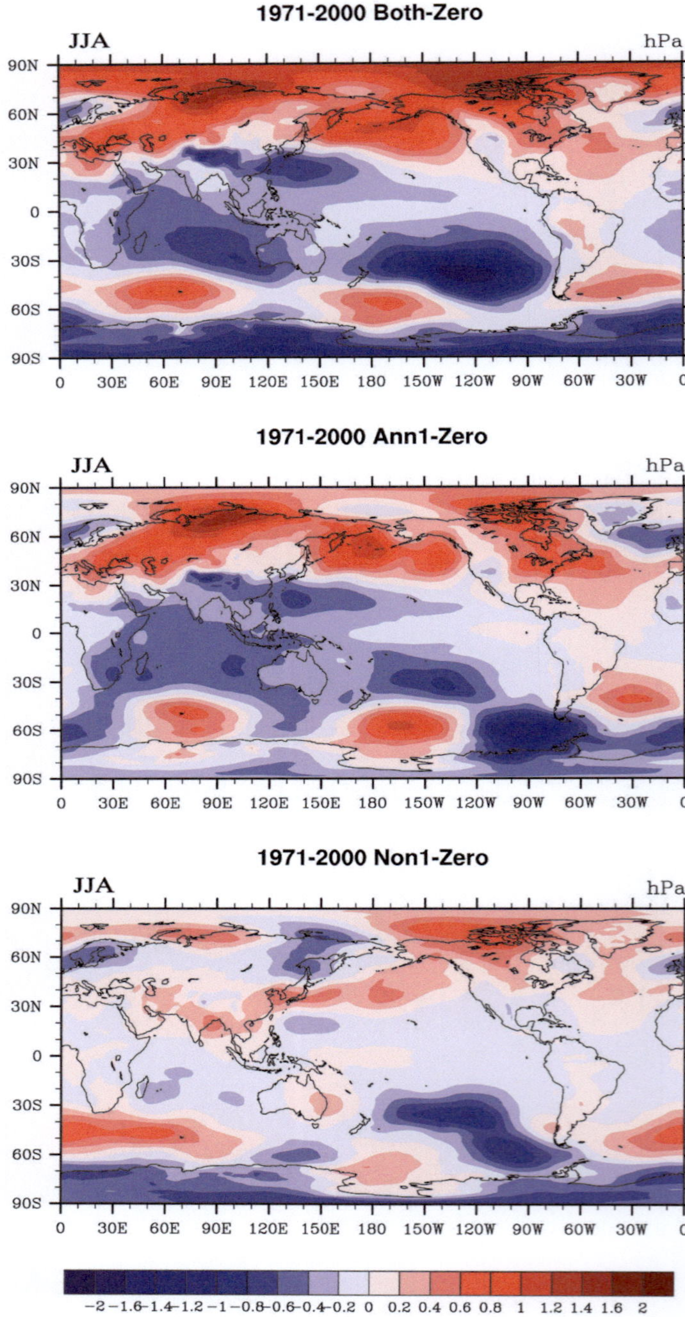

Fig. 5.30 Sea level pressure for June–August averaged over 1971–2000, simulated by Both minus Zero (*upper*), Ann1 minus Zero (*middle*), and Non1 minus Zero (*lower*) (unit: hPa)

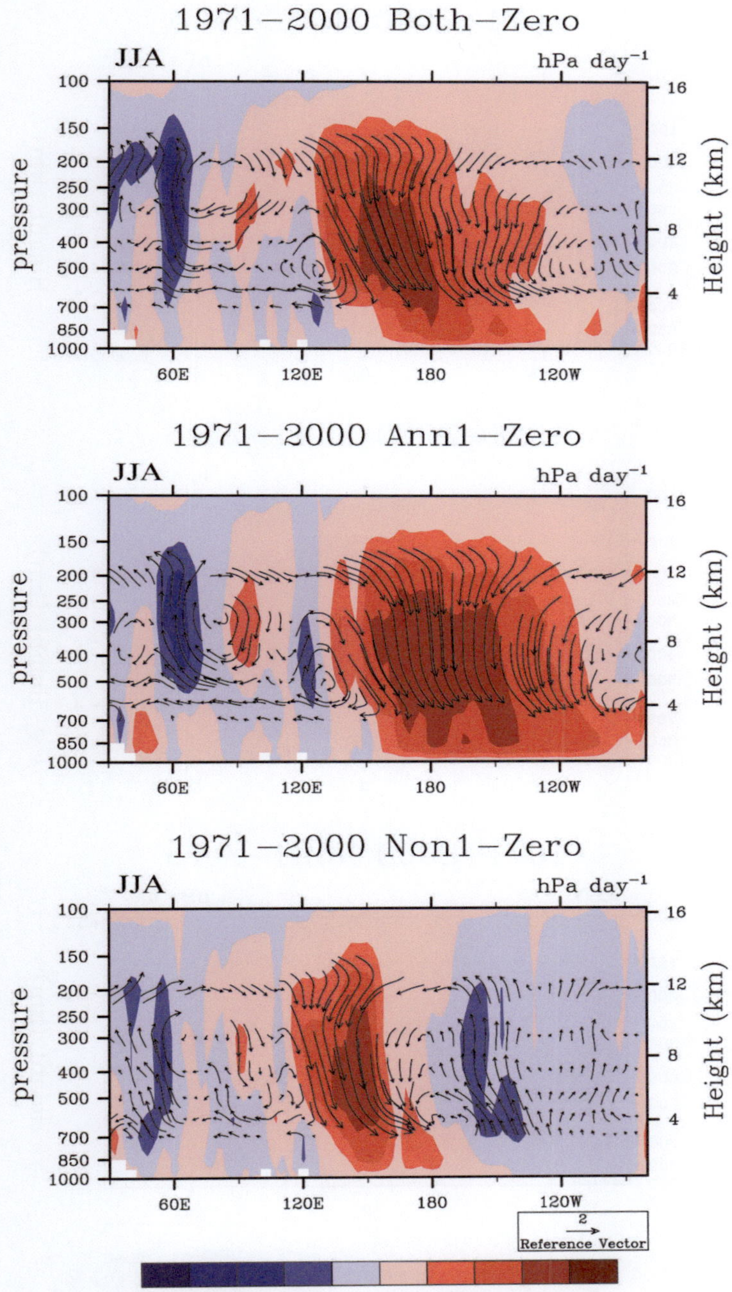

Fig. 5.31 5°S-5°N averaged zonal circulation for June–August averaged over 1971–2000, simulated by Both minus Zero (*upper*), Ann1 minus Zero (*middle*), and Non1 minus Zero (*lower*). The shades denote the latitude-height section of vertical velocity (unit: hPa/day)

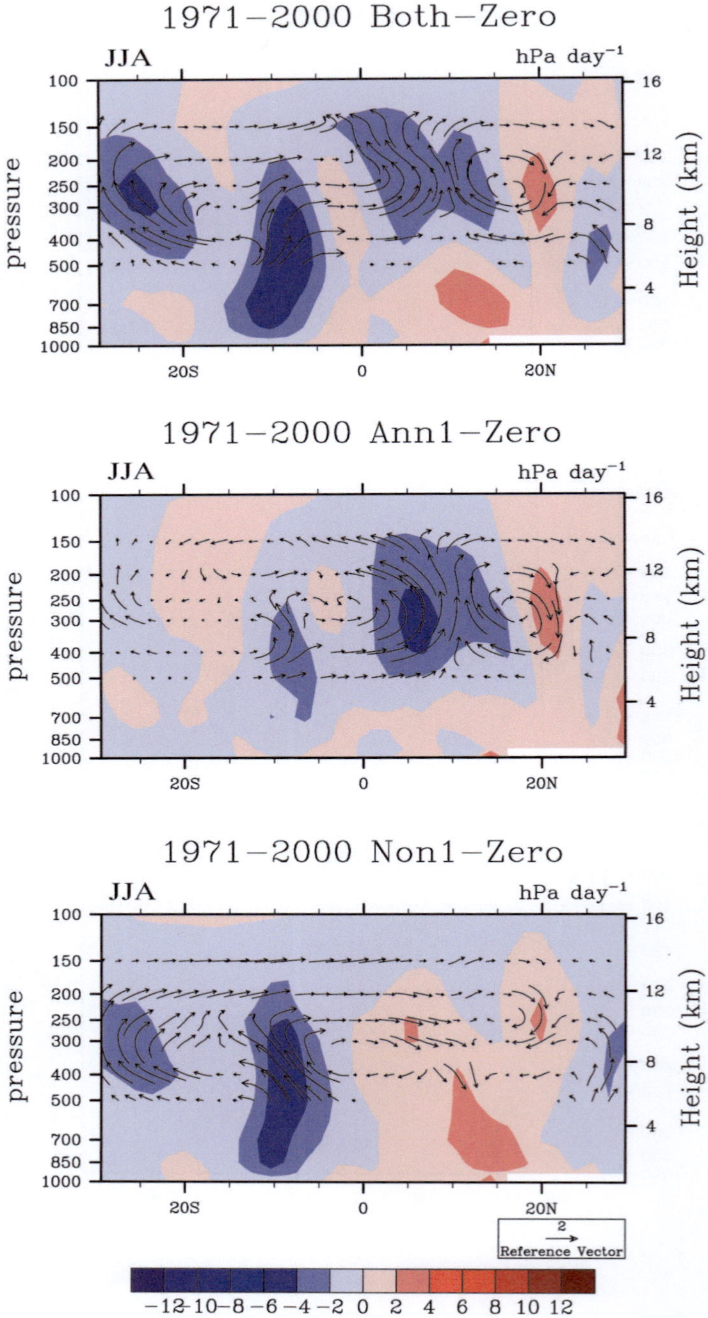

Fig. 5.32 60°E-90°E averaged meridional circulation for June–August averaged over 1971–2000, simulated by Both minus Zero (*upper*), Ann1 minus Zero (*middle*), and Non1 minus Zero (*lower*). The shades denote the latitude-height section of vertical velocity (unit: hPa/day)

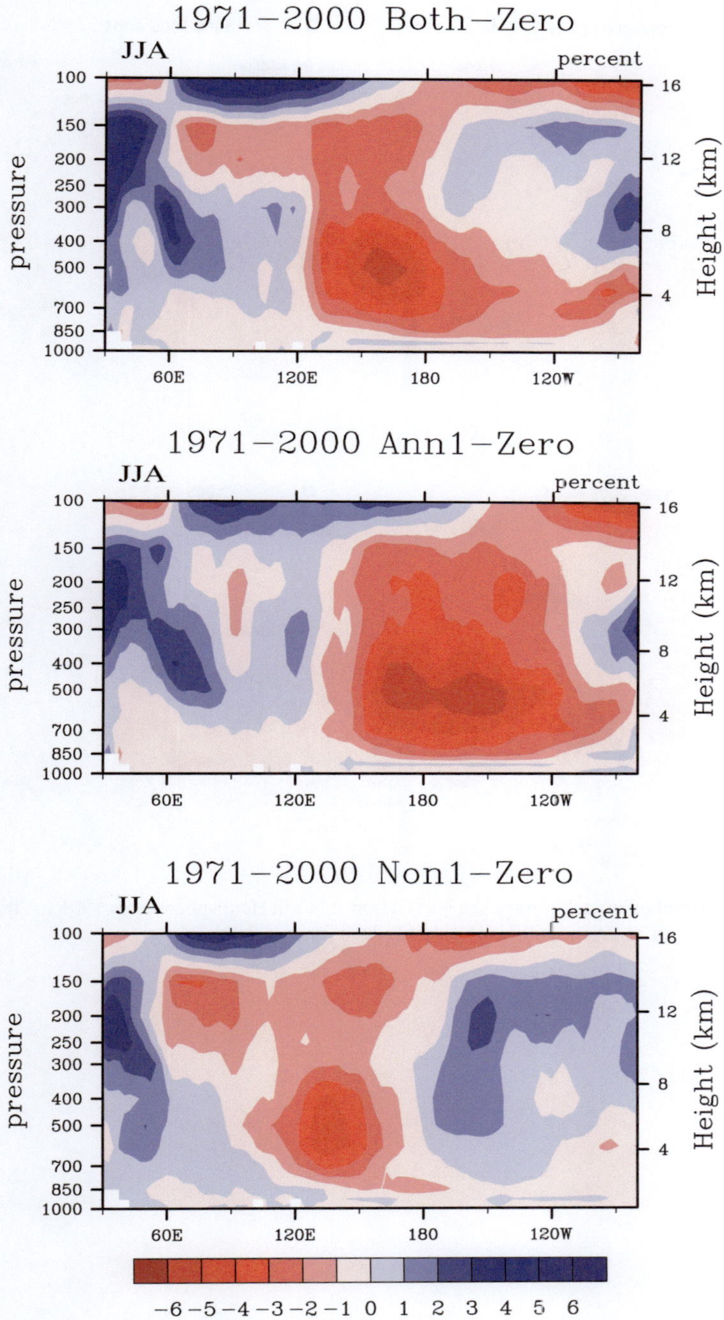

Fig. 5.33 5°S-5°N averaged relative humidity for June–August averaged over 1971–2000, simulated by Both minus Zero (*upper*), Ann1 minus Zero (*middle*), and Non1 minus Zero (*lower*). The shades denote the latitude-height section of vertical velocity (unit: percent)

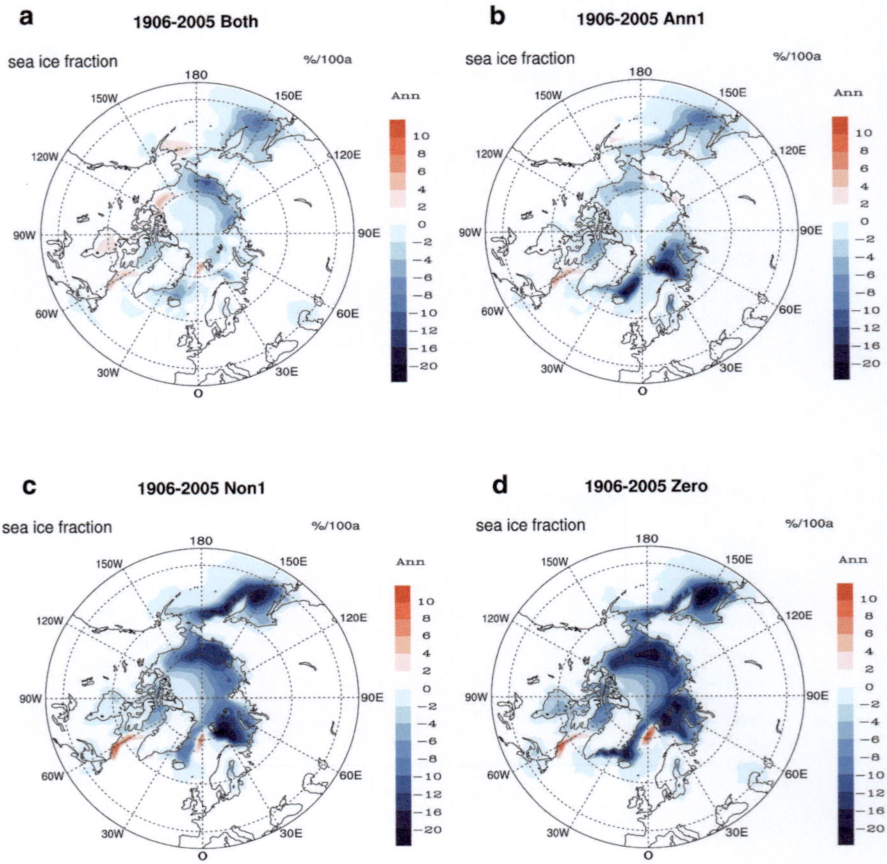

Fig. 5.34 Trend of annual average sea ice fraction at North Hemisphere from 1906 to 2005, simulated by (**a**) Both, (**b**) Ann1, (**c**) Non1, and (**d**) Zero (unit: %/100)

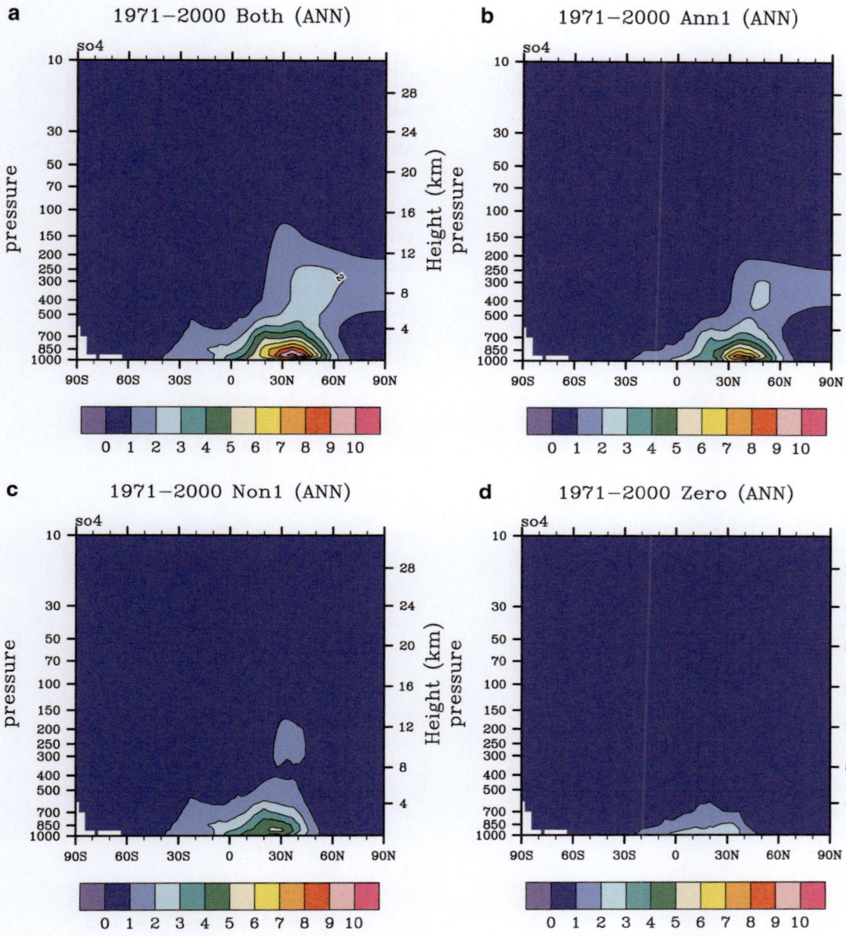

Fig. 5.35 Latitude-depth section of annual average sulfate aerosol concentration averaged over 1971–2000, simulated by (**a**) Both, (**b**) Ann1, (**c**) Non1, and (**d**) Zero (unit: 10^{-6} g/m^3)

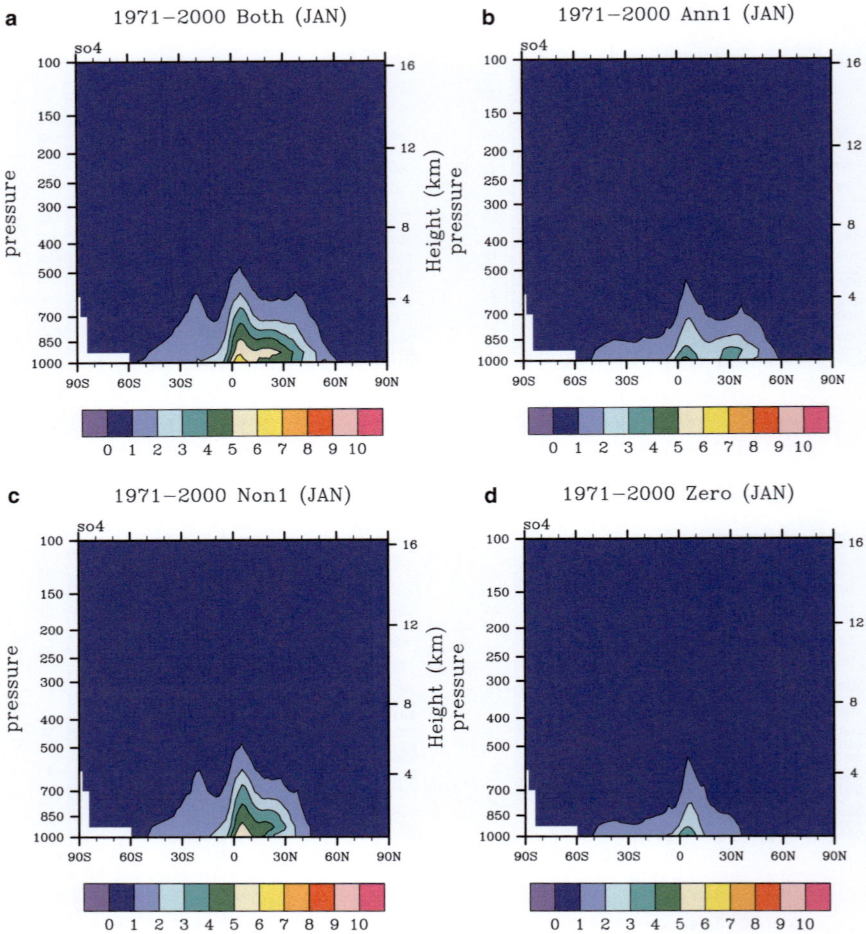

Fig. 5.36 Latitude-depth section of sulfate aerosol concentration in January averaged over 1971–2000, simulated by (**a**) Both, (**b**) Ann1, (**c**) Non1, and (**d**) Zero (unit: 10^{-6} g/m^3)

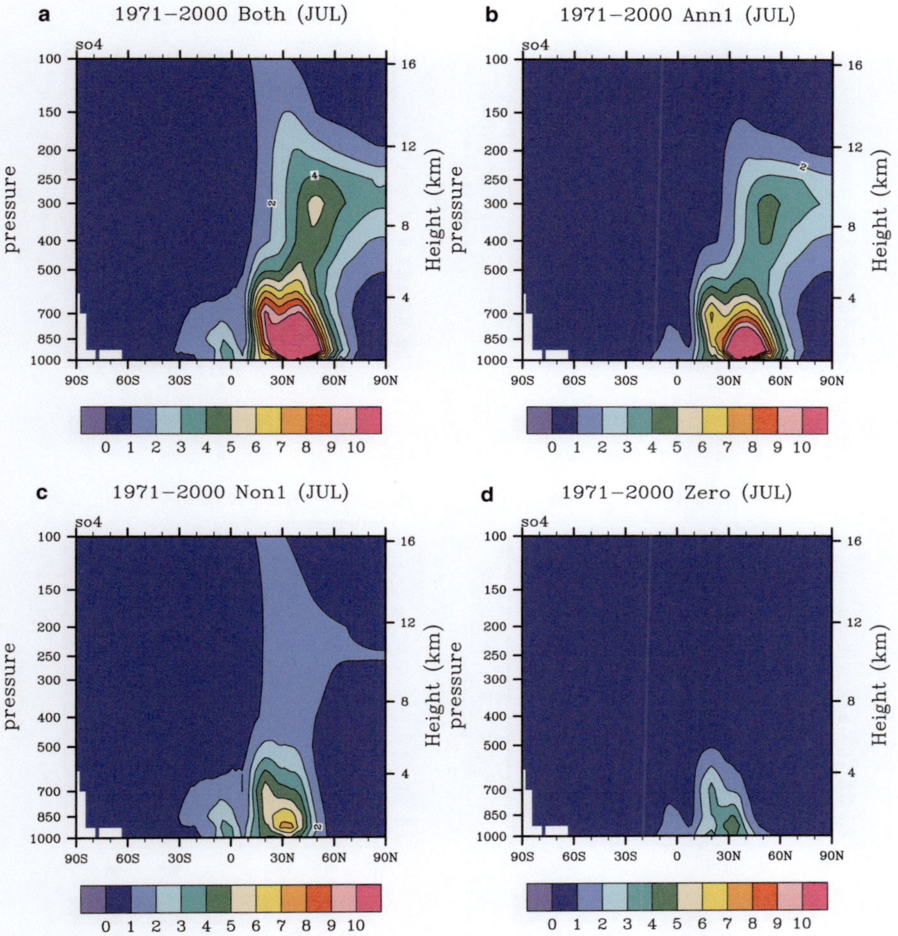

Fig. 5.37 Latitude-depth section of sulfate aerosol concentration in July averaged over 1971–2000, simulated by (**a**) Both, (**b**) Ann1, (**c**) Non1, and (**d**) Zero (unit: 10^{-6} g/m^3)

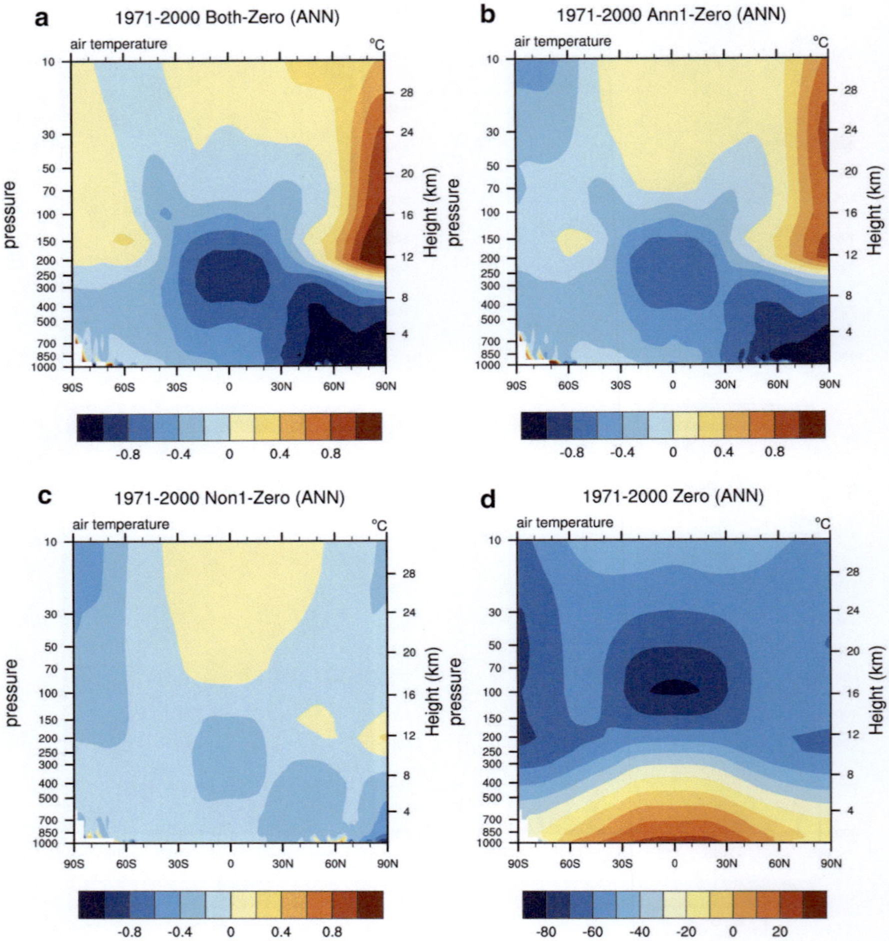

Fig. 5.38 Latitude-depth section of annual average air temperature averaged over 1971–2000, simulated by (**a**) Both, (**b**) Ann1, (**c**) Non1, and (**d**) Zero (unit: °C)

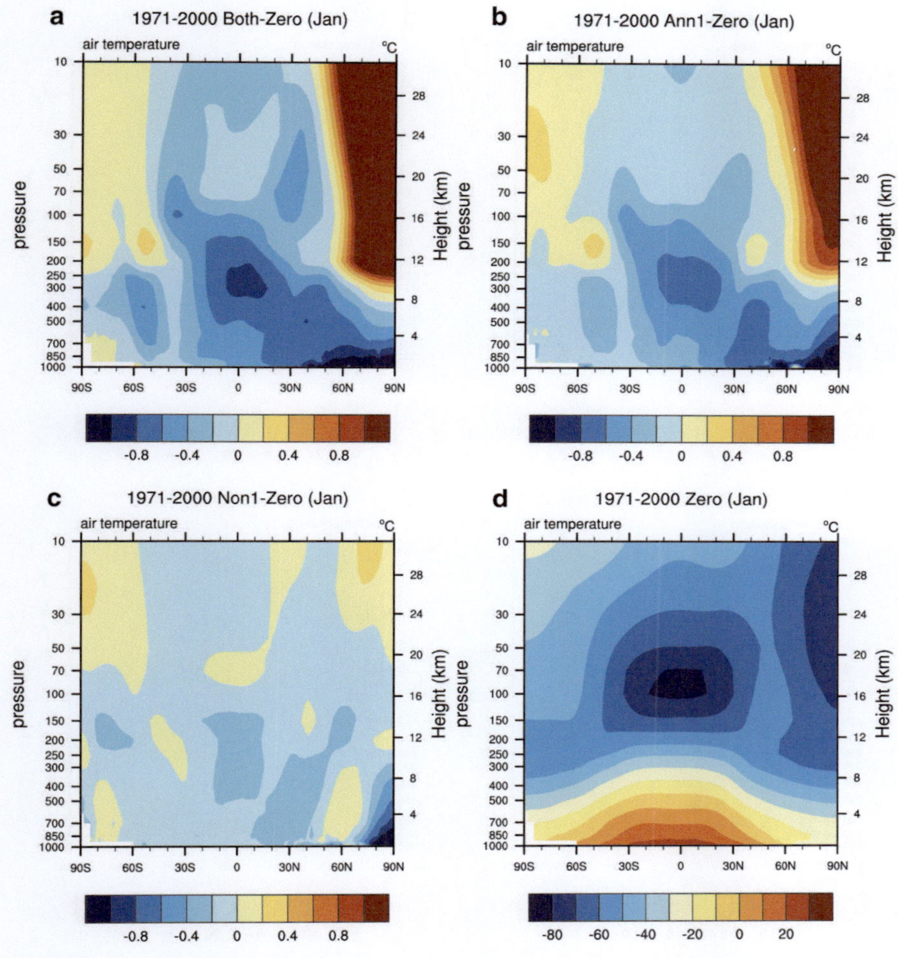

Fig. 5.39 Latitude-depth section of air temperature in January averaged over 1971–2000, simulated by (**a**) Both minus Zero, (**b**) Ann1 minus Zero, (**c**) Non1 minus Zero, and (**d**) Zero (unit: °C)

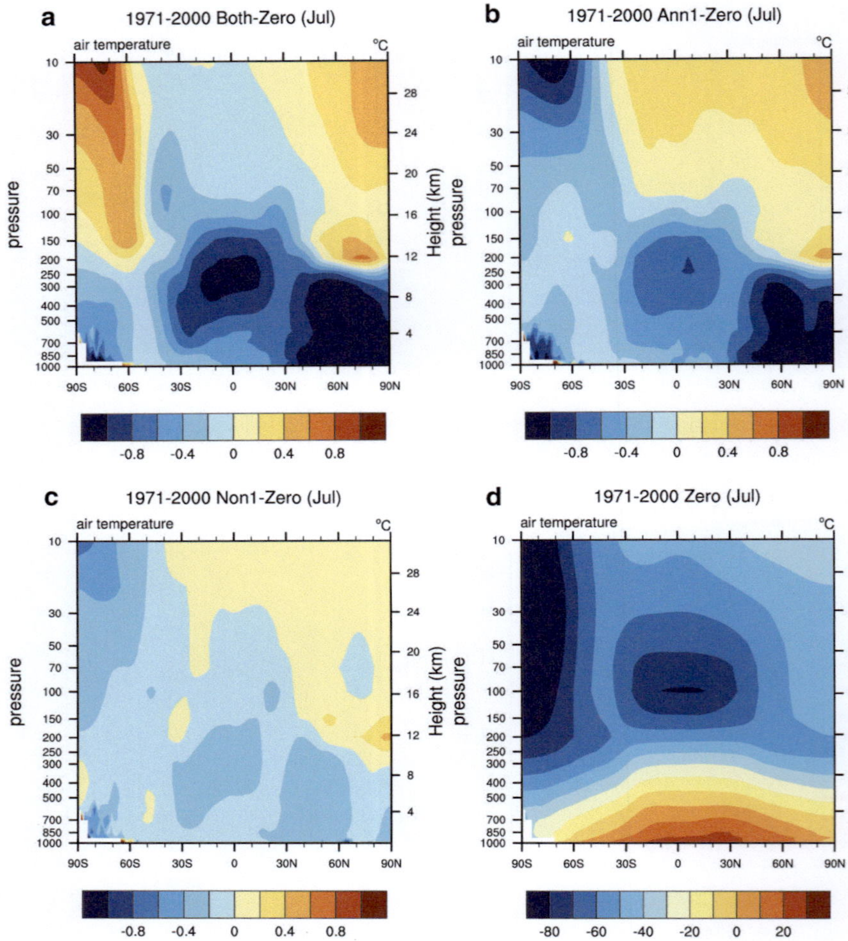

Fig. 5.40 Latitude-depth section of air temperature in July averaged over 1971–2000, simulated by (**a**) Both minus Zero, (**b**) Ann1 minus Zero, (**c**) Non1 minus Zero, and (**d**) Zero (unit: °C)

References

Smith SJ, Pitcher H, Wigley TML (2001) Global and regional anthropogenic sulfur dioxide emissions. Glob Planet Chang 29:99–119

Smith SJ, Conception E, Andres R, Lurz J (2004) Historical sulfur dioxide emissions 1850–2000: methods and results. Technical report, Pacific Northwest National Laboratory, Joint Global Change Research Institute

Smith SJ, van Aardenne J, Klimont Z, Andres RJ, Volke A, Delgado AS (2011) Anthropogenic sulfur dioxide emissions: 1850–2005. Atmos Chem Phys 11:1101–1116

Chapter 6
Transferred Responsibility for Historical Climate Change Induced from Carbon Trade Between Developed and Developing World

Abstract This chapter shows the results of attribution experiments which assess the influences of transferred carbon emissions along with international trade on the historical responsibility of developed/developing countries and on the potential mitigation ability of the Kyoto Protocol. These numerical experiments are performed based on NCAR/CESM1_0_2 model for the period of 1990–2005.

Keywords Transferred emission • Responsibility • Kyoto Protocol

W. Dong et al., *Atlas of Climate Change: Responsibility and Obligation of Human Society*, Springer Atmospheric Sciences,
DOI 10.1007/978-3-662-48444-9_6

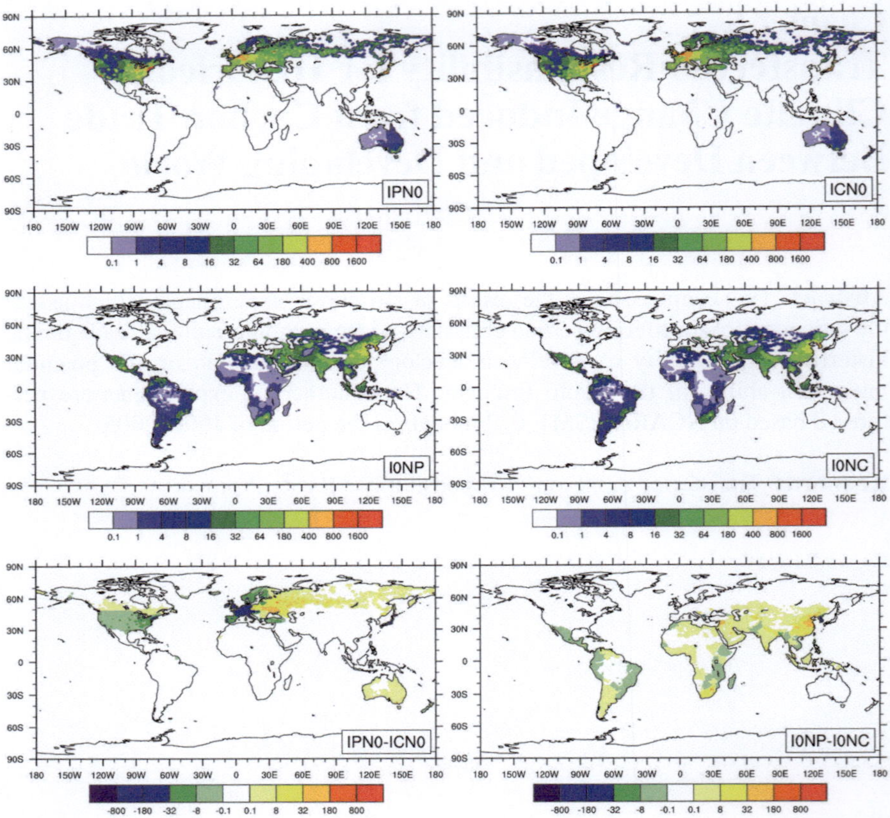

Fig. 6.1 Cumulative CO_2 emission fluxes from 1990 to 2005 for experiments IPN0, ICN0, I0NP, and I0NC. The differences between IPN0 and ICN0, I0NP, and I0NC are also shown here. The spatial resolution of 0.9°×1.25° is used to match that of CESM model (unit: $KgCO2/m^2 \bullet 16y$)

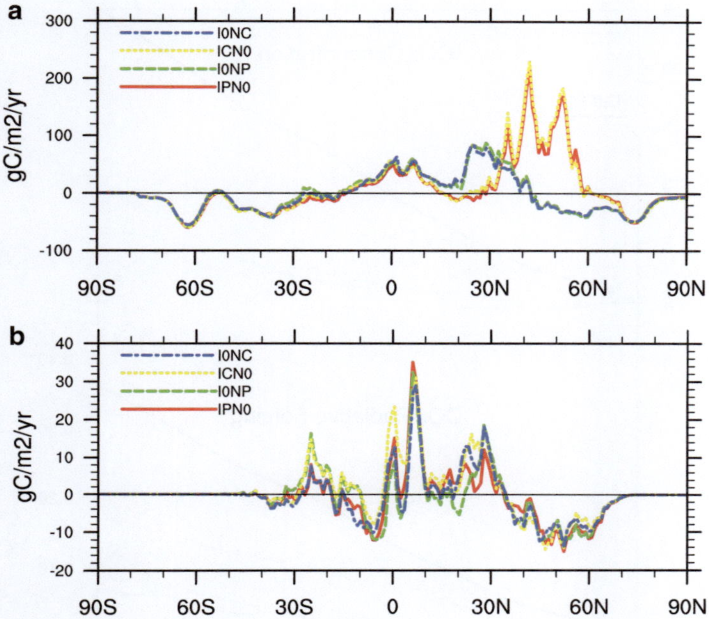

Fig. 6.2 Zonal averaged (**a**) CO_2 emission fluxes (unit: gC/m^2/y) and (**b**) land-air CO_2 fluxes (unit: gC/m^2/y) from 1990 to 2005 for experiments IPN0, ICN0, I0NP, and I0NC

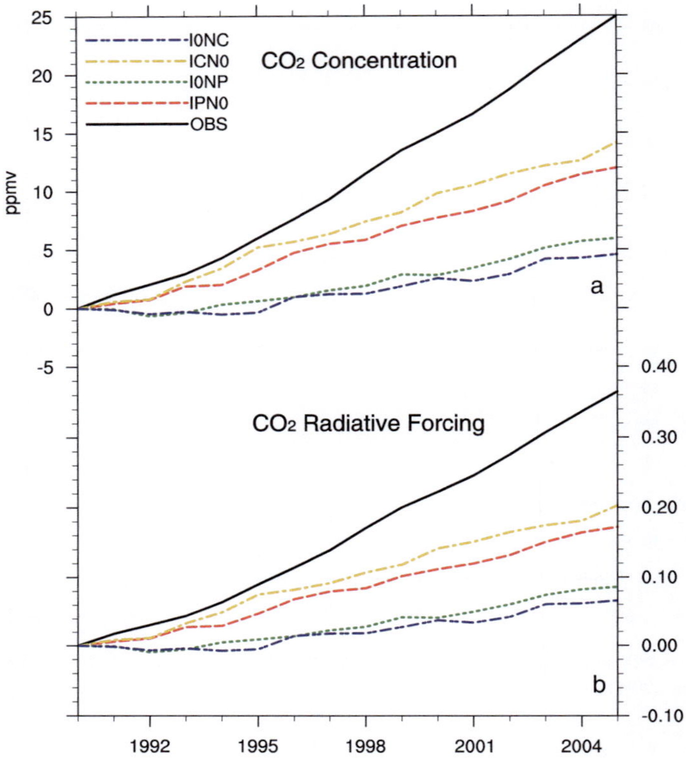

Fig. 6.3 Observed and simulated (**a**) CO_2 concentration (unit: ppm) and (**b**) radiative forcing of CO_2 (unit: W/m^2) under IPN0, ICN0, I0NP, and I0NC experiments

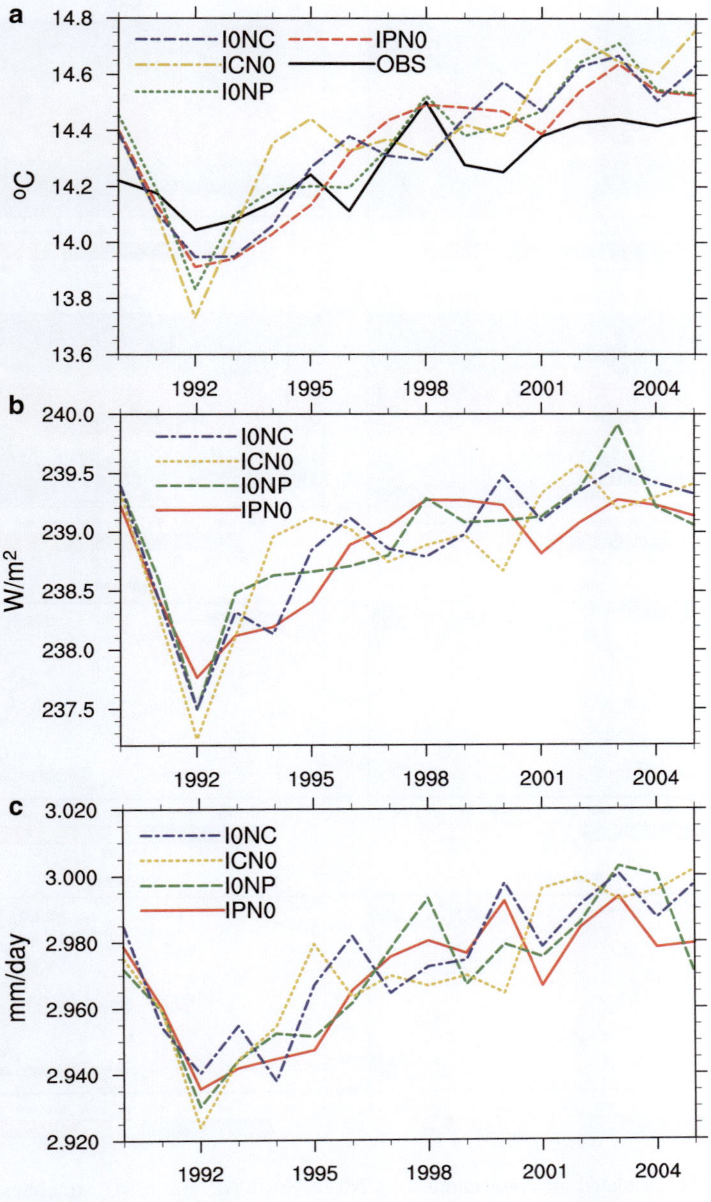

Fig. 6.4 Observed and simulated global mean (**a**) air temperature (unit: °C), (**b**) net long-wave fluxes (unit: W/m^2), (**c**) precipitation (unit: mm/day) under IPN0, ICN0, I0NP, and I0NC experiments

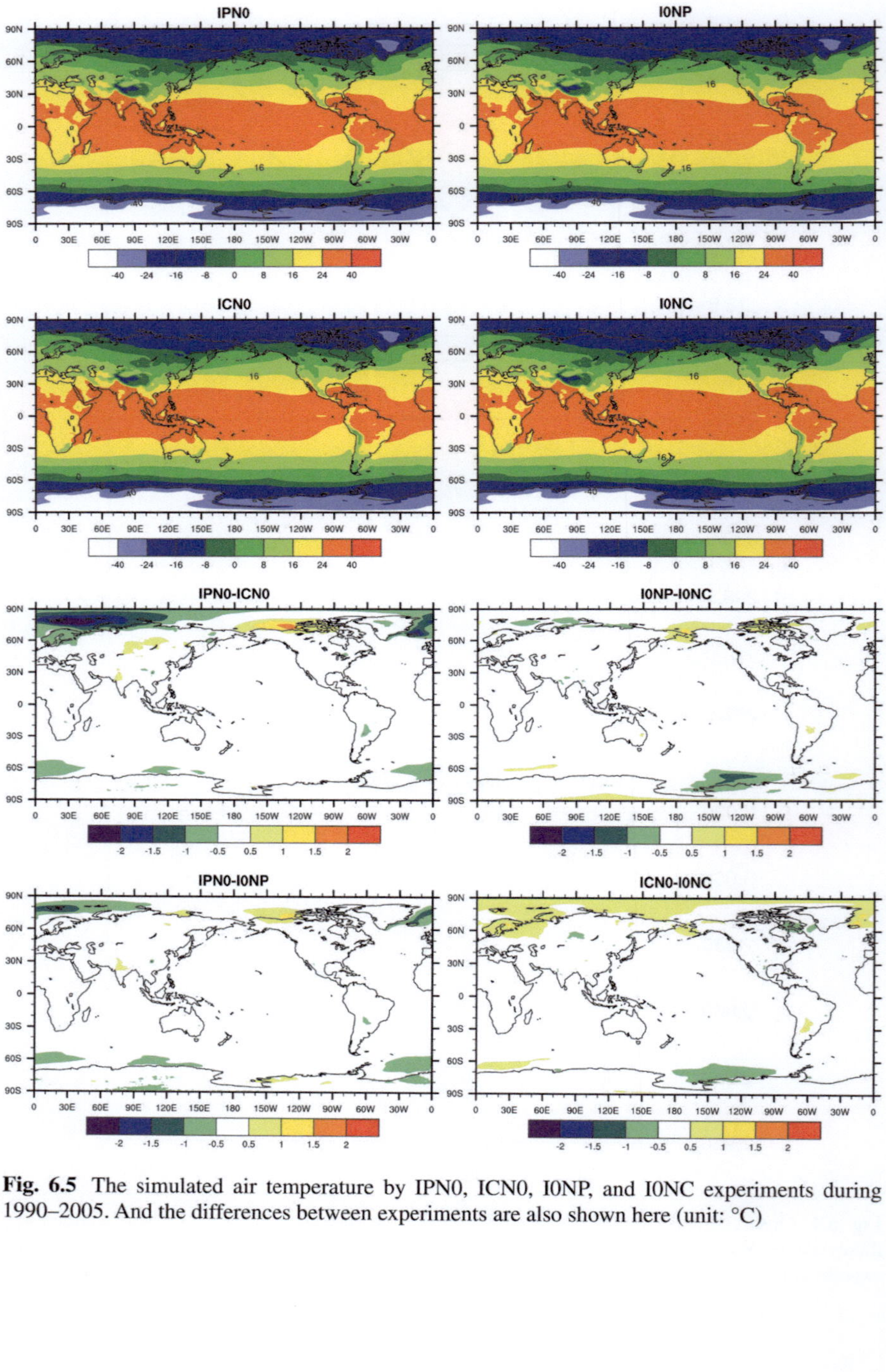

Fig. 6.5 The simulated air temperature by IPN0, ICN0, I0NP, and I0NC experiments during 1990–2005. And the differences between experiments are also shown here (unit: °C)

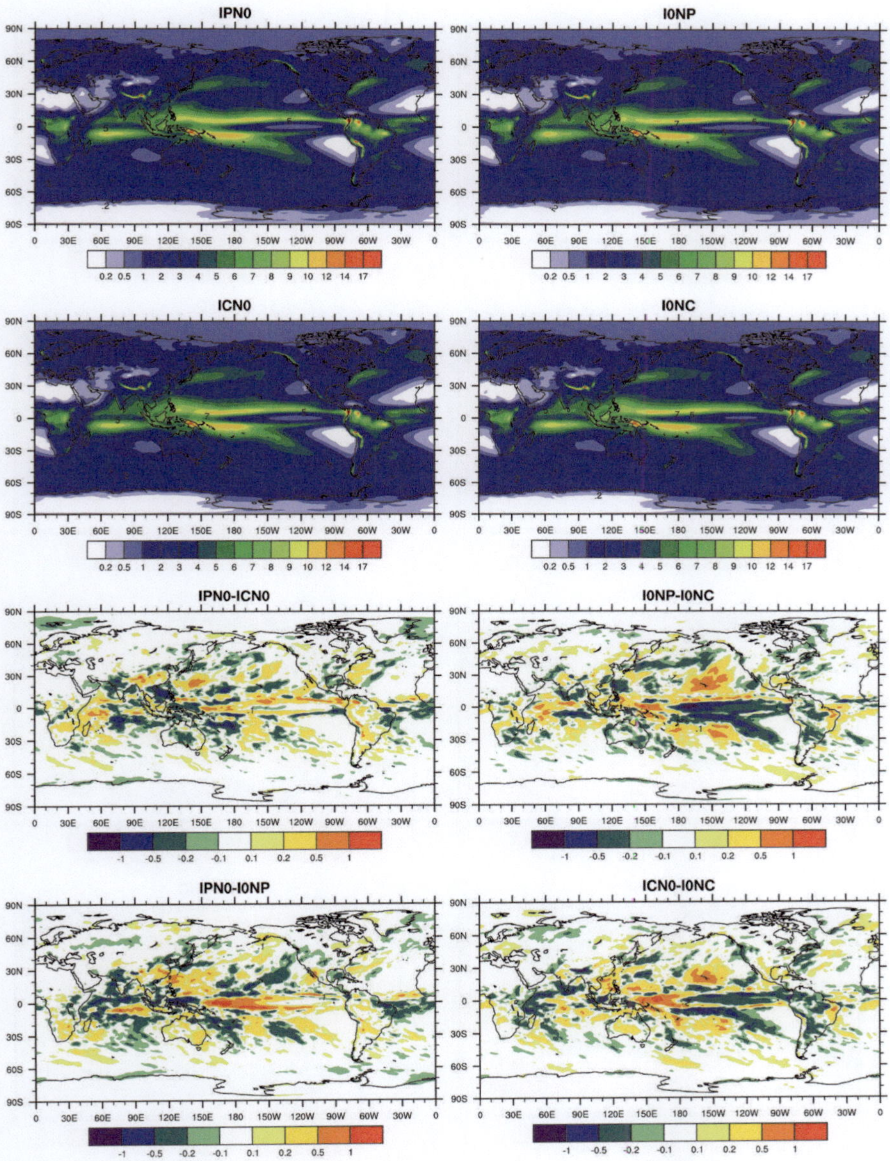

Fig. 6.6 The simulated precipitation by IPN0, ICN0, I0NP, and I0NC experiments during 1990–2005. And the differences between experiments are also shown here (unit: mm/day)

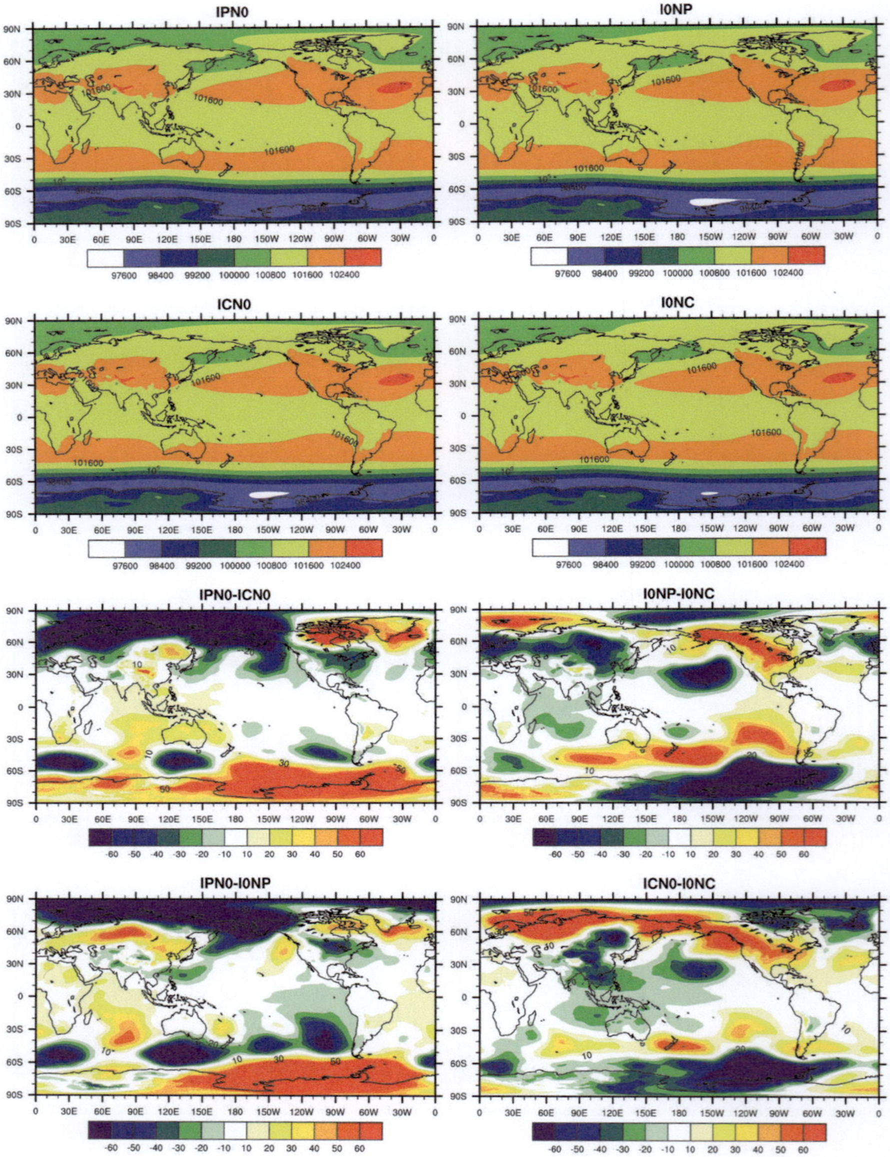

Fig. 6.7 The simulated sea surface pressure by IPN0, ICN0, I0NP, and I0NC experiments during 1990–2005. And the differences between experiments are also shown here (unit: Pa)

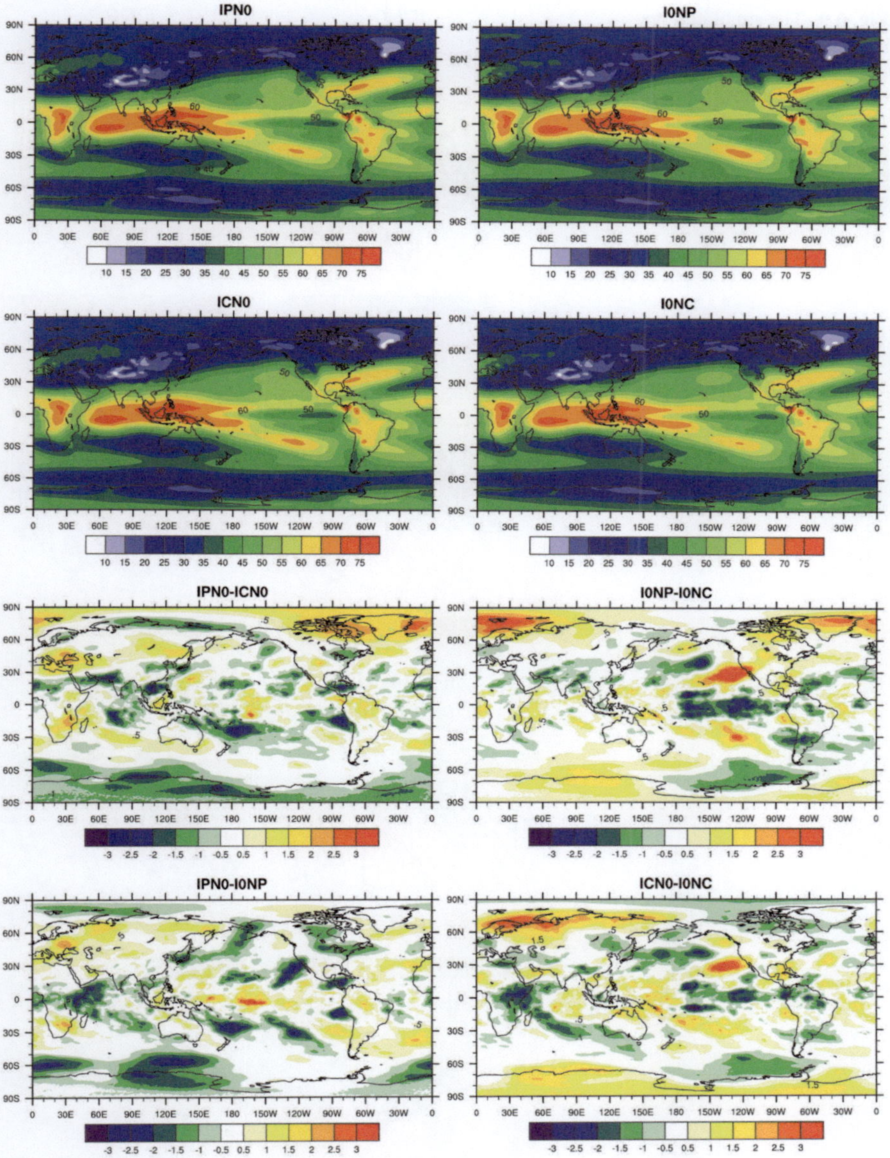

Fig. 6.8 The simulated relative humidity by IPN0, ICN0, I0NP, and I0NC experiments during 1990–2005. And the differences between experiments are also shown here (unit: %)

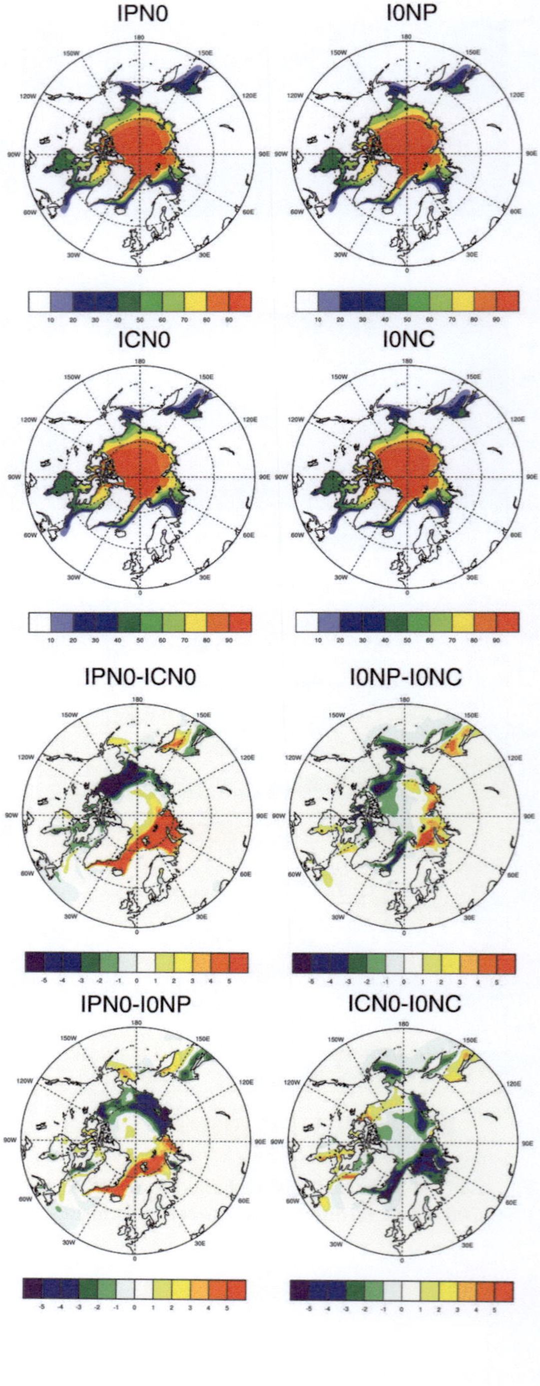

Fig. 6.9 The simulated sea ice concentration by IPN0, ICN0, I0NP, and I0NC experiments during 1990–2005. And the differences between experiments are also shown here (unit: %)

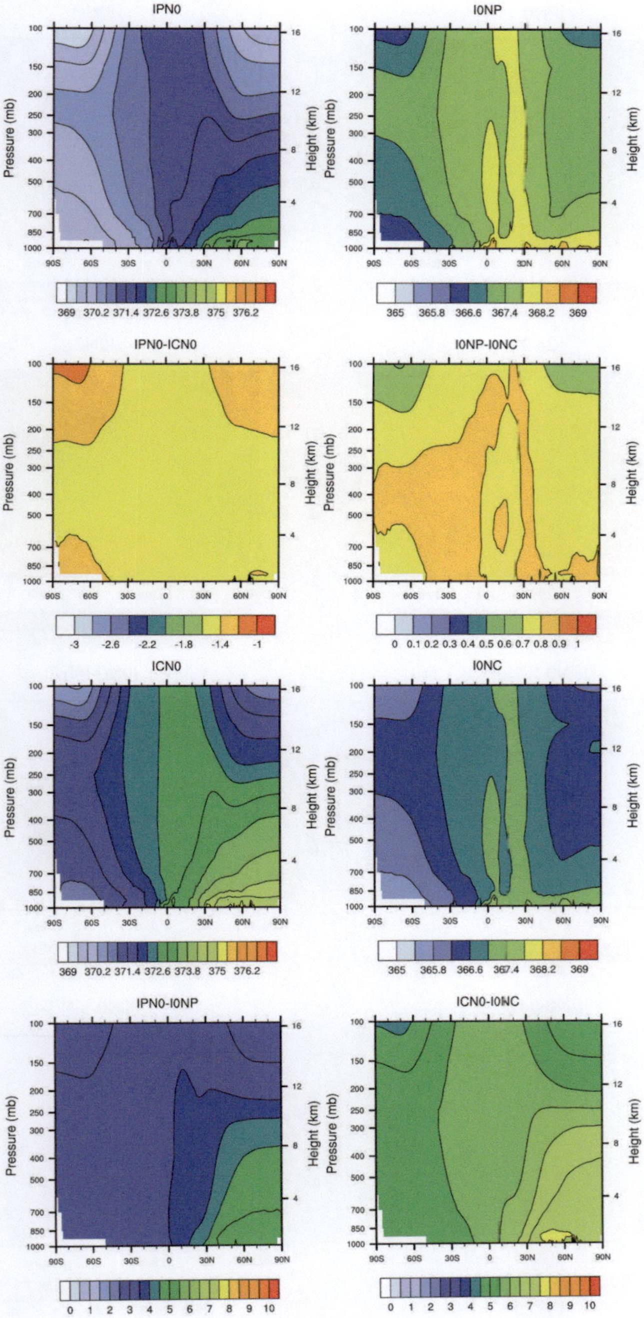

Fig. 6.10 The simulated latitude-depth section of atmospheric CO_2 concentration by IPN0, ICN0, I0NP, and I0NC experiments during 1990–2005. And the differences between experiments are also shown here (unit: ppm)

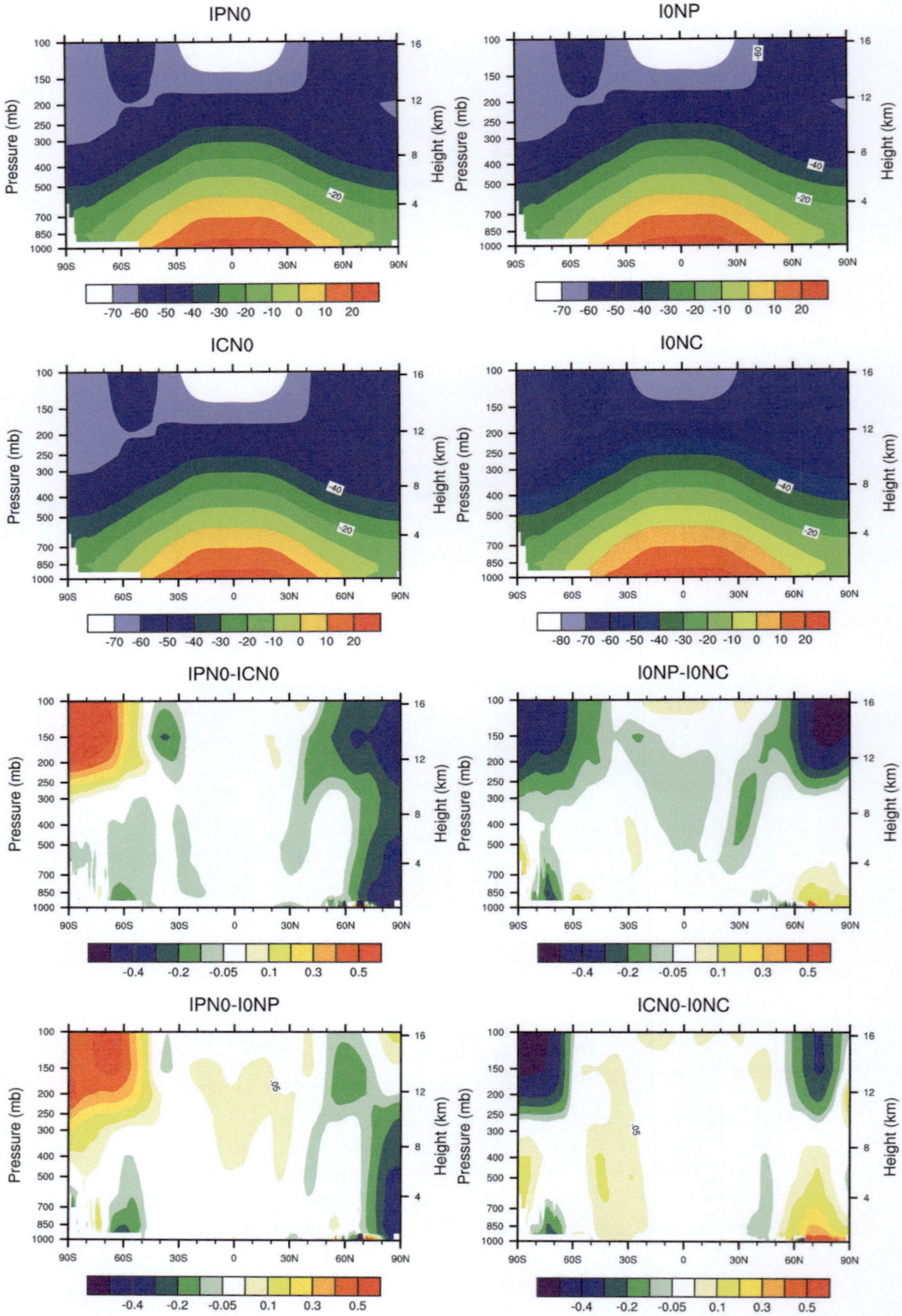

Fig. 6.11 The simulated latitude-depth section of air temperature by IPN0, ICN0, I0NP, and I0NC experiments during 1990–2005. And the differences between experiments are also shown here (unit: °C)

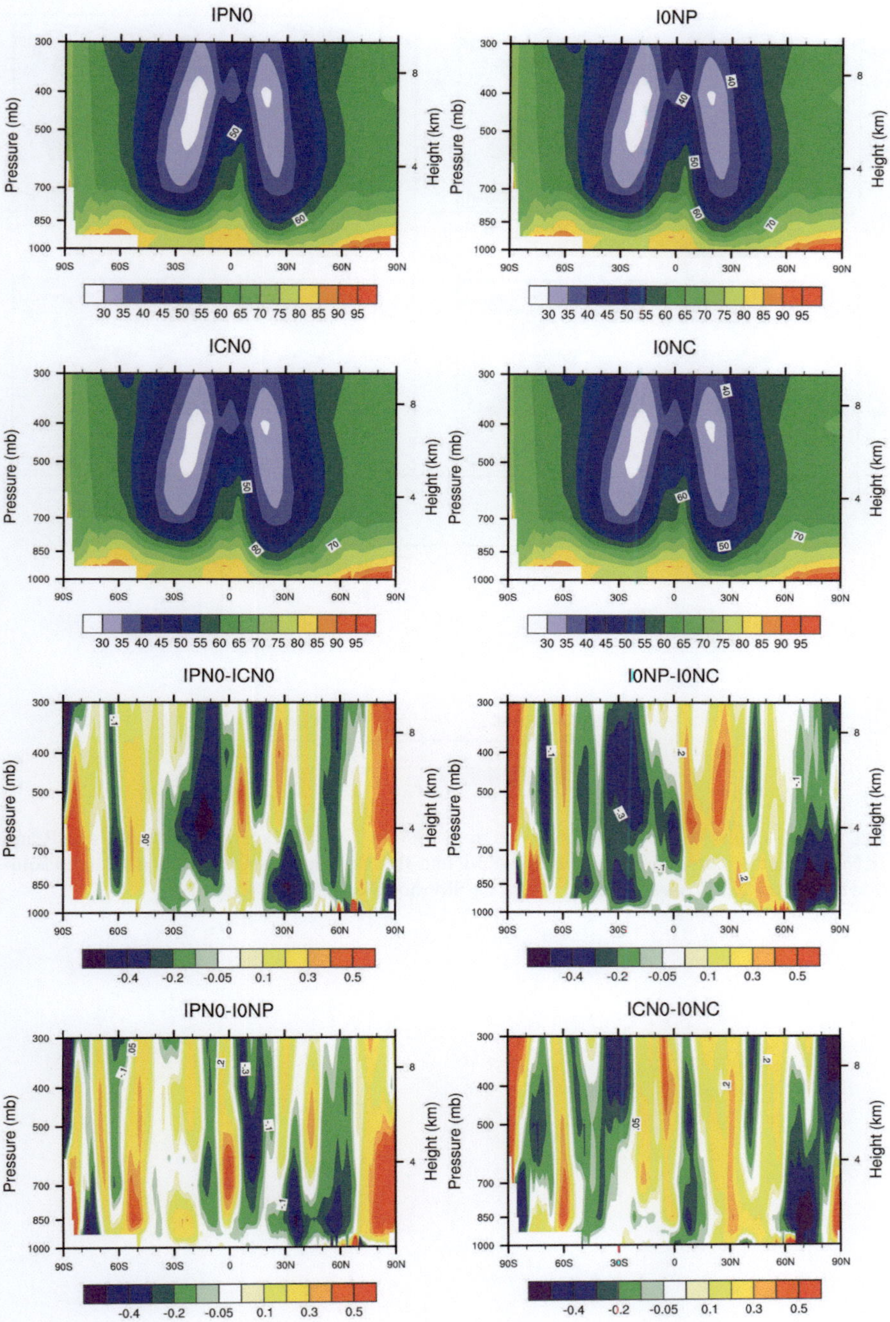

Fig. 6.12 The simulated latitude-depth section of relative humidity by IPN0, ICN0, I0NP, and I0NC experiments during 1990–2005. And the differences between experiments are also shown here (unit: %)

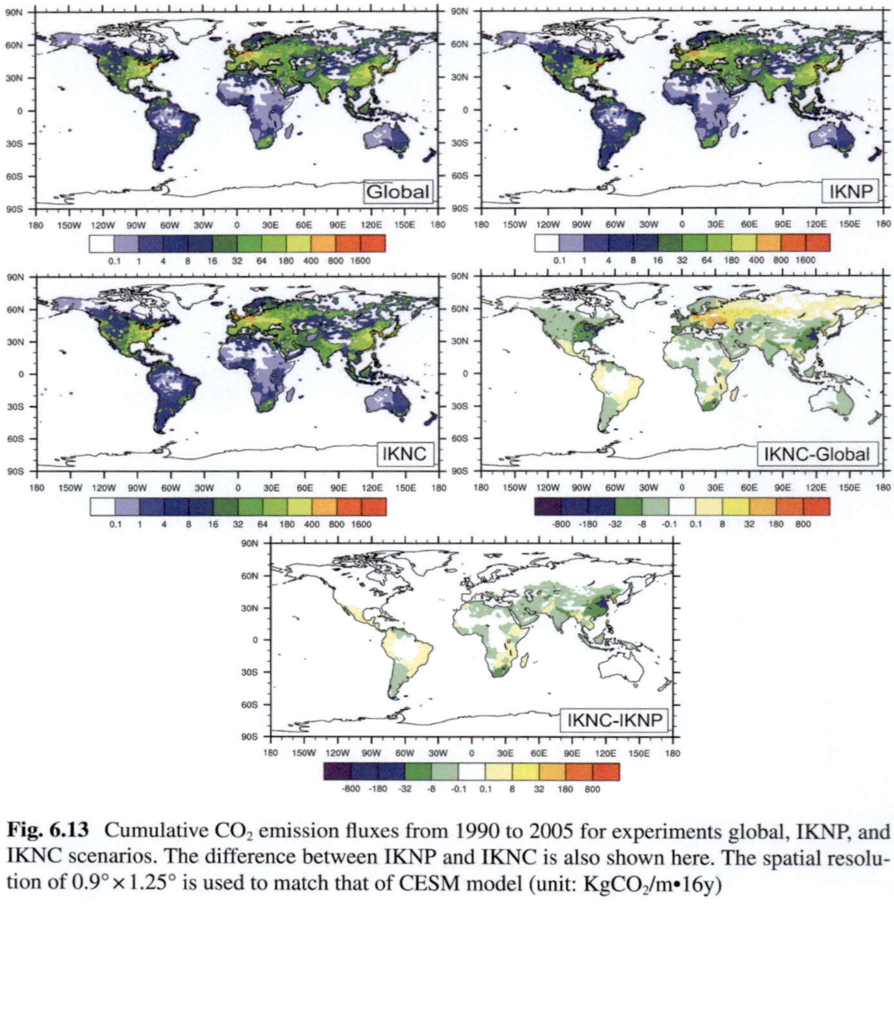

Fig. 6.13 Cumulative CO_2 emission fluxes from 1990 to 2005 for experiments global, IKNP, and IKNC scenarios. The difference between IKNP and IKNC is also shown here. The spatial resolution of $0.9° \times 1.25°$ is used to match that of CESM model (unit: $KgCO_2/m \cdot 16y$)

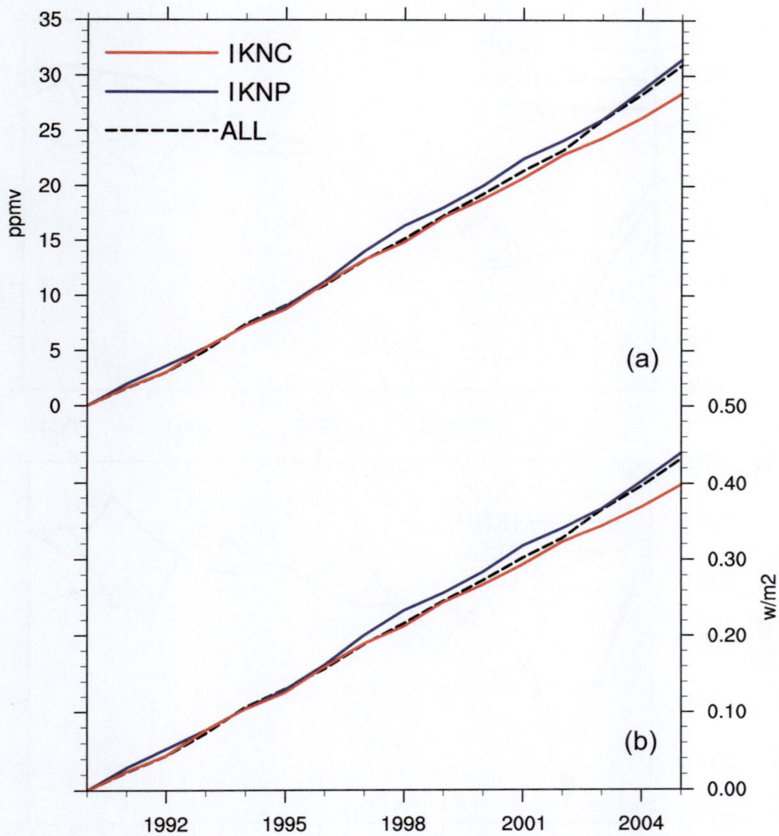

Fig. 6.14 Observed and simulated (**a**) CO_2 concentration (unit: ppm) and (**b**) radiative forcing of CO_2 (unit: W/m^2/) by global, IKNP, and IKNC experiments

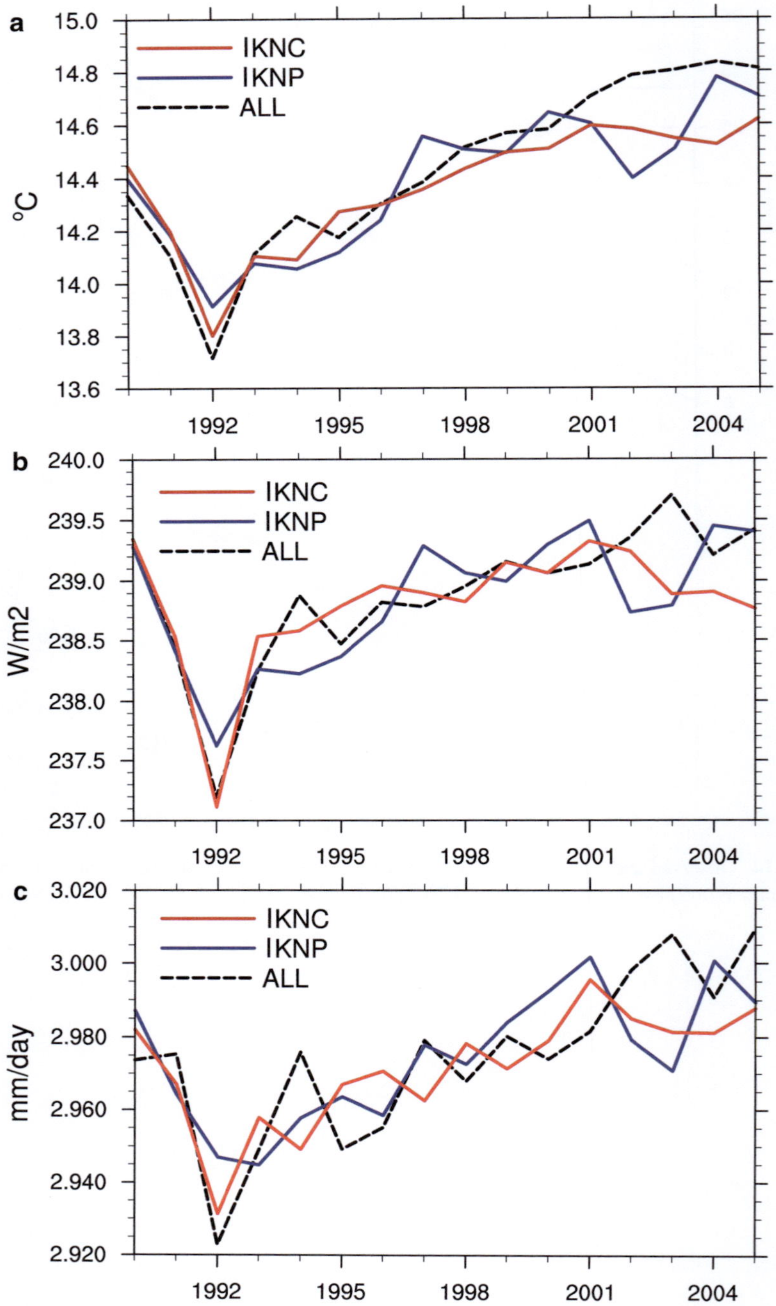

Fig. 6.15 Observed and simulated global mean (**a**) air temperature (unit: °C), (**b**) net long-wave fluxes (unit: W/m^2), (**c**) precipitation (unit: mm/day)

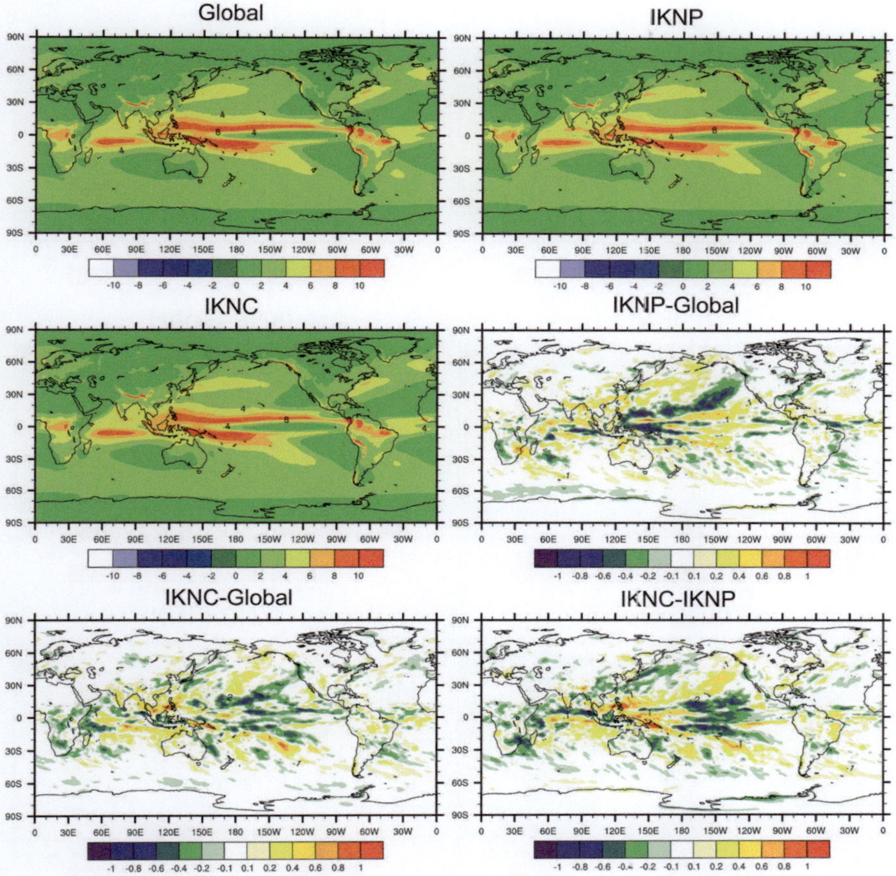

Fig. 6.16 The simulated air temperature by global, IKNP, and IKNC experiments during 1990–2005. And the differences between experiments are also shown here (unit: °C)

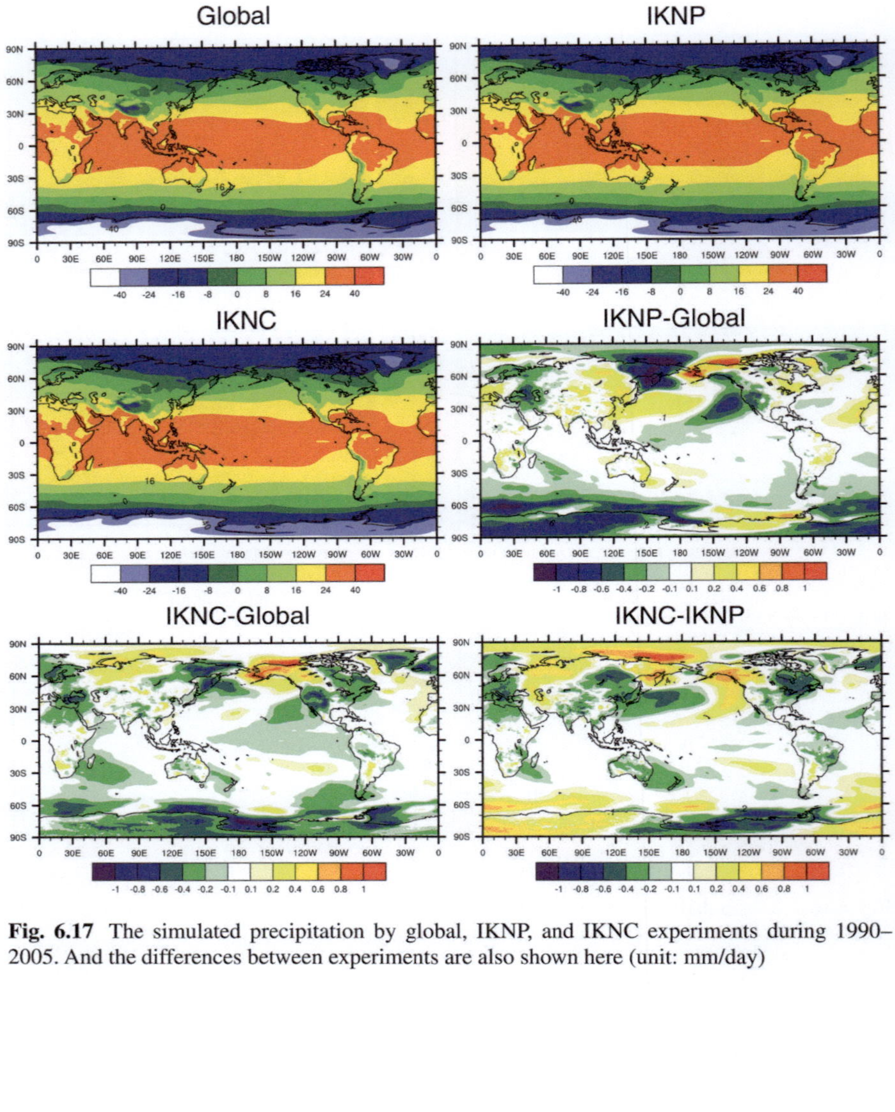

Fig. 6.17 The simulated precipitation by global, IKNP, and IKNC experiments during 1990–2005. And the differences between experiments are also shown here (unit: mm/day)

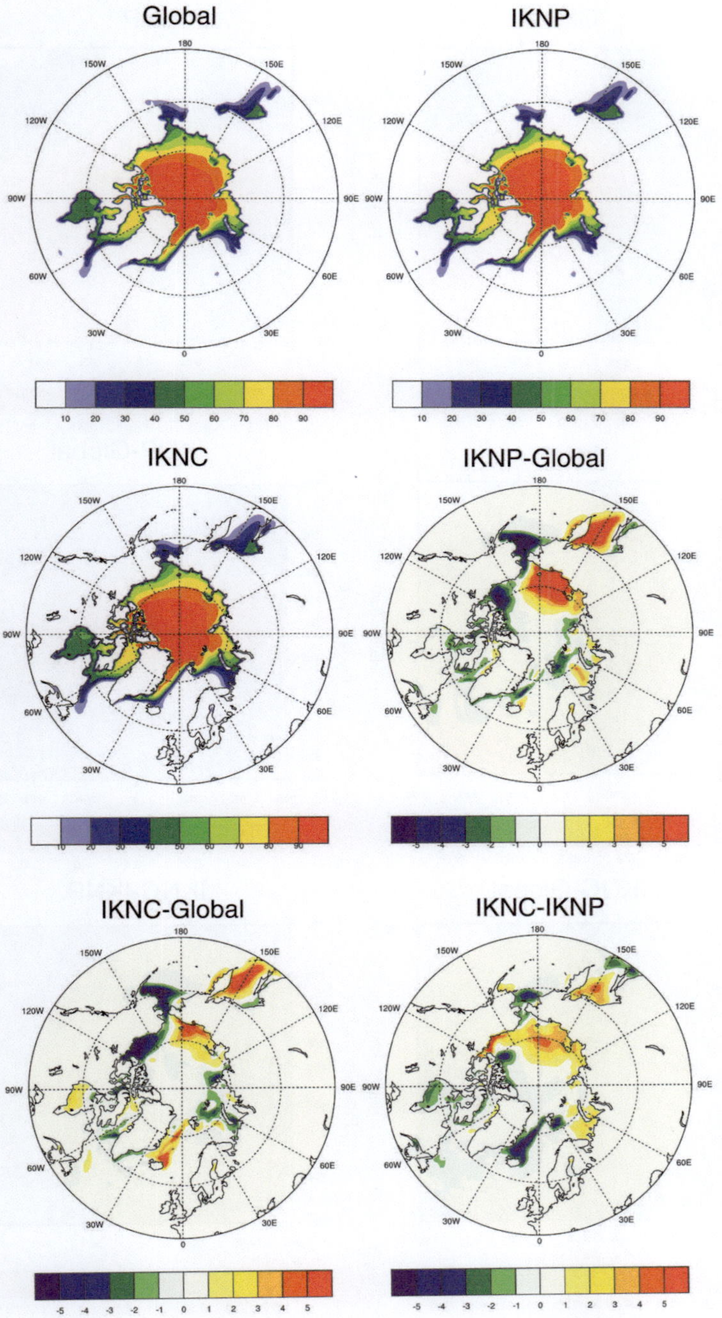

Fig. 6.18 The simulated sea ice concentration by global, IKNP, and IKNC experiments during 1990–2005. And the differences between experiments are also shown here (unit: %)

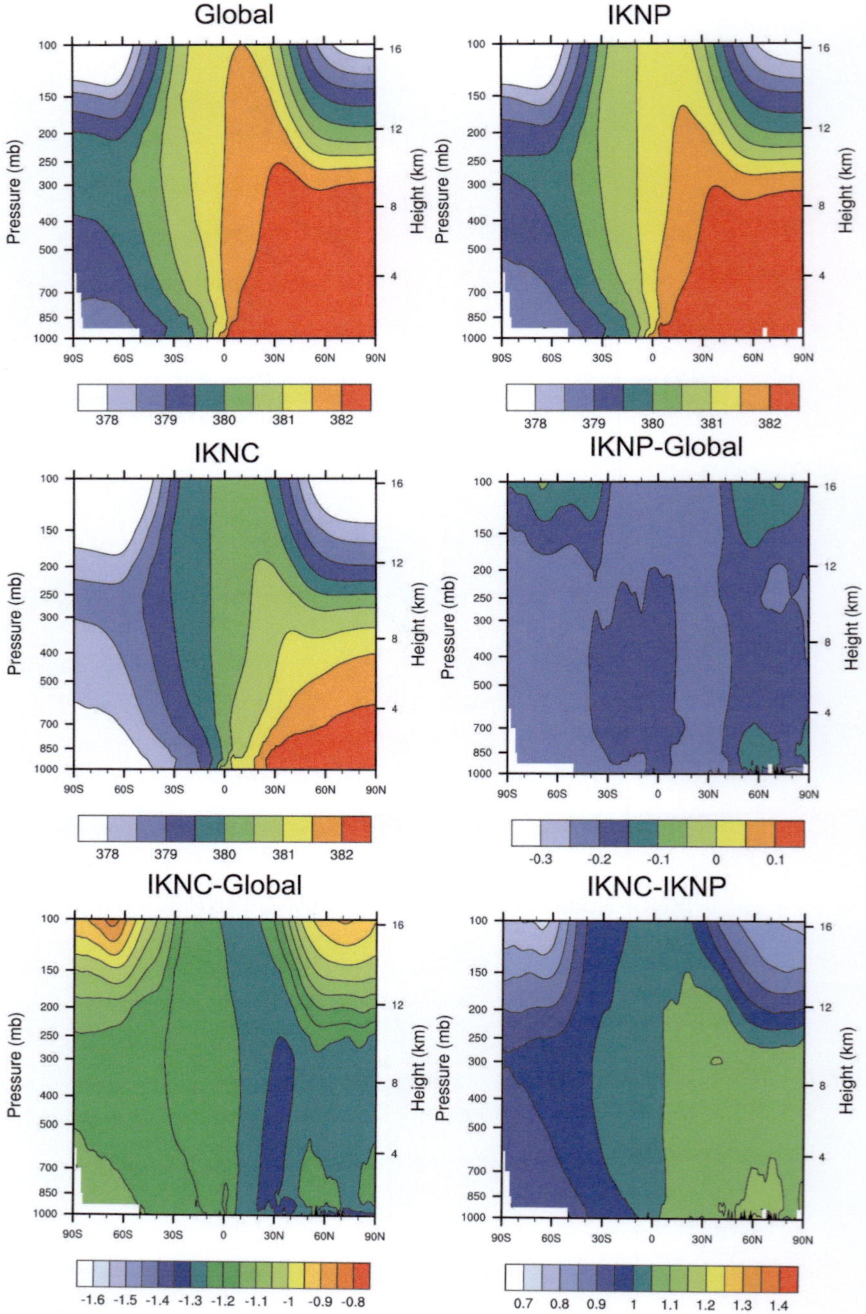

Fig. 6.19 The simulated latitude-depth section of atmospheric CO_2 concentration by global, IKNP, and IKNC experiments during 1990–2005. And the differences between experiments are also shown here (unit: ppm)

MIX
Papier aus verantwortungsvollen Quellen
Paper from responsible sources
FSC® C105338

If you have any concerns about our products,
you can contact us on
ProductSafety@springernature.com

In case Publisher is established outside the EU,
the EU authorized representative is:
Springer Nature Customer Service Center GmbH
Europaplatz 3, 69115 Heidelberg, Germany

Printed by Libri Plureos GmbH
in Hamburg, Germany